CONCRETE TECHNOLOGY AND DESIGN

Volume 5

Natural Fibre Reinforced Cement and Concrete

CONCRETE TECHNOLOGY AND DESIGN

A series highlighting the importance of the inherent properties of concrete materials and their influence on design. Each volume covers a particular aspect of concrete technology or of design, and contains contributions from internationally acknowledged experts. The series will prove very useful to practising engineers and architects, researchers, and advanced students in civil and structural engineering and materials technology.

Series Editor: R.N. Swamy, PhD, CEng, FICE, FIStructE, FASCE

Volume 1: New Concrete Materials

Volume 2: New Reinforced Concretes

Volume 3: Cement Replacement Materials

Volume 4: Ferrocement: Current and Potential Applications

CONCRETE TECHNOLOGY AND DESIGN

Volume 5

Natural Fibre Reinforced Cement and Concrete

Editor

R.N. Swamy, BEng, MEng, PhD, CEng, FICE, FIStructE, FASCE, FIE(Ing)
Department of Mechanical Engineering
University of Sheffield, UK

Blackie

Glasgow and London

Blackie and Son Ltd
Bishopbriggs, Glasgow G64 2NZ
7 Leicester Place
London WC2H 7BP

First published 1988

British Library Cataloguing in Publication Data

Natural fibre reinforced cement and concrete.
1. Construction materials: Fibre reinforced cement. 2. Construction materials: Fibre reinforced concrete
I. Swamy, R.N. II. Series
624.1′833

ISBN 0-216-92493-6

Phototypesetting by Thomson Press (India) Ltd., New Delhi
Printed in Great Britain by Bell & Bain (Glasgow) Ltd

Preface

Nature will continue to be bountiful to mankind so long as we do not wantonly consume or destroy the vast and rich resources that it so generously provides. Natural and vegetable plants and fibres have a unique and irreplaceable role in the ecological cycle, and their natural abundance, plentiful supply and relative cheapness are matched by the ease and readiness with which these resources can be swiftly replenished. Such materials can therefore provide a compatible and competent alternative reinforcing medium to the construction industry, not only for cement matrices, but also for soil cement and ferrocement materials.

When two chapters in the second volume of this series were devoted to the use of natural fibres and bamboo reinforcement in cement and concrete matrices, there was strong and confirmed evidence that these new composites had a wide range of applications in the developing countries. However, health legislations against asbestos (and the consequent reduction in world production and demand for asbestos) has necessitated the widespread development and use of asbestos replacement materials (chiefly fibres of vegetable origin), for the developed countries.

Attempts to use wood fibre as a constituent of cement and plaster-based building materials date back almost 100 years, while cellulose has been used as an asbestos substitute for nearly 50 years. Throughout this period methods of refining cellulose fibres, and replacing asbestos either fully or partly with paper pulp, have been studied extensively. Refined kraft wood fibres have since been identified as the best pulp source to impart strength, stiffness and toughness to cement composites. Flat sheets, corrugated roofing, moulded products and non-pressure fibre pipes are today recognized as the major uses for a wide range of wood fibres applied as reinforcement to cement and concrete.

By comparison with many synthetic organic fibres, wood fibres import high strength and elasticity, possess fire and impact resistance, and exhibit durability and long-term resistance to weathering. They are readily processed in the manufacture of both autoclaved and air-cured products, and can be produced via industrialized production processes as well as by simple manual methods. There is adequate field experience now to enable safe design and construction using natural fibre reinforced cement and concrete products from both automated production processes and village-level small-scale industries. Although natural fibre cement technology is now fully developed in many countries, there are still several industries which use mixtures of wood fibres with synthetic and mineral fibres and/or fillers.

The main aim of this new volume is to appraise fibre reinforcement technology and highlight its global potential. There are eight chapters in this volume, written by authors noted for their blend of research activities and practical experience. These chapters cover, in a comprehensive and authoritative manner, the complete spectrum of natural fibre cement composites—including the physics and chemistry of fibres, fibre cement composite process technology, properties of the composites and their long term stability—and provide a full discussion of the comprehensive range of products, uses and applications. An especially critical assessment of the durability of the fibres and the fibre cement composites is provided also, while careful consideration has been given to appraising manufacturing technology, both at a completely automated level and at the village level.

Natural fibre cement composites have special relevance to developing countries in view of their low cost, savings in energy and their ready application to the rapid development of a country's infrastructure. Bamboo is one of the fastest-growing, highest-yielding and easily renewable of natural resources, and one chapter is devoted to a comprehensive review of its properties and uses as reinforcement. Similarly, a wide variety of natural fibres exist in the Far East and in Latin America, and these have been extensively reviewed in two chapters. Reference is also made to the utilization of disintegrated newsprint in cement and gypsum matrices for the building industry.

The potential for natural fibre cement composites is tremendous. It is hoped that this book points the way for future research, development and application, and that these composite materials will find a permanent place in the building industry.

RNS

Contents

Contributors

V. Agopyan

Instituto de Pesquisas Tecnologicos do Estado de São Paulo S/A, São Paulo, Brazil

Dr Agopyan is Head of the Building Materials Group, Building Technology Division, Instituto de Pesquisas Tecnologicas (IPT) do Estado de São Paulo, Brasil. He is also part-time Assistant Professor in the Department of Civil Construction of the Escola Politécnica of the University of São Paulo. He has published many papers on glass reinforced gypsum, vegetable fibre reinforced materials, the use of by-products in building and durability of materials.

R.S.P. Coutts

CSIRO Division of Chemical and Wood Technology, Australia

Dr Coutts is principal research scientist at the Division of Forestry and Forest Products, CSIRO, Australia. His research interests since 1964 have included organometallic chemistry, interfacial science, fibre composites and inorganic chemistry. He has been author or co-author of some 60 scientific papers, three reviews and a textbook. Currently he is directing his research interests into asbestos fibre replacement, inorganic coatings for building products and wood fibre reinforced composites. He is a fellow of the RACI, a member of APPITA and is on the editorial board of the *International Journal of Cement Composites and Lightweight Concrete.*

Z. Fördös

Aalborg-Portland, Aalborg, Denmark

Zoltan Fördös is Research Engineer at the Cement and Concrete Laboratory of Aalborg Portland Cement Co., Denmark. He received his MSc degree from the University of Veszprém, Hungary, and his special interests in materials research are silicate chemistry, fibre reinforcement and durability. He is a member of several committees in Denmark, and of Nordic and RILEM Technical Committees.

Hans-Erik Gram

Swedish Cement and Concrete Research Institute, Stockholm, Sweden

Dr Gram received his MSc degree in Civil Engineering at the Royal Institute of Technology, Stockholm, in 1977. In the same year he took up consultancy and research work at the Swedish Cement and Concrete Research Institute. His research work on the durability of natural fibres in concrete led to a DSc degree in 1983. He is now project leader for a research project entitled 'Concrete – New Material Combinations'. He is a member of FIP Commission on Concrete.

R. Pama

Asian Institute of Technology, Bangkok, Thailand

Dr Pama is vice-president for development of the Asian Institute of Technology and associate director of the International Ferrocement Information Center. He obtained his Bachelor of Science in Civil Engineering from the Mapua Institute of Technology, Philippines; his Master of Engineering from the Asian Institute of Technology, Bangkok, and his PhD from the University of St Andrews, Scotland. He was with the teaching staff at St Andrews, before joining the AIT as assistant professor and subsequently professor in the Structural Engineering and Construction Division. He has been actively involved in the organization of conferences, workshops and seminars, and has authored three books and 44 technical papers and edited three volumes of conference proceedings.

S.S. Rehsi

Central Building Research Institute, Roorkee, India

Dr Rehsi is Assistant Director at the Central Building Research Institute, Roorkee (U.P.), India, and is responsible for research and development work on cement, concrete and utilization of industrial wastes at the Institute. He is a member of the Indian Ceramic Society, the Indian Concrete Institute and the Indian Standards Institute Committees. He has published over 100 research papers, and is the author of a book on building materials from Indian fly ashes.

L. Robles-Austriaco

Asian Institute of Technology, Bangkok, Thailand

Mrs Austriaco is senior information scientist, International Ferrocement Information Center, and Editor of the *Journal of Ferrocement*. She obtained a Bachelor of Science in Civil Engineering from the Mapua Institute of Technology, and a Master of Engineering from the Asian Institute of Technology. Mrs Austriaco has most recently served as information scientist, Asian Information Center for Geotechnical Engineering, Asian Institute of Technology, and lecturer in the School of Housing, Building and Planning, Universiti Sains Malaysia, Malaysia. She is a member of the Philippine Institute of Civil Engineers and the American Society for Engineering Education. She has authored numerous publications in ferrocement and fibre reinforced materials.

Asko Sarja

Concrete and Silicate Laboratory, Technical Research Centre of Finland, Espoo, Finland

Asko Sarja received his MScTech degree at Oulu University in 1967 and his DTech degree at Helsinki University of Technology in 1979. For three years subsequently he worked at the Bridge Design Office of the Water Resources and Waterways Administration in Finland. Since 1970 he has been involved in research at the Technical Research Centre of Finland. Since 1978 he has been acting as Director of the Concrete and Silicate Laboratory.

1 Wood fibre reinforced cement composites

R.S.P. COUTTS

Abstract

Many of the recent texts on fibre reinforced cements and concretes make little or no reference to wood fibres as a reinforcement. It will be shown that the fibre cement sheeting, corrugated roofing and fibre pipe industries, which constitute a global and multi-billion-dollar activity, could in the near future rely mainly upon the use of a range of wood fibres as reinforcement.

It is the intention of this chapter to describe in some depth the properties of wood fibres and how they are obtained from trees; the methods of incorporating such fibres into cements and mortars, the properties obtained from these wood fibre reinforced cement (WFRC) composites and their applications as commercial products, especially as the main alternatives to asbestos reinforced cement products.

Although publications concerning the theoretical considerations of WFRC composites are few in number, some discussion of bonding the microstructure will contribute to the understanding of how wood fibres are effective in improving the mechanical and physical properties of cementitious matrices.

1.1 Introduction

It has been stated that roughly 30 M tonnes of asbestos cement products were produced per annum at the start of this decade, with more than half of this production coming from the Free World.[1]

During the 1960s and 1970s, considerable public outcry arose against the use of asbestos-containing materials, because under certain circumstances the users and fabricators of such products could develop various forms of terminal health problems.[2–5]

Despite a declining demand for asbestos, due to legislation against its use in western countries, world production and consumption of asbestos are no lower than they were a decade ago.[1] Developing countries like Zimbabwe and China as well as the USSR are expanding the use of asbestos and taking over the role of Canada as world suppliers of asbestos. It has been projected that world production by the year 2000 could be 7.5 M tonne, whereas total production in 1985 was 4.1 M tonne.

Thus we find the type and rate of asbestos replacement in the world is determined more by economic and political pressures than by considerations of health; for these reasons asbestos may continue to be used for some years in developing nations of the world and in Soviet countries.

In the developed countries of the Free World, replacement of asbestos by alternative fibres is of paramount importance. Although the search for asbestos replacements has been going on for many years, the main surge of activity dates from about the mid-1970s.[6]

1.2 History and development of WFRC composites

Although patents from the last century refer to the use of wood fibre as a component of building materials made from cements and plasters,[7–9] interest in wood fibres as a reinforcement for fibre cement has mainly taken place in the last 10–20 years.

Unfortunately the large fibre cement manufacturing companies are the real custodians of the history of fibre cement development and, from the obvious gap in the literature, they have released very little information about the use of wood fibres in cement.[10]

James Hardie and Coy Pty Ltd started manufacturing asbestos cement products in Australia in 1917.[11] After establishing manufacturing plants around Australia, the company extended production to New Zealand in 1938. International expansion continued to Malaysia in 1964 with the formation in 1966 of United Asbestos Cement Berhad, a joint venture with James Hardie Industries, Turner and Newall, European Eternit companies and 51% Malaysian ownership. In 1970 Indonesian production started with P.T. Harflex Asbes Semen. By 1977 James Hardie Industries had 29 plants in Australia, New Zealand, Indonesia and Malaysia employing 6500 people.

James Hardie and Coy Pty Ltd took an active interest in the use of cellulose, as an economic asbestos substitute, in fibre reinforced cement in the early to mid-1940s. This work was intensified during the post-World War II years when there was a worldwide shortage of asbestos fibre. An investigation was conducted at Camellia, NSW, by Heath and Hackworthy[12] to discover whether paper pulp could be used to replace asbestos completely or partially in asbestos cement sheets. Fibres studied included bagasse, groundwood, wheat straw, cement bags and brown paper. The experimental autoclaved sheets showed brown paper (kraft) was the best of the pulp sources, giving greatest strength to the composite material. However, when asbestos supply was reinstated, this work was discontinued.

Renewed interest in wood fibres began almost inadvertently in 1960.[13,14] In those days, the asbestos fibre board, containing 15% asbestos, was made between steel interleaves. James Hardie's was believed to be the only group in the world which at that time was steam-curing its sheets. To make a cheap board as an alternative interleaf, boards were made up with half the asbestos replaced by wood fibres. It was found, however, that this material was a better product than the material they were selling. This board became the first generation 'Hardiflex', and full production started in 1964. From the 1960s onwards their products have contained no more than 8% asbestos, which was about half the amount the rest of the industry was using.

Attempts to further reduce the asbestos content by adding more wood fibre were unsuccessful due to the ineffectiveness of these fibres, compared to asbestos, in trapping the cement particles during formation of the sheet.

James Hardie entered into collaborative research with CSIRO in 1978[15] to study, among other things the refining of cellulose fibres in an attempt to overcome the difficulties of retaining the cement in the wood fibre reinforced cement sheet. By May 1981 the new generation of asbestos-free cement products – Hardiflex II – was being commercially manufactured. This autoclaved product was totally reinforced by refined kraft wood fibres.[16,17]

In Europe during 1975–77, Cape Industries had marketed boards reinforced with 5% cellulose and high levels of mineral fillers called 'Supalux', 'Monolux' and 'Vermiculux', mainly for fire-resistant use. 'Masterboard' and 'Masterclad' were more dense and stronger and used for external cladding.[18] In 1976 Sweden banned asbestos cement products, and all asbestos cement production was closed down. Other Scandinavian countries were forced to use alternatives, and A/S Norcem in Norway and OY Partek AB in Finland decided to form a joint development under the name of NOPA.[19] The material produced was called 'Cellcem', and contained cellulose fibres mixed with other fibres. Manufacture of the products 'Internit' and 'Pernit' started in Norway in 1977. In 1979 Finland started to produce 'Minerit'.

In 1983 James Hardie Industries and Cape Industries of the UK formed a joint venture, Fibre Cement Technology.[20] The objective was to market the new technology they had developed, to manufacture asbestos-free fibre cement building products, to interested companies throughout the world.

It was stated in 1985 that the UK manufacturers had replaced asbestos in about 50% of the fibre cement sheeting products.[6] James Hardie Industries by this time had totally replaced asbestos fibre from its range of building products, which included flat sheet, corrugated roofing and moulded products, throughout Australia and New Zealand. Part of the Malaysian production by the company was also free of asbestos. The Indonesian interests had been sold in 1986 for financial reasons.

As well as flat sheet products, James Hardie Industries had become a world leader in injection moulded fibre cement products and non-pressure fibre cement pipes, all based on wood fibre as the reinforcement material. The first experimental production of WFRC pipe was undertaken at the Brooklyn factory in September 1980.[13] Commercial production began in Western Australia at the Welshpool factory in July 1984. The last asbestos pipes made by James Hardie were manufactured in March 1987.

At the present time there is considerable activity in the patent literature concerning the use of wood fibres or mixtures of wood fibres with other synthetic fibres. This is taking place throughout Europe and Japan, and companies such as Dansk Eternit-Fabrik A/S, Cape Boards and Panels Ltd, Amfu Ltd., TAC Construction Materials Ltd, Rockwool AB, Turner and Newall Ltd, Matsushita Electrical Works and others are involved. This area of research will experience a literature boom in the next few years.

1.3 The nature of wood

Excellent texts are available which deal with the intricate details of wood growth, wood structure, chemical composition and durability of wood[21,22] as well as the processes of pulping and refining or beating.[23–25]

Cells of various shapes and sizes are present in all plant material, but those plants which are considered to be 'wood-like' always contain a complex organic compound called *lignin*. Lignin is nature's adhesive which in wood bonds the main reinforcing elements, more commonly called fibres, together to form such a unique natural composite. Wood possesses strength, toughness and stiffness sufficient to withstand the forces of nature in the forest, or, more importantly, the strains and stresses imposed on timber (or lumber) when used in construction. An understanding of the way in which wood is structured will help in appreciating the way wood fibres can be obtained from wood and employed as an abundant, renewable, cheap, yet effective reinforcing material in modern composite materials such as paper, hardboard, medium-density fibreboard and plaster or cement fibre boards.

1.3.1 *Wood structure*

Wood cells are formed in the cambium layer, a very narrow layer which separates wood and bark. The first layer of the cell wall is comprised of a random network of *microfibrils*, which are complex chains of *cellulose* molecules, and is called the *primary wall*. Inside this is the *secondary wall*, which may be divided into three separate layers: the outer layer, referred to as S_1; the middle layer, S_2; and the inner layer, S_3. The S_2 layer is usually much thicker than either S_1 or S_3. The microfibrils in the secondary wall occur in ordered patterns (Figure 1.1). It should be noted that in the S_2 layer the microfibrils are aligned at a comparatively small angle to the longitudinal axis of the fibre (this will be discussed in section 1.4 with respect to tensile strength).

The membrane at the outside of these layers and which separates the cells from each other is called the *middle lamella*. When the formation of the cell wall is complete, impregnation with lignin begins in the middle lamella and gradually extends through the cell wall.

The newly formed cell on the inner side of the cambium may become one of four types of wood tissue: *parenchyma cell*, *fibre*, *tracheid* and *vessel*. Each type of tissue serves one or more special functions: the parenchyma cells conduct and store food and water; the fibre's main role is to provide mechanical

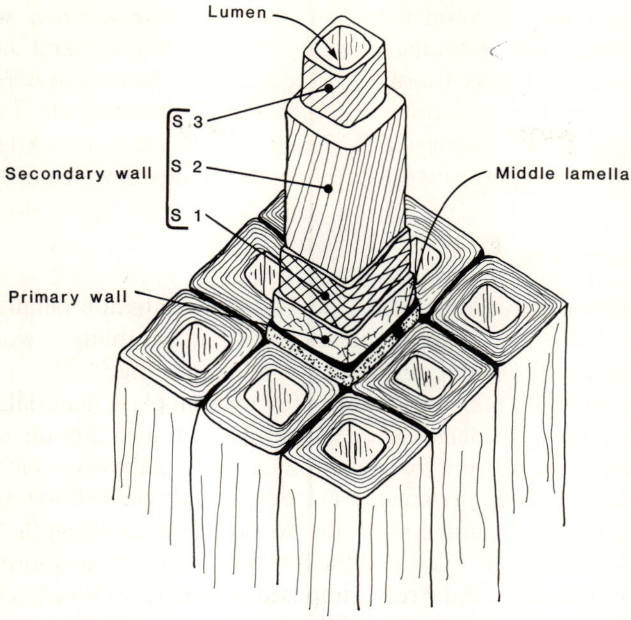

Figure 1.1 The structure of wood.

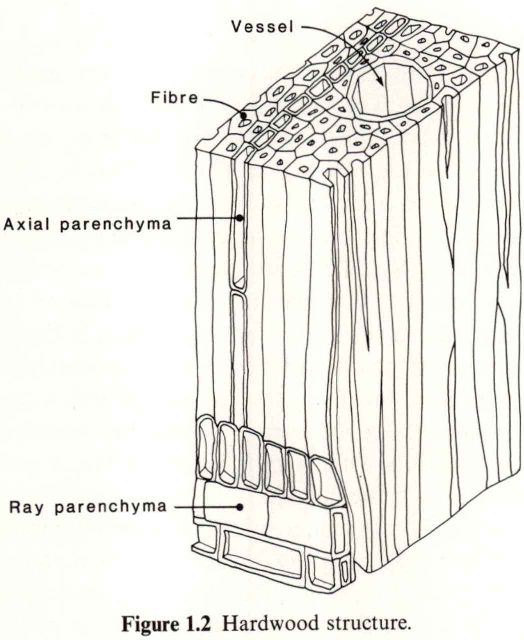

Figure 1.2 Hardwood structure.

support; tracheids and vessels conduct water and dissolved mineral salts from the roots to the leaves as well as providing mechanical support. Vessels (or pores) occur only in hardwoods and are their main distinguishing feature (Figure 1.2).

The names *hardwood* and *softwood* are based mainly on timbers produced in the northern hemisphere. More correctly, the softwoods or coniferous types (pines, firs and spruces) are called *gymnosperms*, and the hardwoods (gums, oaks and ashes) are all classed as *angiosperms*. The hardwood–softwood grouping has little meaning on a world scale, as some hardwoods (such as balsa wood) are extremely soft. Softwoods are much less complex in structure than hardwoods and their cells are of simpler form (Figure 1.3).

Parenchyma cells are rectangular with thin walls. Tracheids are elongated cells with rounded ends and walls, whereas fibres are long pointed cells. Tracheids, which are not common in hardwoods, will be referred to simply as 'fibres'. The properties of 'fibres' will be discussed in section 1.4.

The increase in size of a tree varies with the seasons and climatic conditions. During late autumn and winter, there is little or no growth, but growth is at a maximum in spring. The *spring* or *early wood* usually has wood cells with larger diameter and thinner walls than those cells formed during periods of slow growth–*summer wood* or *late wood* (see Figure 1.3).

As will be seen later in this text, fibres with thick cell walls have different papermaking characteristics compared with thinner-walled fibres. Conseq-

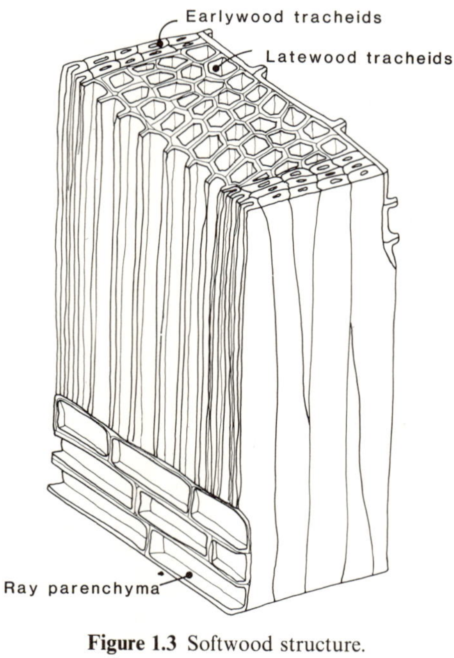

Figure 1.3 Softwood structure.

uently, the ratio of late wood fibres to early wood fibres can have a big influence on the ability of a given wood fibre source to act as a reinforcing element.

1.3.2 *Chemical composition and durability*

The chemical components which constitute most of the mass of wood are cellulose, *hemicelluloses* and lignin. Cellulose is made up of thousands of glucose molecular units joined in long chains, and represents 40–45% of wood. It contains 44.4% carbon, 6.2% hydrogen and 49.4% oxygen, and is relatively unaffected by alkalis and dilute acids.

Lignin occurs in the form of very complex chain polymers of phenolic building blocks consisting of about 65% carbon, 6% hydrogen and 29% oxygen, and is virtually impossible to dissolve without first breaking it down into simpler substances. Lignin comprises between 22 and 30% of wood and varies in chemical composition and amount in different species, being somewhat more abundant in softwoods than in hardwoods.

Hemicelluloses act as a matrix for the cellulose microfibrils, are of relatively low molecular weight and are soluble in alkalis. The hemicelluloses are converted to carbohydrates by the action of acids. They occur as about 20% of the weight of softwoods and between 15 and 30% of the weight of hardwoods.

All the chemical components of wood are formed from the sugars produced in the leaves by photosynthesis. The water obtained from the roots is combined with carbon dioxide from the atmosphere, by complex biochemical processes dependent on chlorophyll which is able to absorb from sunlight the energy needed, to form simple sugars. When these sugars are conveyed from the leaves to the wood cells, they are changed into the much more complex cellulose, hemicelluloses and lignin.

In addition to cellulose, hemicelluloses and lignin, wood also contains a wide range of compounds known as *extractives* which can be extracted by solvents. These materials, usually concentrated in the heartwood, contribute to colour, density, durability, flammability and moisture absorbency, and include polyphenols, oils, fats, gums, waxes, resins and starches.

Extractives may constitute from 5% to as much as 30% of the dry wood mass, varying with species, sapwood or heartwood, rate of growth and season. They can be extracted from wood by a variety of inert solvents, such as water, alcohol, acetone and ether; thus they are not part of the structure of wood.

Over the last three decades, numerous research groups have reported that wood–cement composite boards experience inhibition or retardation to normal hydration and strength development of portland cement. Hardwoods[26] are generally more inhibitory to cement setting than are softwoods.[27] Various workers have reported that hemicellulose, starches, tannins, sugars and certain phenols all affect the setting of cement.[28,29]

Inhibition of the setting of portland cement by finely ground wood has been measured quantitatively by recording the time required for such a mixture to attain a maximum temperature in a Dewar flask.[30] Many researchers have used this technique in systems containing wood chips, wood slivers or sawdust. More recently, commercially available wood pulp fibres were studied, and it was confirmed that those fibres with more extractives present were less desirable as a reinforcement for cement.[31]

1.3.3 *Pulping of wood*

Until about fifty years ago, papermaking, the main use of wood fibre, was based essentially on the use of softwoods because their fibres were longer and easier to felt than those of hardwoods. Australian research workers then developed satisfactory methods for converting the eucalypts of lighter weight into pulp, and such species have provided the basis of the paper industry in Australia.[32,33] Interest in the utilization of short-fibre hardwood pulp has developed worldwide, and Japanese paper manufacturers have been large importers of Australian eucalypts in chip form.

There are a number of ways of producing pulp which will be summarized briefly in the following sections.

1.3.3.1 *Groundwood pulp.* Mechanical separation of the fibres is the least expensive. Billets of wood are pressed against a revolving grindstone immersed in water. The fibres are torn and weakened, but virtually all the wood is converted into pulp, with yields as high as 97%.

1.3.3.2 *Mechanical pulp.* Mechanical pulping in its various forms has been claimed as the pulping method of the future.[34] This method can solve the problems of limited wood supply, of restricted capital, of environmental restrictions and of efficient utilization of mixed forests. In its expansion from groundwood pulping mechanical pulping has advanced into the domain of chemical pulping blurring what was once a clear line of distinction – there is now refiner mechanical pulp, thermomechanical pulp, chemimechanical pulp and chemithermomechanical pulp.

Chemimechanical pulping has the potential to bring pulp and papermaking to areas of the world where it has not been practical before. It can be used in countries where wood supply is not sufficient to supply a chemical pulp mill. It is ideally suited for developing countries where capital is costly and limited, skilled workers are in short supply, markets are small and undemanding, wood species are highly mixed and chemicals are expensive. For a modest investment in the range of A$ 15–25 million, a chemimechanical pulp mill can be constructed to produce about 70 tonnes per day, an economically viable size in most developing countries.

Mechanical pulping is not without its weaknesses. Means must be found to reduce the energy consumption, increase the strength of the pulp and improve light stability. The last factor is important in paper manufacture, but not in the use of the pulp as a reinforcement in cement systems etc.

It has been proven in Australia that softwood (*Pinus radiata*) forests can be used to produce single-furnish thermomechanical pulp for newsprint. While this is still not fully acceptable for paper production in the southern United States, it most certainly should be for mills in North America whose main raw material is spruce.

Mechanical pulps are the pulps of the twenty-first century and they will represent the greatest growth in the paper industry.[34] The objective will be to produce high-yield pulps (> 80%), with high drainage rates and good bonding strength. As kraft pulp prices increase, there will be new opportunities for mechanical pulps to enter new markets, in particular WFRC materials.

1.3.3.3 *Chemical pulps.* Chemical pulping methods dissolve the lignin which binds the fibres together in the wood. The chips are treated with chemicals under conditions of high temperature and pressure and give yields of 45–60% of pulp. The chemical changes that occur are very complex and will not be discussed in this text.

One of the oldest chemical methods is the *sulphite process* which uses an aqueous solution of calcium bisulphite (also sodium, magnesium or ammonium sulphites or bisulphites) as a pulping agent. It is used to produce high-quality bleached paper, but is more limited in the wood species that can be used than with the kraft process. The chips are cooked under pressure in an acidic solution containing the pulping agent and free sulphur dioxide. The lignin is sulphonated and the compounds formed dissolve in the acidic solution.

The *kraft process* is used to produce a major portion of the world's chemical pulp. Sodium sulphide and sodium hydroxide are the active ingredients which attack the lignin. The process operates under alkaline conditions, and penetration of the wood is more rapid. It is applicable to all species and so has a wide industrial appeal, but it creates considerable effluent disposal problems and releases an unpleasant odour into the surrounding district.

In kraft mills, the chemical recovery systems, which include such items as multiple effect evaporators, recovery boilers, causticizing equipment, together with chemical plant and pulping and bleaching equipment, may account for a large part of the total capital cost (of the order of A$300 million for a mill making 500 tonnes per day of bleached pulp).

The alkaline conditions cause less harm to the cellulose than the acidic conditions of the sulphite process, since (as seen in section 1.3) cellulose is more easily hydrolysed by acids than by alkalis. Chips from young trees require less alkali and give better pulp for better fibre properties.

The use of chemical pulps in WFRC products removes the problem of 'extractives' which, as will be discussed, interfere with the cure of autoclaved products resulting in materials of low performance (see section 1.5.4).

1.3.4 *Refining or beating*

No attempt is made to describe in detail the design and operation of the machinery used in these processes, as texts have been published.[23–25]

Refining or beating can be defined as the mechanical treatment of pulp carried out in the presence of water, usually by passing the suspension of pulp fibres through a relatively narrow gap between a revolving rotor and a stationary stator, both of which carry bars or knives aligned more or less across the line of flow of the stock. The term 'beating' is usually applied to a batch treatment of pulp suspension, whereas 'refining' is used when the stock is passed continuously through one or more refiners in series.

It should be pointed out that the refining of chemical pulp does not produce the same effects as it does on mechanical pulp. Chemical pulps (section 1.3.3) are relatively pure cellulose, with the hydroxyl groups accessible. In mechanical pulps the hydroxyl groups are blocked by the presence of lignin. The refining of mechanical pulp is really a completion of the process of disintegration of fibre bundles down to individual fibres.

Changes observed in fibre structure as a result of the mechanical action on the fibrous material can depend on the type of refiner, the refining conditions used, the fibre type (hardwood or softwood) and the pulp (mechanical or chemical). However the main effects which are observed can be classified into four areas:

(i) Internal fibrillation or delamination
(ii) External fibrillation of the fibre surface
(iii) Fines formation
(iv) Fibre shortening.

Internal fibrillation effects (i) are difficult to observe under a microscope, but they can be considered by analogy with a piece of rope. Rope is a helical wrap of strands which themselves are helical wraps of fibres. If one twists a rope in the direction of the helical wrap the rope becomes stiffer; likewise, if the twist is in the opposite direction the rope unwinds (or delaminates) to open up the structure and becomes 'floppy'; such is the case with internal fibrillation. The main effect of internal fibrillation is to increase fibre flexibility and swelling. The fibres may also undergo excessive curling and twisting.

External fibrillation (ii) is easily observed by scanning electron microscopy (Figure 1.4). The fibrils or fibrillar lamellae attached to the fibre surface can vary widely in size and shape (but the process is again like unravelling a piece of rope at its surface).

The last stage (iii) of external fibrillation is the peeling off of the fibrils from the fibre surface with the formation of fines. Depending on the forces acting on the fibre during refining, more or less of the fibrils will be removed from the surface of the fibre.

Fibre shortening (iv) is the other primary effect attributed to refining. An indication that fibre shortening has occurred is the change observed in particle size distribution, which is a result of the cutting action of the blades or discs in the machinery on the single fibres.

Refining plays an important role in producing a large surface area for fibre-

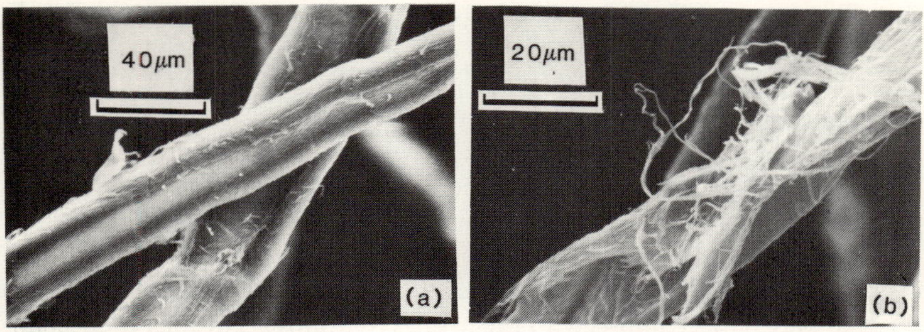

Figure 1.4 (*a*) Unbeaten fibre of *P. radiata*, compared to (*b*) externally fibrillated fibre.

to-fibre or fibre-to-matrix (in the case of composites) bonding and, more importantly, can assist in controlling the drainage rates of processing liquids during the fabrication of products. This is one of wood fibre's main advantages compared to synthetic fibres such as glass, steel, etc., in asbestos replacement.

1.4 Properties of wood fibres

The significance of the properties of the wood fibres will be demonstrated in the next section dealing with wood fibre reinforced cement composites (section 1.5). There we will appreciate the importance of fibre size, morphology, surface charge, etc., in the manufacturing processes of fibre cements. Likewise, the mechanical properties of wood fibres such as tensile strength, fracture toughness, etc., will be shown to provide a unique system when used as a reinforcement with cements or mortars.[15–18]

1.4.1 *Physical properties of wood fibres*

1.4.1.1 *Physical dimensions.* The physical dimensions of fibres are exceedingly important in the application of fibres as reinforcement, be it in paper, medium-density fibre board or in fibre reinforced cement. For this reason, certain dimensions and other properties of a typical softwood (*P. radiata*)[35] and a typical hardwood (*Eucalyptus regnans*)[33] are compared in Table 1.1. Examination of these figures immediately shows the following differences:

(i) Hardwood fibres are much shorter and narrower than softwood fibres.
(ii) Hardwood fibres generally have a higher relative cell wall thickness than do earlywood softwood fibres. This implies hardwood fibres are stiffer and have greater resistance to collapse.
(iii) Hardwood pulps always contain a large amount of thin-walled vessels.
(iv) With softwoods there is a distinct difference between early wood and late wood. The difference is much less noticeable in the case of hardwood fibres (Figures 1.2 and 1.3).

Table 1.1 Some typical fibre properties for softwood[35] and hardwood[33]

Fibre property	*P. radiata*	*E. regnans*
Fibre length (mm)	3.5 (2–6)	1.0 (0.5–2.5)
Fibre width (μm)	—	20 (10–23)
Early wood	45	—
Late wood	13	—
Cell wall thickness (μm)	—	4
Early wood	3	—
Late wood	5	—
Type of cells (%)		
Fibres	>90	65
Vessels	—	18
Parenchyma	<10	17

(v) The number of hardwood fibres per unit mass is always much larger than that of softwood fibres. In one gram of pulp there may be of the order of seven or eight times as many *E. regnans* as *P. radiata* fibres.[36]

1.4.1.2 *Surface potential and surface area.* The Hatschek process is the major manufacturing method for fibre cement products and, as will be seen (section 1.5.2), is very much akin to a papermaking machine. The retention of cement particles by asbestos (or its replacement) and the drainage properties of the sheets during the dewatering stage are paramount importance, hence fibre surface potential and surface area are of considerable significance.

Zeta-potential – the electrokinetic charge on a colloidal particle – has been discussed in many areas of science. When the particle is a wood fibre it has been postulated that the zeta-potential plays a major role in the process of papermaking. The zeta-potential is a controlling parameter in filler and fines retention, drainage and pulp flow behaviour.[37]

Current work in the field of paper science[38,39] shows quite dramatically that zeta-potential is not a measure of surface charge, and cannot be used for comparison of the surface charge of even very similar materials, let alone two materials as distinct as asbestos and wood fibre. In this discussion it suffices to say that in the case of cellulosic fibres, the present evidence indicates that dissociation of ionic groups is the main source of the charge. At very high pH values, dissociation of the hydroxyl groups can contribute to the particle charge, but at low pH values the charge must be due to dissociation of carboxyl groups or sulphonic acid groups, depending on the pulp yield and on whether a sulphite or sulphate pulp is used. A cellulosic fibre dispersion is negatively charged at all pH values. When cationic hydroxylated metal species, such as occur in cement particles, are present, charge reversal may occur, as has been established in the case of wood fibres in the presence of papermaker's alum.[37]

1.4.2 *Mechanical properties of wood fibres*
The use of wood fibres in paper, paperboard, and fibre reinforced composite materials has created a need for better characterization of fibre mechanical properties. Most of the pioneering work in wood fibre testing was carried out by researchers from the paper industry, and much of the data on fibre properties interpreted from tests carried out on paper sheets. The information on single fibres is limited, although new techniques in recent times are providing more data as time passes.[40]

1.4.2.1 *Modulus of elasticity.* Recent studies by Dutch workers[41] have suggested that the upper limit for the modulus of elasticity for the cellulose chain would be 150 GPa. Page and coworkers[42] have shown that the maximum value of modulus of elasticity, for single fibres of softwood, varies from approximately 80 GPa to 18 GPa as the fibril angle varies from 0° to 50°. Due to defects and inhomogeneities within the cell wall structure, the practical average value is much lower. Jayne[43] reports average moduli for softwood fibres ranging from 18 GPa for early wood to 42 GPa for late wood in Douglas fir. The above results were for fibres dried under conditions which restricted twisting and shrinkage. The influence of restraint during drying is even more obvious from the work of Spiegelberg[44] which showed increased elastic moduli when fibres were dried under load (an average fibre modulus of 97 GPa). Hardacker[45] dried fibres on a Teflon-coated plate to eliminate any restraint on drying, and reported moduli of 9 GPa for early wood and 22 GPa for late wood. Thus the elastic moduli of individual fibres separated by chemical means may range for 10–100 GPa depending upon fibril angle, drying restraints and imperfections.

1.4.2.2 *Tensile strength.* Like elastic modulus, the tensile strength of single fibres is dependent on the helical angle of the S_2 layer and the presence of defects. Page and co-workers[45–47] have reported a value of 2000 MPa for defect-free black spruce fibres with a zero fibril angle; this figure should be considered as a maximum value. A more realistic value would be in the range 500–1000 MPa for selected fibre fractions carefully prepared to minimize fibre damage. Jayne[43] found values for tensile strength between 350–1000 MPa.

Considerable data from zero-span tensile testing of paper sheets have been equated to tensile values for single fibres. It would appear from all the available evidence that the likely average tensile strength of fibres commercially prepared from reasonable quality wood by normal methods would be not less than 700 MPa. Strengths of 800–900 MPa might be anticipated if the wood supply is carefully selected to contain a large proportion of late wood. This is not likely to be readily achieved in Australia or New Zealand where the main softwood species is young *P. radiata.*[40]

1.4.2.3 *Fibre flexibility.* The flexibility of a fibre is of great importance during the preparation stage of a composite material and also during the process of composite failure. Certainly wet fibre flexibility has been known to be important for paper manufacture, but the property was seldom measured quantitatively.

Tam Doo and Kerekes[48,49] have recently developed a single-fibre flexibility test method which eliminates the disadvantages of earlier methods. This has enabled them to measure the effect of wood species and pulping conditions on fibre flexibility and confirm quantitatively what has long been known in a qualitative way. They showed that cedar fibres are flexible ($1.33 \times 10^{-2}\,\mathrm{N\,m^{-2}}$), Douglas fir fibres are stiff ($6.8 \times 10^{-2}\,\mathrm{N\,m^{-2}}$) and Southern pine pulp fibres are very stiff ($11.7 \times 10^{-2}\,\mathrm{N\,m^{-2}}$). It was also shown that mechanical pulps are 20–30 times stiffer than chemical pulps from the same species. By chemically treating mechanical pulps, a marked decrease in stiffness can be noted.

1.4.2.4 *Work of fracture.* The energy needed to form a fracture surface is called work of fracture, and is related to fracture toughness by consideration of the area of the surface formed during fracture.

Gordon and Jeronimidis,[50] during their research on the work of fracture of several species of wood, found that weight for weight, the strength and stiffness of timber along the grain compares well with the best engineering metals, as does its work of fracture across the grain. To explain their findings they searched for a novel energy-absorbing mechanism in the structure of wood.

Earlier work by Page *et al.*[46] showed that when single wood fibres were tested in tension, some of them showed a pseudo-plastic behaviour which curiously resembled that of a ductile metal, even to the existence of a distinct yield point on the load-extension curve. Furthermore, although cellulose fibrils seldom show breaking strains more than 2 or 3%, some of the fibres showed elongations as high as 20%. These workers explained the properties in terms of the structure of wood fibres (see section 1.3.1). Wood fibres are hollow tubes composed of layers of fibrils wound in a steep spiral, and so behave as spirally wound fibre reinforced composite tubes. Under axial tensile strain, it has been demonstrated that such structures can fail by buckling due to induced shear stresses. This failure mechanism can be shown to produce composite materials with work of fracture comparable to those of ductile metals, i.e. 40 $\mathrm{kJ/m^2}$.

1.4.3 *The effect of moisture on wood fibres*

We have seen that cellulose is the primary component of the cell wall and the crystalline microfibrils are the elements which give the fibre its tensile strength; however, there are disordered zones which are believed to play a significant role with respect to mechanical properties.

Hemicellulose and lignin act as matrix materials in wood fibres and are generally considered to be amorphous. They are hygroscopic thermoplastic substances, and so environmental conditions such as humidity and temperature have a strong influence on them and hence the mechanical properties of wood fibres. With increasing moisture content, the torsional stiffness of fibres decreased by about 50% (between 25% and 90% RH at room temperature).[51] This softening is related to the softening of the hemicelluloses in the cell wall. In the longitudinal direction this effect (up to 90% RH) results in a drop in modulus of only 11%, although the hemicellulose modulus is reduced by a factor of a thousand. This is clearly a reflection of the fact that the stiff cellulosic microfibrils are preferentially aligned along the fibre.

In many cases, a maximum in fibre tensile strength and tensile modulus is found at low moisture contents. When relatively dry, fibres show low tensile properties due to a poorer stress distribution within their structure. Such an effect is not restricted to wood fibres but also occurs in synthetic fibres.

When single fibres are immersed in water, the relative longitudinal stiffness drops considerably i.e. by about 70–80% from the value at 50% RH. This has been explained as being due to softening of the disordered zones of the cellulose microfibrils and also to a reduction in cell wall cohesiveness leading to a slippage between fibrils. The relative decrease in modulus of pulp fibres, when immersed in water, is almost independent of yield.

1.5 Wood fibre reinforced cement composites

The amount of data available on WFRC products, in the scientific literature, has been limited until recent times due to the fact that manufacturing interests have been responsible for much of the preliminary work and for commercial reasons have retained the knowledge in-house, or locked away in the patent literature.

Ideally one should commence by defining the system and then using the theoretical principles of fibre reinforcement to predict the various properties. However, it has been found difficult to relate the current theories of reinforcement to the wood fibre reinforced cement systems under examination (section 1.5.1).

By recording the mechanical (section 1.5.4) and physical (section 1.5.5) properties of WFRC composites and examining their microstructural behaviour when loaded to failure (section 1.5.6), considerable confidence can be obtained to justify the commercial utilization of such building materials (section 1.6) without a complete theoretical understanding of the working system.

Concrete, mortars and cements are attractive for use as materials of construction in that they are cheap, durable and have acceptable compressive strength and stiffness. Unfortunately they possess inferior tensile strength and are brittle. The inclusion of fibre reinforcement can improve the tensile strength (or more importantly the flexural strength) and fracture toughness, while modifying properties of elastic modulus, density, hygroexpansivity and durability.

Much has been written about the enhancement of engineering properties of cement-based products by the inclusion of fibres as reinforcement. Excellent texts and reviews[52–54] on the basic concepts of fibre reinforcement of brittle matrices are available, which deal with synthetic fibres such as steel, polypropylene, glass and more recently Kevlar, carbon and high modulus polyethylene as well as a host of new synthetic fibres which are currently being developed.

It is interesting to note that the use of wood fibres predates the use of all of the above-named fibres as reinforcement for cement matrices.[7–9] Not only does wood fibre predate them in use, but as a viable commercial alternative to asbestos fibre it far outranks them in performance.[6,55] In many cases wood fibres have been mixed with other man-made fibres to enable such synthetic fibres to be used on conventional fibre cement manufacturing equipment.[56,57]

A multitude of theories have been derived based on the reinforcement of cement and concrete by fibres[52–54] such as glass, steel, etc. Unfortunately there has been a tendency to extrapolate these theories into areas of natural fibres such as asbestos, bamboo, wood fibre and an immense range of different vegetable fibres. At times consideration has been given to certain deviations from ideal behaviour, but the number of assumptions needed to deal with such fibre systems sheds considerable doubt on the validity of using theoretical equations derived for systems containing synthetic fibres which have constant fibre length, fibre diameter, chemical composition, density, fibre shape and fibre surface area.

Although doubt has been cast on the acceptance of current theory for the WFRC system, there is still much to be gained by considering certain basic findings of the theories of fibre reinforcement with respect to WFRC materials.

The difficulty of examining WFRC systems arises from the fact that wood fibre properties vary greatly from species to species, from tree to tree within a species (due to location, climate, soil conditions etc.) and even within the same tree (late-wood and early-wood fibres, section 1.3). Pulping methods and refining can yield reinforcing fibres from the same source with widely varying properties. One must also consider the effects of matrix formulation, curing conditions and test conditions (section 1.5.6.1) when comparing studies of similar wood fibre reinforcement.

In an attempt to cover completely the available literature of WFRC materials, the theoretical studies that have been carried out on the WFRC system will be considered.

1.5.1 *Theoretical principles of fibre reinforcement*

If the maximum benefit of composite materials as engineering materials is to be achieved, it is necessary to understand their potential for bearing loads. Linear elastic fracture mechanics (LEFM), the discipline concerned with failure by crack initiation and propagation, is often used to assess the suitability of conventional engineering materials such as metals and concrete in load-bearing situations. Its use in explaining composite behaviour is

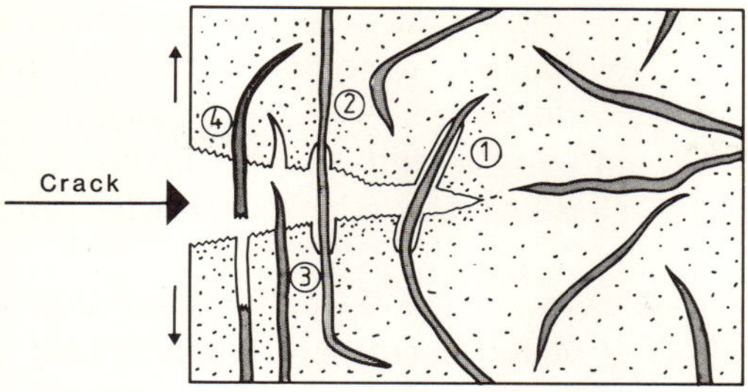

Figure 1.5 Schematic representation of crack travelling through a fibre reinforced matrix.

debatable.[58,59] Nevertheless, the use of fracture mechanics in fibre reinforced cements and concretes has received considerable attention.

Failure in a fibre composite emanates from defects in the material. These may be broken fibres, flaws in the matrix and debonded fibre/matrix interfaces. Figure 1.5 shows a schematic representation of a cross-section through a fibre reinforced matrix. The diagram shows several possible local failure events occurring before fracture of the composite. At some distance ahead of the crack, which has started to travel through the section, the fibres are intact. In the high-stress region near the crack tip, fibres may debond from the matrix (e.g. fibre 1). This rupture of chemical bonds at the interface uses up energy from the stressed system. Sufficient stress may be transferred to a fibre (e.g. fibre 2) to enable the fibre to be ultimately fractured (as in fibre 4). When total debonding occurs, the strain energy in the debonded length of the fibre is lost to the material and is dissipated as heat. A totally debonded fibre can then be pulled out from the matrix and considerable energy lost from the system in the form of frictional energy (e.g. fibre 3). It is also possible for a fibre to be left intact as the crack propagates. This process is called crack bridging.

From this simplistic approach we are immediately made aware of the importance of fibre to matrix bond strength, frictional stress opposing pull out, tensile strength of the fibre, fibre length and fibre content.

1.5.1.1 *Strength and modulus.* Fibre reinforced materials are often analysed by the 'rule of mixtures' for calculating strength and modulus of composites.

The rule for composite strength (σ_c) can be stated as:[52]

$$\sigma_c = \sigma_m v_m + \eta_1 \eta_2 \sigma_{fu} v_f \qquad (1)$$

where v_f and v_m are the volume fractions of fibre and matrix, respectively, σ_{fu} is the tensile strength of the fibre, σ_m is the stress on the matrix when the fibres fail, and η_1 and η_2 are factors allowing for fibre orientation and fibre length respectively. There is a need for critical fibre volume ($v_{f\,crit}$) to be exceeded, i.e. $v_f > v_{f\,crit}$, for equation (1) to apply.

When the matrix cracks, the composite may carry a lesser or greater load than the matrix, depending on fibre content. Figure 1.6 shows load/deflection

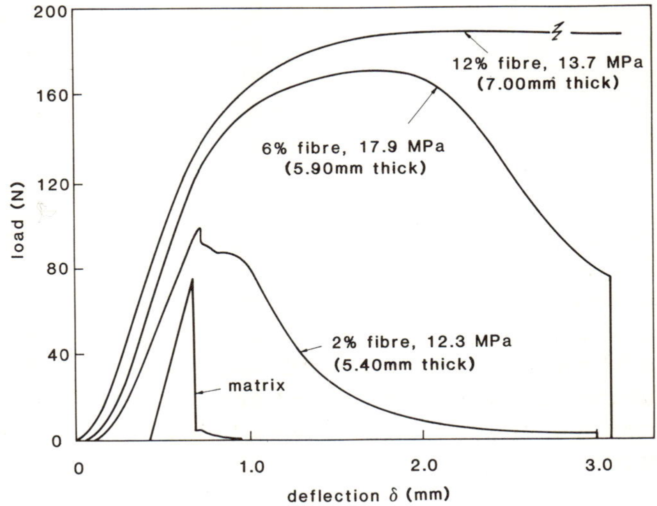

Figure 1.6 Load–deflection curves for a range of WFRC composites tested in flexure.

curves for a range of fibre composites tested in flexure. The 'critical volume' of fibre needed to exceed matrix strength is much lower in the flexural case than for tensile strength tests, but it does indicate the ability of fibres to carry the load after the matrix has cracked. For economic reasons the lowest value of $v_{f\ crit}$ is desired. Theoretical calculations often show that very low values of v_f are needed to achieve strengthening; however, factors such as poor bonding or fibre orientation make higher values necessary for practical applications. When the critical volume of fibre is exceeded it is possible to achieve multiple cracking, and the brittle matrix can respond as a pseudo-ductile material (see Figure 1.6).

For elastic modulus E_c of the composite, the rule is written as:

$$E_c = E_m v_m + \eta_1 \eta_2 E_f v_f \qquad (2)$$

where E_c, E_m and E_f are the moduli of elasticity for composite, matrix and fibre respectively.

The rule of mixtures assumes that fibres are aligned in the direction of stress, there is no effect from Poisson's ratio, and before cracking of the composite, fibres are fully bonded to matrix, i.e. equal strains in fibre and matrix (however, cement matrix has a strain at failure between 50 and 500×10^{-6}, whereas wood fibre has a strain at failure $> 20\,000 \times 10^{-6}$)

Andonian *et al.*[60] attempted the first theoretical analysis of the mechanical properties of WFRC composites using the rule of mixtures for random fibre-reinforced composites, taking into consideration only fibre pull-out. The calculations gave close predictions (of tensile and flexural strength and elastic moduli) for the experimental results they obtained.

Laws[61] questions the use of the rule of mixtures for predicting the tensile strength of fibre reinforced brittle matrices; the matrix is brittle and fails at a strain very much lower than the failure strain of the fibres, and it is widely accepted that the tensile strength should depend on the fibre contribution alone and is given simply by

$$\sigma_c = \eta_1 \eta_2 \sigma_{fu} v_f \tag{3}$$

Swift and Smith[62] have used this equation to explain flexural strength for sisal reinforced cement.

The rule of mixtures assumes that fibres are aligned in the direction of the stress, and although η_1 allows for various orientations to compensate for a degree of randomness, it cannot be determined as a precise value. Likewise, η_2 is determined by the length of fibre in relation to the critical length (l_c). The critical length is defined as twice the length of fibre embedment which will cause fibre failure during pullout.

$$\eta_2 = \frac{l}{2l_c} \quad \text{when } l < l_c$$

or

$$\eta_2 = l - \frac{l_c}{2l} \quad \text{when } l > l_c$$

Again, in practical terms the precise value of η_2 is rarely known, because the determination of l_c with any accuracy is very difficult.

The critical fibre length as stated above can be shown to equate to equation (4) if the shear stress (τ_s) developed at the interface is uniform.

$$l_c = \frac{\sigma_{fu} \cdot r}{\tau_s} \text{(where } r \text{ is fibre radius)} \tag{4}$$

Andonian *et al.*[60] calculated the critical length of *P. radiata* fibres in WFRC composites to be between 18 and 23 mm, and as the measured length of the fibre is about 3.5 mm, fibre fracture was not possible. Coutts and co-workers[63,64] had observed fibre fracture during WFRC composite failure and concluded that the 'apparent critical fibre length' must be less than 3.5 mm.

The conflicting reports on the predominance of fracture or pull-out of wood fibres from a cement matrix prompted Morrissey *et al.*[65] To study a model system consisting of sisal slivers embedded in cement and protruding from one end of the cement matrix. About 200 such samples were tested under tension. It was found that the slivers did not behave in the manner predicted for uniform cylindrical fibres. After a break of the elastic bond between the fibre and the matrix, the fibre started to pull out. The force resisting pullout was not proportional to the length of embedment but was dominated by the highest local resistances, present due to fibre shape. As pullout proceeded, anchor spots developed and the force required rose or fell in an apparently random manner.

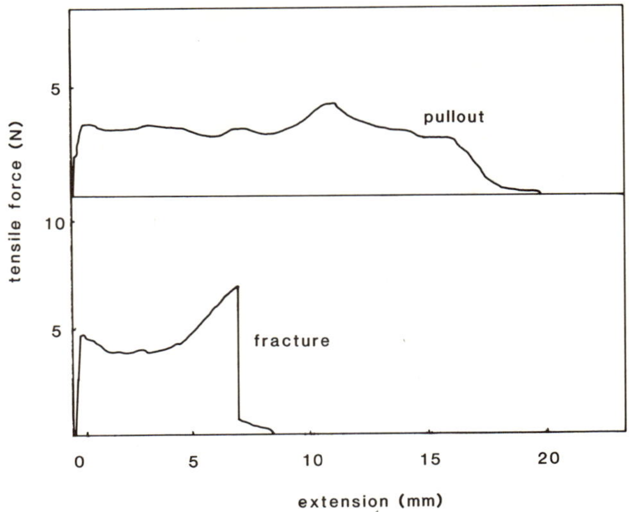

Figure 1.7 Tensile load–extension curves for different failure modes of sisal slivers embedded in cement.

When the anchorage was too strong to be dislodged by the maximum force the fibre could carry, tensile fracture of the fibre occurred (Figure 1.7).

There was a '*critical length*' of embedment, which for the sisal slivers was approximately 30 mm. When the embedment was shorter, fibres tended to be pulled out, (Figure 1.7) and when it was longer they tended to break. This 'critical length' is not the length for which a uniformly distributed frictional stress reaches its critical value under the maximum sustainable tensile load, as is commonly assumed (equation 4), but it corresponds to the length around which the probability of a local strong anchorage becomes high.

The 'critical length' of 30 mm for sisal slivers corresponds to an aspect ratio of 110 ± 50, which is comparable with the aspect ratio of *P. radiata* fibres. This would support the observation that *P. radiata* woodpulp fibre cement composites frequently experience fibre fracture during failure.

Numerous studies of steel fibre and glass fibre reinforced cements and concretes have indicated that the major factors affecting flexural strength are the fibre volume and aspect ratio of the fibre. Higher values of either lead to higher values of flexural strength.[52]

Without considering the theoretical equations leading to the above conclusions, it is interesting to note that in cellulose fibre reinforced cement composites an increase of fibre volume (not surprisingly) leads to increased flexural strength up to about 8% by mass of fibre, at which stage efficient packing of the fibres becomes difficult and strength starts to drop (Table 1.2).

With respect to the aspect ratio of the fibres it has been noted by Coutts and Warden[66] that air-cured WFRC samples reinforced with softwood fibres

Table 1.2 Effect of aspect ratio and fibre content

Fibre content	Flexural strength (MPa)				Fracture toughness (kJ/m^2)			
	P. radiata		NZ flax		*P. radiata*		NZ flax	
(% by mass)	RH†	Wet	RH	Wet	RH	Wet	RH	Wet
2*	15.6	11.5	17.4	12.8	0.31	0.37	0.19	0.24
4*	16.6	11.7	20.5	16.9	0.57	0.88	0.34	0.71
6*	19.9	12.4	21.4	16.6	1.15	1.81	0.48	0.94
8*	23.1	14.8	23.2	18.0	1.86	3.15	0.84	1.74
10*	21.3	12.9	23.4	17.3	1.92	2.97	1.19	2.15
12*	21.7	10.5	17.7	—	2.09	2.88	2.59	—
	P. radiata		*E. regnans*		*P. radiata*		*E. regnans*	
2**	14.9	10.1	10.6	8.6	0.41	0.64	0.25	0.33
4**	20.2	11.9	14.2	10.5	0.64	1.52	0.51	1.00
6**	25.1	13.3	20.9	10.4	1.40	3.72	1.06	1.61
8**	30.3	14.8	20.3	8.4	1.93	4.51	1.37	1.49
10**	29.2	13.1	20.1	9.6	2.28	4.60	1.46	1.83
12**	27.6	10.4	20.6	9.3	2.25	3.60	1.68	1.79

*Autoclaved mortar $^\dagger 50 \pm 5\%$ RH,
**Air-cured cement 22 ± 2°C

($l/d = 80$–100) when compared to similar samples reinforced with hardwood fibres[67] ($l/d = 50$–60) displayed higher flexural strength at the same fibre content by mass (see Table 1.2). Conflicting with this observation is the fact that autoclaved samples of mortars reinforced with softwood fibres[68] and New Zealand flax[69] ($l/d = 200$) show very similar flexural strength, but, more importantly, lower fracture toughness (see next section). The reason that NZ flax composites do not produce higher strength properties[69] is related to the fact that most of the fibres (tensile strength approximately 300 MPa) are broken, and so σ_{fu} is a limiting factor and not l/d. When tested wet, the longer NZ flax fibres produce stronger samples than the short *P. radiata* fibres, due to the fact that more of the long fibres can be loaded up to failure. In keeping with this, we find the very short *E. regnans* fibre reinforced materials are weak when tested wet or at RH test conditions ($50 \pm 5\%$ relative humidity, 22 ± 2°C, because the short fibres cannot be loaded to failure. In the above argument it is proposed that the interfacial bond is constant, as all fibres were prepared by the kraft pulping method.

1.5.1.2 *Fracture toughness.* We have seen in Figure 1.6 that the post-cracking ductility imparted to the composite by fibre addition can be considerable. Gordon states[70]: 'The worst sin in an engineering material is not lack of strength or lack of stiffness, desirable as these properties are, but lack of toughness, that is to say, lack of resistance to the propagation of cracks'. The origin of fracture toughness in WFRC composites was claimed to come mainly from fibre pullout (85% to 90%),[60] although later studies by these same workers[71] considered fibre fracture must be of importance in agreement with other researchers in the field.[63,64,73]

Toughness measurements can be conducted in several ways;[74] impact testers such as Charpy or Izod which involve stored energy in a pendulum, calculating the area under stress–strain curves, and the use of fracture mechanics involving stress intensity or similar parameters, or more practical tests such as dropped balls or weights, etc. Some of these techniques are more suited to particular composites, but all have limitations which render elusive the well-defined material properties useful to engineers and material scientists.

Mindess and Bentur[75] studied the fracture of WFRC products and found that saturated samples were weaker and more compliant than air-dry specimens. It was found that the wet samples were completely notch-insensitive, while the air-dry specimens may be slightly notch-sensitive. Notch sensitivity is a requirement for the application of linear elastic fracture mechanics to cement composites, and so these workers concluded that LEFM could not be used for WFRC materials.

Mai and Hakeem[76,77] have studied the slow crack growth of WFRC composites for both dry and wet materials, using a double-cantilever beam system. The results were analysed using K-solutions, and compliance measurements within the framework of LEFM. It was concluded that LEFM concepts can be used for WFRC products.

The use of LEFM requires a linear elastic homogeneous matrix. The introduction of fibres into a matrix, to achieve ductility, must by their very nature result in a heterogeneous material with a non-linear stress–strain curve after matrix cracking.

The fracture toughness results given in Table 1.2 for a range of WFRC composites have all been obtained by the method of measuring the area bounded by the load/deflection curve. As the materials are non-linear and non-elastic, this approach is useful in that the total recorded energy will include the contributions of the work of fracture of the matrix, the debonding and frictional slipping of the reinforcement, and any strain energy released by fibre fracture.

For fibre cements and concretes the matrix work of fracture is virtually that of the unreinforced cement or concrete and is less than 50 J/m^2. Therefore it is assumed that any improvement in composite toughness will depend on whether the fibres bridging the crack are able to support the load previously carried by the matrix, and on whether the fibres break or pull out of the matrix.

Fibre pullout is the most common mode of failure for steel and glass fibre cements and concretes. This is because, to incorporate sufficient fibre into the formulations without fibre tangling, the fibre aspect ratio cannot be high enough to exceed the critical fibre length.

It has been recorded that wood fibres can be loaded into the matrix material with fibre volumes in excess of $v_{\mathrm{f\ crit}}$ (Figure 1.6) and can be fractured during failure, hence the 'apparent critical length' is exceeded. This behaviour does result in relatively high levels of fracture toughness (Table 1.2) but does not lend itself to a conventional form of theoretical analysis. The explanation of such properties is discussed in terms of fracture mechanics and bonding in sections 1.5.4.2 and 1.5.6.2.

Recently, Hughes and Hannant[78] have studied the reinforcement of Griffith flaws in WFRC in an attempt to explain their behaviour, and concluded that stresses developed are significantly higher than would be expected using the rule of mixtures theory.

It is hoped that as commercial production of WFRC becomes a global activity, more interest will be shown in the theoretical explanation of how this system works and hence in developing ways to optimize its use in design.

1.5.2 *Fabrication processes*

The production of asbestos fibre cement products is a mature industry – it has been suggested[55] that over 1100 sheet machines and 500 pipe machines are in production around the globe, with total capital investment in excess of A$4.6 billion.

The Hatschek process (or wet process) is the most widely used method of production. The process was discovered at the beginning of this century and remains basically the same today (Figure 1.8). The manufacturing techniques are closely related to conventional heavy paper- and boardmaking processes. An aqueous slurry of asbestos and cement, about 7–10% solids, is supplied to a holding tank which contains a number of rotating screen cylinders. The cylinders pick up the solid matter, removing some of the water in the operation. An endless felt band travels over the top surface of the cylinders, and picks up a thin layer of the formulation from each cylinder. The laminated ply then travels over a vacuum dewatering zone which removes most of the remaining water. The dewatered formulation is then wound up on a steel calender or assimilation roll until a product of desired thickness is formed. The material is further compressed by pressure rolls, which are in contact with the assimilation roll.

For sheet production, the layer built up on the assimilation roll is automatically cut off, and drops on to a conveyor to be transferred to a stack for curing. If corrugated roofing is to be made, the flat sheet is taken off to a corrugating station where the sheets are deposited on oiled steel moulds for shaping.

Pipe machines are similar to the Hatschek process but usually have one or two vats in series, and the pressure imposed on the mandrel by the press rolls is much greater than for sheet production, so as to form a dense product. The machine may stop while the mandrel carrying the pipe is set to one side for pipe withdrawal. The process is often referred to as the Mazza process.

The Magnani (or semi-dry) process can be used to prepare pipes and corrugated sheet. This process has the advantage that it can provide a greater

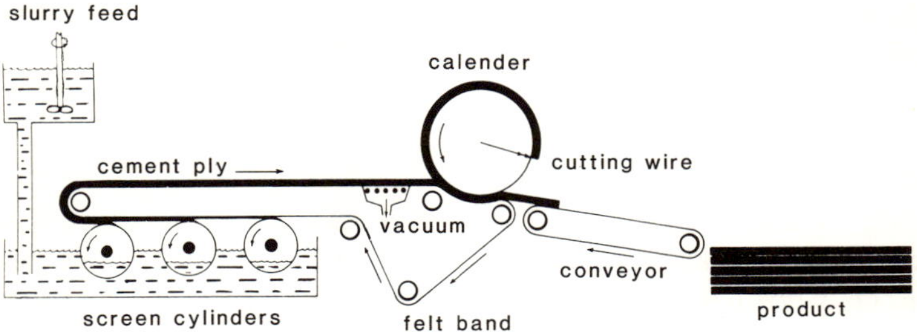

Figure 1.8 The Hatschek process.

thickness of material at the peaks and troughs of the corrugations, and so increase the bending strength. This profile also enables overlapped sheets to fit more closely together in use. The thick slurry (about 50% solid) of this process flows uniformly and directly on to a felt conveyor which passes over numerous vacuum boxes to dewater the formulation. In the case of corrugated roofing the felt is compressed over a corrugated former by a shaped roller. Pipe formation is similar to the Mazza process.

Injection moulding is now tending to replace the hand moulding of green sheets (from the Hatschek process) for the manufacture of special fittings for use with corrugated sheets. A slurry of 40–50% solids is pumped into a permeable mould and then subjected to pressures in excess of 20 atmospheres in a hydraulic press, via a rubber diaphragm. The mix is dewatered by this process of pressure filtration, and then has sufficient green strength for the product to be demoulded by means of a suction lifting pad, and transferred to a pallet for curing. The operation is very fast.

From the brief descriptions of the main manufacturing processes, certain facts become evident when an alternative fibre is sought to replace asbestos for use in these mechines. The replacement fibre must be such that the drainage properties of the mix are not altered, and the fibre must be able to be uniformly dispersed through the matrix.

The formulation of the matrix, and hence the cure of the product, has varied from country to country and even between companies within those countries. The autoclave curing process has always been favoured in Australia and the USA, and in some factories in Europe for the manufacture of pressure pipes.

In the autoclave process, the matrix is usually a mixture of ordinary portland cement (opc) and fine-ground sand (silica). The product, after an initial pre-cure period in air, is cured in an autoclave in a steam environment, say 8h at 170–180°C. The cured sheets are at full strength after autoclaving and can be dispatched from the factory in a short time.

By contrast the more traditional air-cured products require 14–28 days of air-curing before they can be dispatched, the time delay involving considerable stock inventory. The process is lower in capital outlay, as no high-pressure autoclaves and steam raising plant are required; however, cement is more expensive than silica, and material costs are therefore higher.

1.5.3 *Asbestos replacement*

Asbestos is a general name used to describe several varieties of naturally occurring crystalline fibrous silicate minerals which, possessing unique physical and chemical properties, have over the years contributed greatly to human safety through their use in braking systems and in various fields of fibre protection. The fibres which constitute the minerals are, however, both brittle and very fine and, if not bound tightly together by a 'matrix' such as cement or resin, may readily become airborne. Under these circumstances, the physical form and chemical inertness of the fibres combine to create a major health hazard for workers in industry or others similarly exposed.

Medical research has shown that exposure to asbestos fibres can cause three major diseases.[79] These include increased susceptibility to lung cancer, particularly amongst those who smoke – it has been suggested that the risk for a smoker may be more than fourteen times that for a non-smoker under similar exposure. Asbestosis, the most common disease, is basically the progressive formation of scar tissue on the lung; premature death is the common result. Mesothelioma is a relatively rare form of cancer of the outer surface of the lung or of the intestinal lining; it is always fatal.

During the late 1960s and through the 1970s, the press of the world[1–5] highlighted the threat of asbestos to the population at large, and more stringent standards were adopted concerning the levels of dust containing asbestos.

Over the last two decades, much research has been conducted into the possible replacement of asbestos with various synthetic fibres such as glass, carbon, Kevlar, nylon and polypropylene. However, little published material is available in the open scientific literature concerning the use of natural fibres as a reinforcement, despite the fact that, by comparison with many synthetic organic fibres, wood fibres possess relatively high strength and modulus (stiffness), and have already been used in cement formulations during times of shortage of asbestos fibres.

1.5.3.1 *Asbestos fibre properties.* We are really concerned only with those properties of asbestos fibre that are applicable to fibre cement construction products. We have seen in section 1.5.2 that the preparation of commercial products involves high capital cost equipment, and so any substitute fibre should, ideally, be processable on existing equipment with minimum modification.

Three forms of asbestos are significant in the fibre cement products industry; chrysotile (white asbestos), amosite (grey-brown asbestos) and crocidolite (blue asbestos). The properties of these forms are given in Table 1.3. Chrysotile is the most commonly used asbestos mineral, accounting for about 95% of

Table 1.3 Physical properties of some varieties of asbestos

Property	Chrysotile	Amosite	Crocidolite	Reference
Tensile strength (MPa) (average)	3100 MPa	2500 MPa	3500 MPa	80
Young's modulus	162 GPa	162 GPa	190 GPa	80
Flexibility	Good	Fair	Good	80
Specific gravity	2.6	3.4	3.4	80
Fibre diameter (μm)	0.03–100	—	0.5	81
Fibre length (mm)	1–80	—	<5	81
Surface area	10 m^2/g	—	—	81

total annual production. It is characterized by long, flexible fibres and is found in Canada and the USSR. Amosite has superior heat resistance and is used principally in the manufacture of insulating boards, although minor amounts are used by fibre cement products manufacturers as it has surface charge properties which assist with drainage properties in the production process.

It may be seen from Table 1.3 that the reported tensile strengths of asbestos fibres are very high. Such measurements of strength are usually obtained from tests conducted under rigorous laboratory conditions. A more realistic value for asbestos fibres used for industrial applications would be approximately 700 MPa.

While tensile strength is the most important property of a fibre in its application as a reinforcement, other factors are more important in actually forming the composite material. For instance, the surface area of asbestos is of paramount importance in fibre cement production. The fibre can be opened up or fiberized (akin to the process of refining or beating used with wood fibres, see section 1.3.4).

Associated with surface area is the property of surface charge. Chrysotile has at its surface a predominant layer of hydroxyl sites connected to an adjacent inner layer of magnesium sites. These alkaline sites would be expected to exert a strong surface activity, either by partial solution or by attraction of ions of opposite charge. In contrast, amosite has external bonds of silica sites which are recognized as having a weakly acidic nature. Such sites, being highly insoluble,would be expected to exert a weak surface charge solely by attraction of suitable ions.

The surface charge of asbestos fibres reflects the sedimentation behaviour of asbestos in aqueous suspension.[80] Measured sedimentation volumes are important test parameters for asbestos, as the behaviour of fibres in the asbestos–cement–water system is of considerable practical importance. Chrysotiles, having typically small and dense sedimentation volumes, react with cement to form sediments which possess poor permeability of water, while the amosite fibres under similar conditions form open networks of fibres, giving a relatively high porosity. Thus the drainage rates of amosite–cement–

water systems are some three to four times better than those based on chrysotile, a fact which has bearing on the speed of production of most types of asbestos-cement manufacturing machinery. It is therefore a widespread practice to blend amosite and chrysotile fibres in the manufacture of asbestos-cement products to avoid drainage problems.

1.5.3.2 *Property requirements for asbestos alternatives.* With over four million tonnes of asbestos being consumed each year and with more than two-thirds being used by the fibre cement indsutry, any replacement fibre will have to be available on a massive scale. The important aspects of fibre form and surface characteristics have been mentioned in the discussion on asbestos properties, and any strong similarities with alternative fibres should contribute to compatibility with cement. The billions of dollars of capital investment in sheet machines and pipe machines necessitate the use of existing equipment with the least amount of modification. One of the big disadvantages of asbestos-cement sheeting is its low impact strength. Hence the use of a fibre which may improve this aspect of fibre cement behaviour would be a positive gain. In any new product on the market, the alternative fibres must be seen to be durable in the matrix, with long-term stability to weathering. With a large amount of the market dependent on autoclaved products, due consideration must be given to the behaviour of the fibre under such conditions of manufacture.

Wood fibres have been included in asbestos cement products since the early 1960s in Australian manufacturing procedures. This action was taken to increase the fracture toughness of the product for better transportation and for fabrication properties. Around the globe, other manufacturers have also used hybrid fibre formulations to improve the performance and profitability of their products. The patent literature contains many references to the use of glass, other synthetic inorganic, natural organic and synthetic organic fibres as additional fibre reinforcement to asbestos in a range of fibre cement products. Such fibres have been added to aid the drainage problem associated with short chrysotiles or to improve the impact strength of the product.

The fibres studied possessed advantages and disadvantages with respect to processing or final composite performance. An excellent coverage of the type of fibres studied has been provided by Hodgson[55] and Harper[18]. Some reference to other fibres will be necessary to place the wood fibre work in perspective with the total effort made towards asbestos replacement in fibre cement composites.

We have seen from the description of the common manufacturing processes (section 1.5.2) and the properties of asbestos fibres that a number of prerequisites exist for any successful replacement fibre. These can be listed under a number of headings, as described in Table 1.4.

In Australia and New Zealand, manufacturers are committed to the autoclave process[15] and so the combination of cement's high alkalinity

Table 1.4 Comparison of properties of fibres for possible asbestos alternatives

Fibre	Alkali resistance	Temperature resistance	Processing ability	Strength contribution	Toughness contribution	Cost
Wood pulp (chemical)	1	1	1	1	1	3
Wood pulp (mechanical)	1	1	2	2	3	3
Polypropylene	1	3	3	3	3	2
Polyester	3	3	3	3	3	2
Kevlar	1	1	2	1	1	1
Steel	1	1	3	3	3	2
Glass	3	1	3	3	3	2
Mineral fibre	3	1	3	3	3	3
Carbon	1	1	3	1	1	1

1. High
2. Medium
3. Low

(approximately pH 12) and high temperature (> 170°C) in the processing of the product eliminates most fibres apart from wood pulp, carbon and Kevlar. The cost of the carbon or Kevlar fibres is almost a factor of 10 greater than that of the wood pulp. Kraft wood fibre *P. radiata* costs approximately A$700 per tonne at the time of writing, while the cost of Kevlar and high modulus carbon is > A$7000 per tonne.

If one considers air-curing fibre cement products, there are limitations in the processing ability, e.g. the fibre must readily disperse into a slurry in the water medium and then be formed into a thin sheet for fabrication. The ability of asbestos fibres to perform this operation was discussed earlier, and was shown to be the main reason why asbestos has remained unchallenged for such a long time. The inorganic fibres of steel, glass, etc., tend to be too stiff or dense to perform well in this area, while the synthetic organic fibres lack a surface suitable for bonding and also introduce drainage problems.

1.5.4 *Mechanical properties of WFRC*

During the last decade a number of papers have appeared in which wood fibres are the sole source of fibre reinforcement. These studies have included chemical and mechanical pulps of softwoods and hardwoods in air-cured and autoclaved matrices. It will be seen that, when suitably pretreated by refining, wood fibres can afford a strong, tough and durable fibre cement suitable for commercial production on traditional slurry/dewatered systems. Such WFRC materials can be used for production of flat sheeting, corrugated roofing and low-pressure pipes which traditionally have used asbestos fibre.

Although wood fibres are cheap and readily processed on existing plant, they have the disadvantage of being hygroscopic. The composite properties are altered by absorption of water, and for this reason extensive testing when both wet and dry is required.

1.5.4.1 *Strength and modulus properties.* Fibre reinforced cement products are generally loaded in bending, and so flexural strength has more meaning than tensile strength in the characterization of such materials.

Some early studies of air-cured WFRC containing low levels of thermo-mechanical pulp (TMP) resulted in laboratory samples having flexural strengths little more than matrix strength, although fracture toughness was improved considerably.[82–85]

The interest in the high yield mechanical pulps, as compared to chemical pulps, was due to a number of reasons.[84] Such pulps make a lower demand on forest resources for the production of a given quantity of pulp (about twice as much pulp from the same amount of wood), effluent treatment of disposal is less troublesome, chemical requirements are much lower, and plants are economical at a smaller scale.

More recent studies of high yield pulps[86] have shown than such pulps at loadings of 8–10% by mass of fibre can result in air-cured cement products with flexural strengths greater than 18 MPa at densities of about 1.5 g/cm^3. Such pulps have also been used to assist in the manufacturing of fibre cement boards containing polyacrylonitrile, polypropylene and polyvinyl alcohol fibres.[87]

Coutts and co-workers have demonstrated that, compared to chemical pulps,[68,88] the mechanical pulps perform poorly.[86] Figure 1.9 shows the flexural strength versus fibre content for both kraft and TMP *P. radiata* fibre reinforced cement when air-cured or autoclaved.

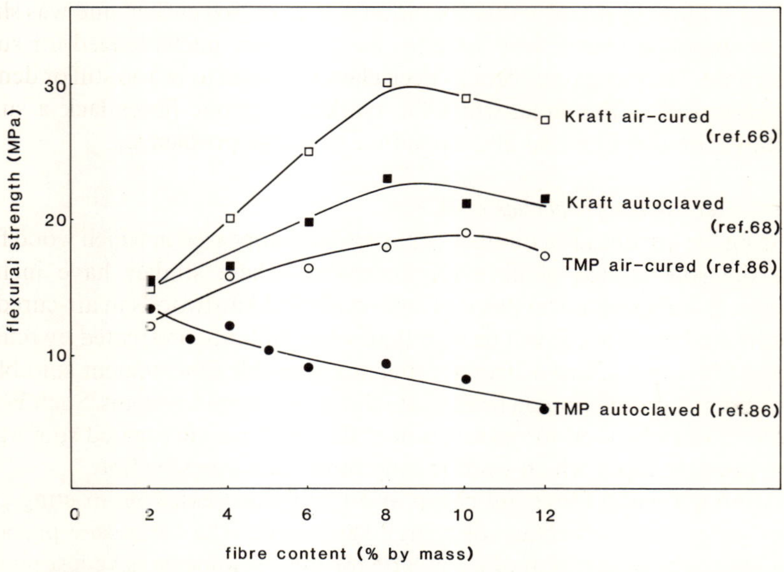

Figure 1.9 Graph of flexural strength *v.* fibre content for various WFRC products.

In the case of TMP pulps, the high temperature of the autoclave ($>160°C$) coupled with the alkaline conditions of the matrix material (pH > 12) results in decreasing strength with increasing fibre content, and has been attributed[28,29] to 'poisoning' of the cement by extraction of polysaccharides and wood acids (see 1.3.2), materials which, in the case of the kraft pulp have been removed during pulp preparation. Air-cured samples are not exposed to such drastic conditions and so setting and strengthening of the cement matrix takes place with less chemical interaction, yet we still note that TMP pulp reinforced cement is weaker than kraft pulp reinforced cement, which when air-cured can have flexural strengths of 30 MPa.[66]

Different chemical pretreatment of the wood chips and different temperature of reaction have been shown to affect the pulp properties of TMP fibres and their performance as reinforcing fibres for cement matrices.[84,86]

Although it was demonstrated in the laboratory that kraft wood fibres were effective as a reinforcement in a cement matrix,[60,83] the pulp performed poorly in asbestos-free formulations prepared on a pilot-plant Hatschek machine because the fibres were unable to form a web capable of retaining cement and silica particles.[15] The open nature of the fibre webs permitted rapid drainage, and in the case of slurries of low concentration, loss of the particulate matrix with a resulting loss of product strength.

Well documented in paper manufacture is the reduction of natural drainage rate of the pulp fibres by beating or refining (see 1.3.4). A collaborative project between James Hardie Industries and CSIRO Australia resulted in data which

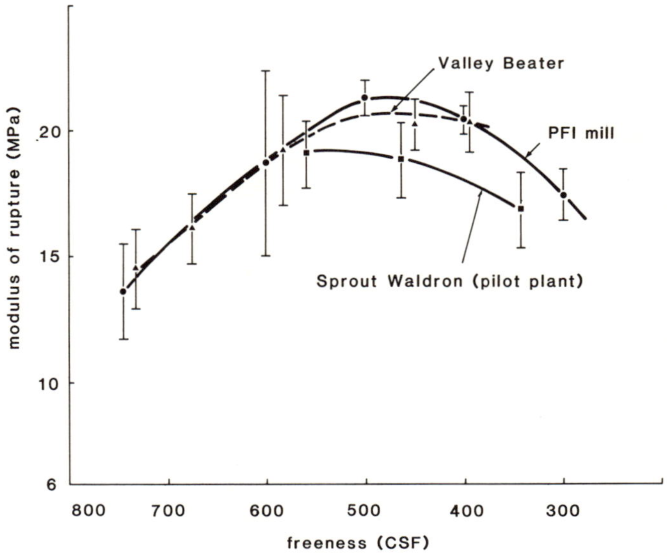

Figure 1.10 Graph of flexural strength *v.* freeness for different refining equipment.

demonstrated the benefit of beating or refining wood fibres for use in WFRC materials.[15–17,38]

A plot of flexural strength versus freeness for autoclaved mortar samples containing 8% fibre by mass prepared in various beaters or refiners[16] is depicted in Figure 1.10. It will be seen that there is a trend for the flexural strength to optimize as the degree of beating is increased. At an 8% fibre loading the optimum freeness is approximately 500 CSF.

It has been shown that a given fibre source (such as *P. radiata*) is affected by the method of pulp preparation (kraft, TMP etc.) and by refining, and depends on the matrix condition (air-cured cement, autoclaved mortar). Different fibre sources also have a major effect on the properties of the fibre cements.

Hensen and Pedersen[89] claimed that pine-wood cellulosic fibres provide a maximum obtainable flexural strength of 16.5 MPa using about 4% by mass of fibre in cement, whereas eucalyptus fibre reinforced cement can achieve about 22.6 MPa at about 5% fibre content. Recent work by Coutts[66] agrees with their findings that at 6% fibre content *E. regnans* provides a WFRC product with a flexural strength about 21 MPa; however, as has already been seen, *P. radiata* reinforced WFRC has a flexural strength value close to 30 MPa (see Figure 1.9). Some earlier work on a limited range of eucalypt fibres[90] (*E. regnans, E. saligna* and *E. grandis*) incorporated in autoclaved mortars, when compared with *P. radiata*, showed little variation in flexural strength, although, as we will see later, this did result in differences in fracture toughness.

The effect of moisture on the strength properties of WFRC composites is of importance. Earlier in WFRC research, variations in test conditions produced variations in test results, and it was evident that standard conditions must be adopted.[16] Figure 1.11 shows the variation of flexural strength at various fibre loadings with change in moisture.[68] Figure 1.12 shows an alternative

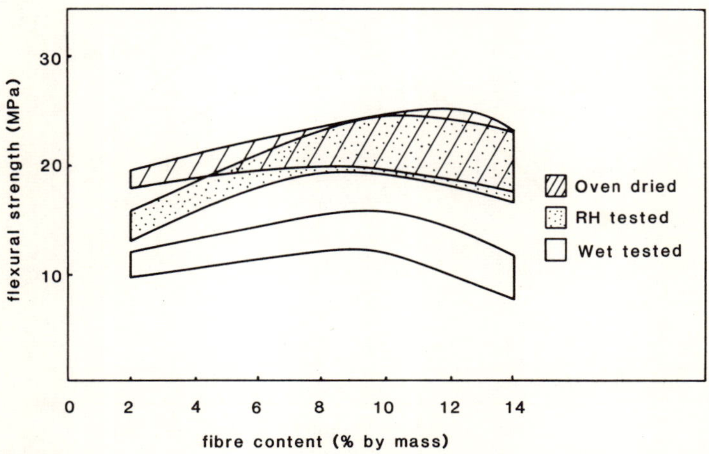

Figure 1.11 Effect of moisture content on the flexural strength of WFRC products of various fibre contents.

representation in which the flexural strength of oven dried (100°C for 24 h) and wet-tested (soaked 48 h in water) samples is given as a percentage of the flexural strength of the same composite formulation when tested at 50% RH and 22°C. This change in strength performance between oven-dried and RH tested samples is greatest for low fibre contents. At fibre contents above 6%, little variation results. On the other hand, wet-tested samples tend to deviate further from the flexural strength value of RH-tested samples as the fibre content is increased.

Tensile strengths of WFRC products have been published by Andonian *et al.*[60], and values ranged from 6.5 to 9.7 MPa as the fibre content was increased from 2 to 10% by mass – this was less than the void free matrix value of 9.7 MPa determined independently from separate experiments. They found flexural strengths ranged from 21–27 MPa for the same fibre range. These latter results are of the same order as those shown in Figure 1.9 (when allowance is made for the fact that their samples were dried at 116°C for 24 h). In agreement with Figure 1.12, they have higher values at lower fibre content than those measured in Figure 1.9 for autoclaved kraft WFRC measured at 50% RH.

Recently Fördös and Tram[73] reported wood fibre cements containing microsilica (> 20% by mass) with excellent strength properties ranging between 25 and 55 MPa. Over the last 10–15 years, microsilica additives have

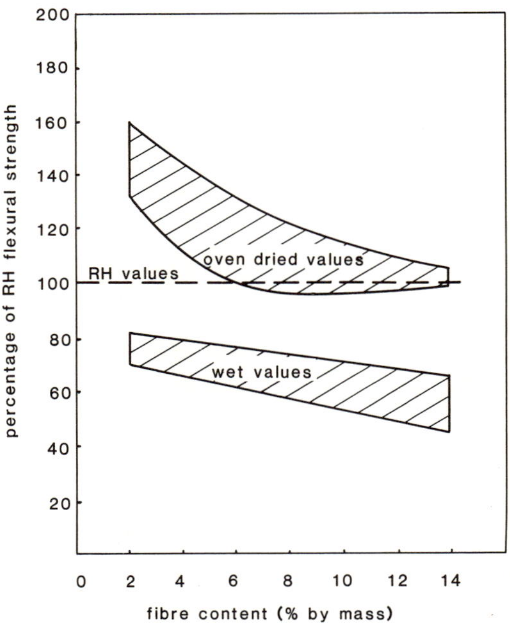

Figure 1.12 Effect of moisture content on flexural strength of WFRC samples tested oven-dry or wet, expressed as a percentage of the same composite tested at 50 ± 5% RH and 22 ± 2°C.

been used increasingly in cement and concrete formulations.[91] The main advantage of such material is as a filler with high silica content made of ultrafine spherical particles. This can result in high matrix strength and dense samples. Fördös and Tram[73] used very high stack compression pressures of approximately 20 MPa, whereas other workers have usually reported pressures of approximately 2–3 MPa.[66–69]

Air-cured WFRC mortars have been studied by Coutts.[92] The use of ground sand as a cheap filler may afford a product with adequate properties for applications such as renders on walls or outdoor garden-ware situations where fibre reinforcement may impart fracture toughness with adequate strength. Although these products have lower flexural strength (approximately 20 MPa at 8% fibre by mass), they have relatively high values of fracture toughness (approximately 3 kJ/m^2) when compared to air-cured WFRC (approximately 2 kJ/m^2).[66]

Few workers have reported the elastic modulus of WFRC materials. Mai and co-workers[60,77] have shown that both tensile and bending moduli are reduced in value from approximately 13 GPa to about 9 GPa as the fibre content is increased from 2 to 10% fibre by mass. In section 1.6 it will be seen that commercially produced samples have higher values of elastic modulus when they have been subjected to high pressures during manufacture.

1.5.4.2 *Fracture toughness.* The mechanisms taking place during composite failure which contribute to fracture toughness are discussed at some length in sections 1.5.1 and 1.5.5. This section will deal mainly with the effects of fibre type, pulping method, refining and test conditions on the values of fracture toughness for given composite formulations. Table 1.5 contains a spread of fracture toughness results showing a range of effects for samples that were prepared under similar conditions.

It is noted that as the fibre content increases from 2 to 8% by mass, there is initially a rapid increase in fracture toughness which starts to taper off at loadings $>$ 12% fibre.

Thermomechanical pulps, when compared with chemical pulps of the same species (e.g. *P. radiata*), tend to have considerably lower fracture toughness values, usually less than half as high. This has been explained in terms of fibre morphology and fibre number.[84,86] It must be remembered that the pulp yield from chemical methods is about half that from mechanical means, and so for a given weight of fibre there is about half the number of mechanical fibres (due to lignin and other non-extracted materials still present in the TMP fibre).

The variation of fracture toughness values between softwood (*P. radiata*) and hardwood (*E. regnans, E. saligna, E. grandis*) fibre reinforced cement is attributed to the fibre length[90] which, as we have seen, is so important during the process of fibre pull-out. Softwood fibres (average 3.5 mm) are longer than hardwood fibres (average 1.0 mm), and hence offer more resistance to pullout during failure of the composite. An interesting observation with respect to the

Table 1.5 Comparison of fracture toughness for various WFRC composites

Fibre type	Condition at test	Fracture toughness (kJ/m²) 2	4	6	8	10	12	Comment	Reference
P. radiata HT-TMP*	50% RH	0.21	0.32	0.56	0.57	0.64	0.68	Air-cured	86
P. radiata LT-TMP**	50% RH	0.25	0.41	0.60	0.8	1.06	1.10	Air-cured	86
P. radiata kraft	RH	0.41	0.68	1.40	1.93	2.28	2.25	Air-cured	66
	wet	0.64	1.52	3.72	4.51	4.60	3.60		
	Oven-dried	0.18	0.33	0.53	0.75	0.97	—	—	60
P. radiata kraft	RH	0.31	0.57	1.15	1.86	1.92	2.09	Autoclaved (700 CSF)†	68
	Wet	0.37	0.88	1.81	3.15	2.97	2.88	Autoclaved (700 CSF)	
P. radiata kraft	RH	—	0.47	0.84	1.30	1.29	1.81	Autoclaved (200 CSF)	68
	Wet	—	0.51	1.02	1.85	1.53	2.53		
E. regnans kraft	RH	0.25	0.51	1.06	1.37	1.46	1.68	Air-cured	67
	Wet	0.33	1.00	1.61	1.49	1.83	1.79	Air-cured	
E. saligna kraft	RH	—	—	—	0.95	—	—	Autoclaved	90
E. grandis kraft	RH	—	—	—	1.36	—	—	Autoclaved	

*High-temperature thermomechanical pulps
**Low-temperature thermomechanical pulps
†Canadian Standard Freeness

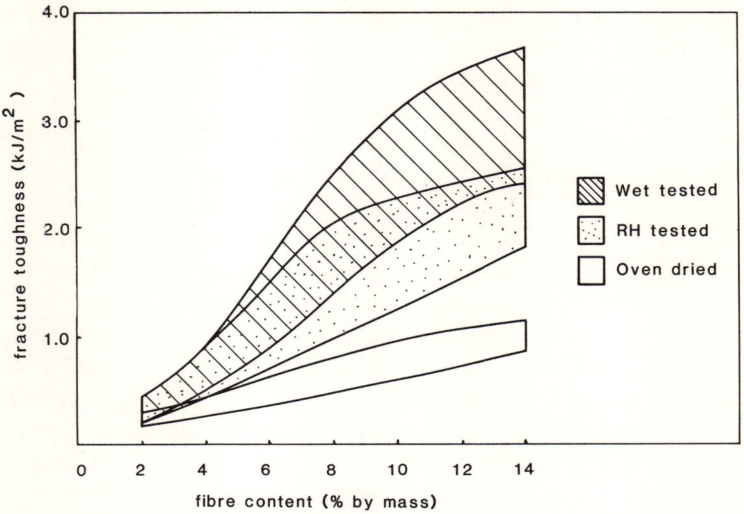

Figure 1.13 Effect of fibre content on fracture toughness of WFRC composites at various moisture contents.

fracture toughness of wet tested samples of hardwood reinforced samples is that there is little change in value over dry-tested samples (see Table 1.5), whereas, in the case of softwood WFRC composites the values often double when comparing dry- and wet-tested samples.

Figure 1.13 shows the fracture toughness of samples of various compositions

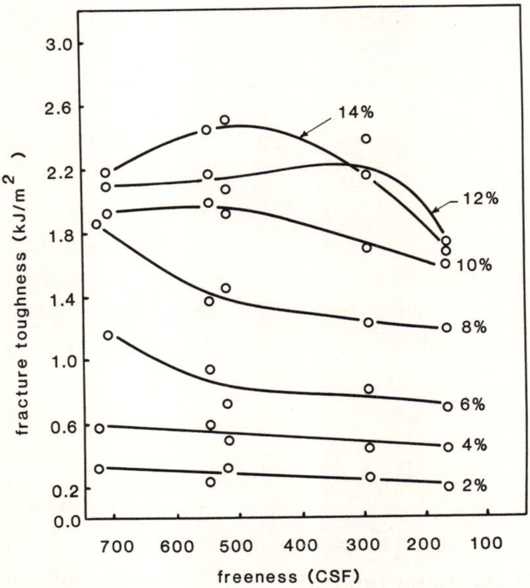

Figure 1.14 Effect of freeness on fracture toughness at various values of fibre content.

(over the complete range of freeness values studied) when tested at different moisture contents. It can be observed that at lower fibre content ($< 6\%$ fibre by mass) there is a general area of overlap between the values of fracture toughness for samples tested wet and those tested at 50% RH. By contrast, oven-dried samples show a dramatically reduced fracture toughness. This variation in toughness or shatter resistance has been attributed to the change in fracture mechanism from one of predominantly fibre fracture in the oven-dried samples to a mixture of fibre pull-out and fibre fracture in the samples at higher RH (see section 1.5.6). A scanning electron microscopic examination of the fracture surfaces of this type of wood fibre-reinforced composite[63,64] has supported this hypothesis, as has a theoretical study by Morrissey *et al.*[65] which considered single sisal slivers embedded in cement.

Figure 1.14 shows a plot of fracture toughness vs fibre freeness at various fibre percentages. In general terms, the toughness of a sample of constant fibre composition tends to decrease as the fibre is further beaten, although at fibre loadings $> 10\%$ an initial increase in toughness with beating is noted. It can also be seen that the brittle matrix rapidly improves in toughness as the fibre content is increased (see Figure 1.13).

The decrease in toughness with refining is due to shortening of the fibres. Hence the fibres offer less frictional resistance during fibre pullout from the matrix.

Mai *et al.*[93] studied fracture toughness due to intermittent interlaminar bonding. The basic principle of the method was to have many layers of perforated papers within the cement matrix so that good matrix bonding resulted through the perforations, and poor bonds existed in regions where the paper separates the matrix. During bending, they hoped to give rise to a sink of fracture energy absorption and hence tough materials. The composite at about 8% paper content (by mass) had a fracture toughness of 0.58 kJ/m^2 which was comparable to their earlier findings for 8% wood fibre reinforced material (0.75 kJ/m^2), but the flexural strength was only 5.7 MPa compared to 24 MPa for the wood fibre composite.[60]

The commercial products prepared by high compaction, such as 'Compressed Sheet II' (see section 1.6), have very high fracture toughness values especially when wet (> 13 kJ/m^2), and do display considerable interlaminar debonding during failure in bending. This interlaminar bond failure is due to the mode of formation of the sheet by winding up on the rolls of the Hatschek machine.

1.5.5 *Physical properties*

The physical properties of WFRC products can have a considerable influence on their acceptability for use in construction. If materials can be made strong and tough yet with low values of density, then workers who handle such products at the building site will prefer them to a similar material which is more dense. At the same time, due consideration must be given to water

Table 1.6 Physical properties of WFRC composites

Fibre type	Matrix	Water absorption (%)					Density (g/cm^3)					Reference
		2	4	6	8	10	2	4	6	8	10	
HT-TMP (*P. radiata*)	Opc (air-cured)	—	15.1	15.5	17.8	25.3	—	1.72	1.54	1.47	1.34	86
LT-TMP (*P. radiata*)	Opc (air-cured)	14.5	16.2	18.3	19.0	19.1	1.89	1.74	1.61	1.52	1.44	86
Kraft (*P. radiata*)	Opc (air-cured)	14.2	16.0	19.2	21.1	23.4	1.90	1.82	1.62	1.55	1.47	66
Kraft (*P. radiata*)	Mortar (air-cured)	15.3	20.9	22.2	24.2	25.4	1.79	1.56	1.54	1.46	1.44	92
Kraft (*E. regnans*)	Opc (air-cured)	17.8	20.5	24.5	25.8	28.0	1.75	1.63	1.50	1.45	1.39	67
Kraft (*P. radiata*)	Mortar (autoclaved)	21.2	25.5	31.9	33.9	33.8	1.65	1.49	1.37	1.31	1.30	68
Kraft (*E. saligna*)	Mortar (autoclaved)	—	—	—	35.7	—	—	—	—	1.27	—	90
Kraft (*E. grandis*)	Mortar (autoclaved)	—	—	—	36.5	—	—	—	—	1.26	—	90
Kraft (*E. regnans*)	Mortar (autoclaved)	—	—	—	36.3	—	—	—	—	1.27	—	90
NZ flax	Mortar (autoclaved)	20.8	27.0	29.6	35.0	37.7	1.62	1.48	1.40	1.31	1.26	69

absorption, for as the density of a product is lowered, this usually results in an increase in void volume with an associated potential increase in water absorption. Thus the load on a structure may be considerably increased should the material become wet, with more than 30% increases in weight occurring in some cases.

Table 1.6 reports properties of water absorption and density for a range of WFRC products containing different fibres, different pulps of the same fibre species and different matrices.

Density can be affected by the type of pulp used. For instance, if the fibre is stiff, like HT-TMP fibres, then packing of the fibres becomes difficult at high fibre content and voids are introduced into the product.[86] This results in a lowering of the density at a greater rate than would be expected for more flexible fibres such as kraft[66] or LT-TMP fibres[86] (LT–low temperature) of the same species of wood which are more conformable and more readily packed together at high loadings. Figure 1.15 depicts variation of density versus fibre content for a range of *P. radiata* WFRC composites. It can be seen that products containing the flexible kraft and LT-TMP fibres are more dense, at a given fibre loading, than the stiff HT-TMP (HT–high temperature) reinforced WFRC.

The matrix material also affects the density of the composite. Figure 1.15 shows that autoclaved kraft WFRC mortar is less dense than air-cured kraft WFRC, but density values for both fall at about the same rate as the fibre content is increased. More recently, a study of air-cured WFRC mortars[92] has

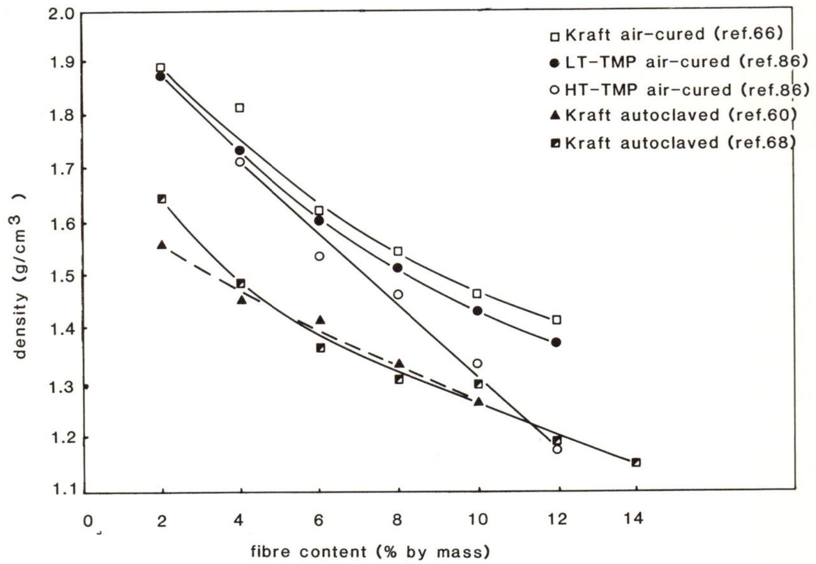

Figure 1.15 Variation of density *v.* fibre content for a range of *P. radiata* WFRC composites.

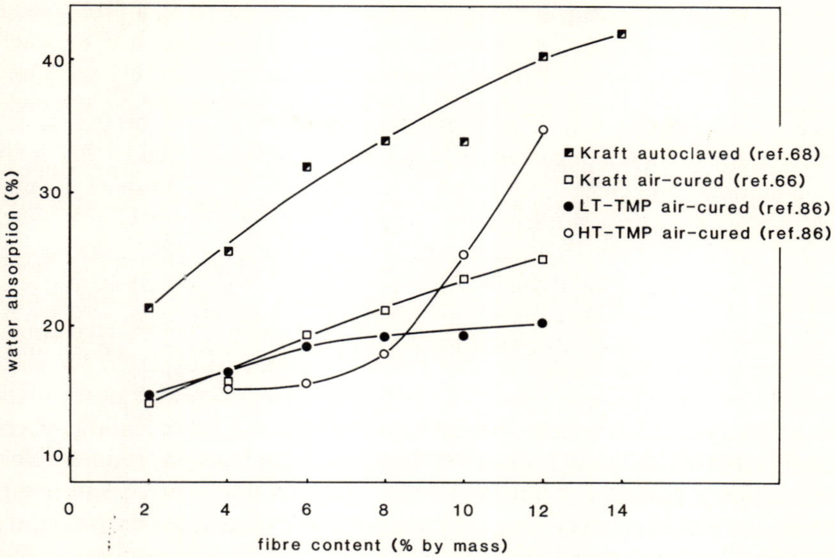

Figure 1.16 Variation of water absorption *v.* fibre content for a range of WFRC composites.

shown that the density of these products falls between those values of air-cured cement and autoclaved mortar (see Table 1.6). All the materials (reported in Table 1.6) were subjected to the same pressure (3.2 MPa) during fabrication. It has been shown by a number of workers that pressure also alters the properties of density and water absorption by reduction of void volume. This is very obvious in commercial production when products are subjected to high compacting pressures compared to laboratory samples.

Water absorption is indirectly related to density in that both are dependent upon the void volume of the sample. Thus in Figure 1.16 we observe that as the fibre content increases the water absorption also increases. The rate of increase is again dependent upon the pulp type, matrix and fibre type (see Table 1.6).

1.5.6 *Bonding and microstructure of WFRC composites*

As well as the physical properties of the fibre and the matrix, a major factor which controls the performance of a composite is the type and arrangement of bonds linking the two materials.

According to composite theory, the interface (that is the region of intimate contact between fibre and matrix) plays the dual role of transmitting the stress between the two phases and of increasing the fracture energy of the composite by deflecting cracks and delocalizing stress at the crack tip.

The interfacial bond itself can be physical or chemical in nature, or a combination of both. Too strong a bond between fibre and matrix results in a brittle material which has strength whereas a weak bond results in a tough material lacking strength. The mechanical performance of a WFRC composite is therefore directly related to the nature and properties of the fibre–matrix interface.

1.5.6.1 *Chemical bonding*. The chemistry and morphology of the matrix material has been well documented[94] and will not be considered further, apart from stating that cement is strongly alkaline ($pH > 12$) and presents metal-hydroxy groups at its surface, such as —Ca—OH, —Si—OH, —Al—OH and —Fe—OH (due to hydration and hydrolysis of silicates, aluminates and to a lesser extent ferrites of calcium that are present in the cement matrix). Cellulosic fibres such as wood fibres contain covalent hydroxyl groups, —C—OH, either phenolic (from residual lignin) or alcoholic (from the cellulose component) and carboxylic groups, $-\overset{\overset{\textstyle O}{\|}}{C}-OH$, due to oxidation of end groups. Hydrogen bonding and/or hydroxide bridges may play a major role in the bonding of WFRC composites. The chemical implications will not be considered quantitatively; it suffices to say that hydrogen bonds may form *between fibres* or *between fibres and matrix*.

Some early work by Coutts and co-workers[83] considered the possibility of using chemical pretreatments of the wood fibres to enhance the bond between fibre and cement. A preliminary screening of a range of coupling agents, inorganic additives, organic polymers and polyelectrolytes was conducted with wood fibres to see if the bond between pretreated fibres and cement could be improved.

A more extensive study of coupling agents was undertaken,[95] in keeping with the use of such materials to assist bonding between inorganic glass fibres and organic polyesters, an area well documented in composite science.

The most acceptable theory for the development of coupling agents is the 'chemical bonding theory', which suggests that the coupling agent acts as a link between fibre and matrix by the formation of a chain of covalent chemical bonds.

The synthesis of a range of titanium derivatives which contained functional groups theoretically capable of reacting chemically with both fibre and matrix was undertaken.[96] These compounds, along with some commercially available coupling agents, were used to treat fibres which were then incorporated into cement for the production of composite materials.[95] More recently, Courtaulds[97] have taken out a world patent on a number of titanium and zirconium compounds to produce similar effects in WFRC composites. Although small improvements in composite mechanical performance have been recorded from the use of pretreated fibres, the costs of such operations are currently considered prohibitive.

The initial hypothesis that some form of coupling agent or additive was needed to achieve bonding stemmed from the general belief that the bond between wood fibre and cement would be weak. It has subsequently been suggested that this may not be the case[72] and that the presence of hydroxyl groups on the surface of both wood and cement may be sufficient for adequate chemical interaction to take place via hydrogen bonding.

Coutts and Ridikas[16] reported the effect of moisture on the mechanical performance of WFRC composites, and showed that as the moisture content of a WFRC test specimen was increased, the flexural strength of a given sample decreased while the fracture toughness increased. These observations have been confirmed by Mai *et al.*[71] This behaviour is common to both air-cured[66] and autoclaved samples[16,68] containing a range of cellulosic fibres, be they softwood,[60,98] hardwood,[67] abaca[99] or NZ flax.[69]

Coutts and Kightly[64,72] suggested that these observations could be supported by the hypothesis that hydrogen bonds and/or hydroxide bridges play a major role in the bonding and hence in the mechanical performance of WFRC composites.

Wet or dry, a wood fibre has about the same tensile strength, but its stiffness is considerably lower when wet. Thus a dried WFRC composite has stiff, highly contorted fibres (see 1.5.6.3) locked into a rigid cement matrix which could be bonded at the interface by a large number of hydrogen bonds or hydroxyl bridged sites. This system when stressed can transfer the stress from the matrix to the fibres via the many interfacial bonds, and hence sufficient stress may be passed on to the fibre, after the matrix has cracked, to cause the reinforcing fibre to fracture under tensile load.

On the other hand, in a moist sample, the hydrogen bonds between fibres or between fibre and matrix are destroyed (by insertion of water molecules between the bridging hydroxyl groups); and, at the same time, the cellulosic fibres are swollen by water absorption and have become less stiff. Under stress this system allows the fibres to move relative to each other or to the matrix. However, due to the pressure of swelling and the highly contorted assemblage of fibres, considerable frictional forces are developed. If the forces are effective over sufficient length of a fibre, they can result in the fibres being loaded to failure; however, the number of fibres that pull out without failure is higher than when the sample is dry, and hence the observed values of fracture toughness for wet samples is higher than for dry samples.

1.5.6.2 *Mechanical bonding.* As well as the chemical bonding aspects of a fibre, we must consider the physical bonding potential. Much of the theoretical data on fibre reinforcement is based on smooth cylindrical fibres of uniform shape and dimensions (section 1.5.1).

Maximum fracture energy is often achieved if frictional energy is dissipated via fibre pullout. Wood fibres are relatively long compared to their diameter and hence have aspect ratios of say 60–200 (depending on whether they are hardwood or softwood, early wood or late wood fibres), but, more importantly, the fibres are hollow and can be collapsed to ribbons and at the same time develop a helical twist along their length (like a cork-screw). When fibres such as these are used to reinforce a brittle matrix, an asymmetrical process will be taking place during pullout (after interfacial debonding has occurred). In classical pullout of straight fibres (glass, steel, etc.) the forces are symmetrically distributed around the fibres; in the case of the contorted wood fibres, the leading edge of the helical fibre can experience considerable compressive stresses resulting in a ploughing action which can damage fibre or matrix resulting in increased fracture surfaces and hence increased fracture energy. Similar observations were recorded by Bentur *et al.*[100] when they studied the physical processes taking place during the pullout of steel fibres of various geometries from portland cement.

The effect of refining or beating wood fibres was discussed in section 1.3.4. It was reported by Coutts and co-workers[16,68] that such mechanical treatment of the wood fibres resulted in improved mechanical performance from autoclaved mortars reinforced with such fibres. This phenomenon can again be discussed in terms of mechanical bonding in that the external surface of the fibre is 'unwound' and the fibrils so formed offer extra anchoring points by which the fibres can accept stresses from the matrix and so become more effective reinforcing elements. Although refining the fibre can assist in mechanical bonding, its main value is in improving the drainage rate and solids retention in the commercial Hatschek process.[15,55]

Mechanical bonding has been discussed in other systems, such as asbestos fibre cement, in which Akers and Garrett showed[101] that asbestos can be fiberized like wood fibres, steel fibres can be kinked, as we have noted in the work of Bentur *et al.*,[100] or polypropylene can be fibrillated into a netlike form, as reported by Hannant and co-workers.[52]

1.5.6.3 *Microstructure of WFRC composites.* The examination of wood–cement composites by scanning electron microscope (SEM) is relatively recent. Ahn and Moslemi[102] examined the manner in which wood particles and portland cement bonded together in cement particle boards and decided that mechanical interlocking plays a significant role. This interlocking is due to crystal growth during the hydration of cement. A similar crystal interlocking effect has been reported for wood fibre reinforced plaster products.[103]

WFRC composites have been examined by different research groups with strong interests in the micromechanical behaviour of such composites. In a number of these SEM studies of WFRC materials, the microstructural features of the fracture surfaces of the broken composites were described, with emphasis being placed on the surface of the fibre and how it had failed.[60,63,64] More recently, interest has turned to the matrix, and in particular, the interfacial region adjacent to the fibre.

In a study of air-cured WFRC composites, containing kraft or CTMP (chemithermomechanical pulp) fibres, Davies *et al.*[63] proposed that the fracture surfaces indicated that both fibre fracture and fibre pullout were taking place during loading to failure under ambient conditions. Andonian *et al.*,[60] studying autoclaved WFRC samples, stated that fibre pullout was the main source of fracture toughness (85–90%) even though the samples had been dried in an oven at 116°C for 24 h. Mindess and Bentur[75] also considered that a pullout mechanism was the main process taking place when commercially produced WFRC samples containing 20% by mass of wood fibres were tested. These workers used only photographic evidence, stating that at 18 × magnification (their highest), lack of focus restricted their observations due to the unevenness of the surface. As the diameter of a wood fibre is approximately 15–40 μm, the observation of end fractures would be difficult by optical means.

More recent SEM studies by Coutts and Kightly,[64,72] using autoclaved WFRC samples, complement the earlier findings with air-cured samples, namely, that failure takes place by mechanisms of fibre fracture and fibre pullout. More importantly, this study showed that the relative importance of these mechanisms is very dependent upon the moisture content of the test sample.[72] Figures 1.17–1.19 show the SEMs of fracture surfaces of WFRC samples obtained from the same formulation when tested dry, wet and at 50 ± 5% RH. Most of the fibres protruding from the fracture surface of a sample preconditioned in an oven at 105°C for 24 h before testing have fractured ends (Figure 1.17*a*). By contrast, Figure 1.18 shows the fracture surface of a sample which has been preconditioned in water for 48 h before testing and indicates considerable fibre pullout, although fibre fracture is still very obvious (see Figure 1.18*b*). Figure 1.19 shows the fracture surface of the sample conditioned at near ambient conditions (50 ± 5% RH and 22 ± 2°C). Both fibre fracture and fibre pullout appear to have taken place. Higher magnification

Figure 1.17 (*a*) SEM showing fracture surface of WFRC preconditioned at 100–105°C for 24 h. (*b*) As (*a*), but higher magnification.

Figure 1.18 (*a*) SEM showing fracture surface of WFRC preconditioned by soaking in water for 48 h. (*b*) As (*a*), but higher magnification.

(Figure 1.19*b*) shows that considerable damage to some fibres of relatively short length has taken place while other fibres have been pulled out of the matrix. Mai *et al.*,[71] on re-examining their earlier work, are now of the opinion that fibre pullout is not the only major source of toughness for WFRC, but indeed that fibre fracture is of considerable importance.

The significance of moisture content, as evidenced in the above SEM work, has been adequately described in the previous section on chemical bonding (1.5.6.1).

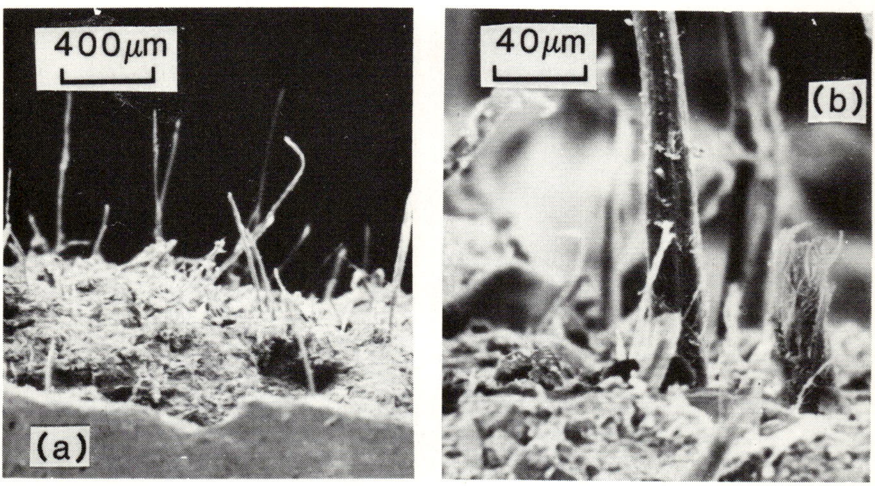

Figure 1.19 (*a*) SEM showing fracture surface of WFRC preconditioned at $50 \pm 5\%$ relative humidity and $22 \pm 2°C$. (*b*) As (*a*) but higher magnification.

Most of the foregoing studies have related to the condition of the fibre after failure and rather little has been said of the most important aspect of the composite – the interfacial region. In studies of air-cured cement matrices reinforced with aggregate particles,[106] steel[107,108] or glass fibres,[109,110] detailed reference has been made to an interfacial region with structure and porosity different to the bulk cement paste. This interfacial zone has, in several cases, been considered to be 50–100 μm thick.

In the cases of aggregate, steel or glass fibre there is a general similarity of behaviour.[106–110] At the surface of the reinforcement a dense layer, consisting largely of hexagonal lamellar crystals of portlandite (calcium hydroxide or CH), forms and replicates the topography of the reinforcement over much of its surface area. At discontinuities in this dense layer, inclusions of hydration products such as calcium silicate hydrate (CSH) gel, ettringite and large crystals of CH occur.

With increasing distance from the reinforcement surface, the proportion of CH decreases, and a relatively porous layer (in comparison to the bulk of the matrix material) forms for some distance before gradually becoming more dense and merging into the bulk matrix.

An SEM study of WFRC composites which focused on the region of the wood fibre–matrix interface,[104,105] revealed that the general characteristics associated with the interfaces of other fibre reinforced cement composites were not observed. Coutts[104] noted that the interface between wood fibre and matrix is usually dense and replicates the surface of the fibre, although discontinuities do occur, especially at higher fibre loadings when normal packing of the fibres becomes more difficult (Figure 1.20). Again, if one looks at the intimate area of contact between fibre and matrix (Figure 1.21), no

Figure 1.20 SEM showing cement surface at interface contains dense matrix with some discontinuities.

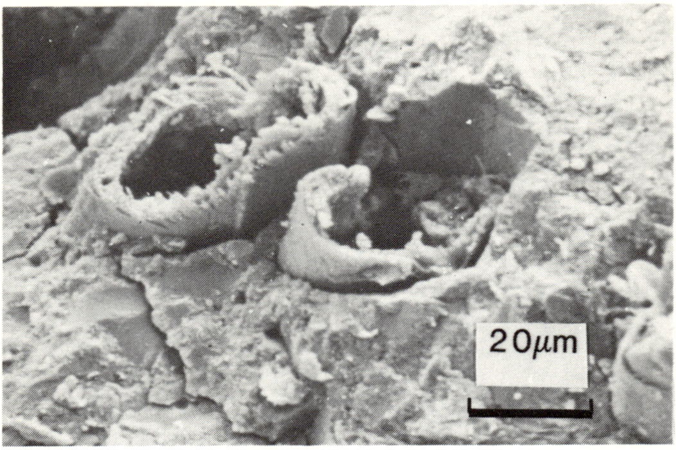

Figure 1.21 SEM shows fractured fibres with dense material from bulk of matrix up to fibre wall.

obvious zone of weak matrix material exists at the interface. It was stated that during fabrication of WFRC composites the samples were placed under pressure to compact the structure and reduce the water–cement ratio. It was hypothesized that, unlike the glass or steel fibres, which are not compressible, the hollow wood fibres are compressed, and after the pressure is removed the water level immediately adjacent to the fibre is lowered even further as the sponge-like fibre draws excess water back into itself. This would reduce the occurrence of voids at the interface of fibre and matrix particles, because less free water would be present, and produce a homogenous dense matrix material from the bulk of the material right up to the point of contact with the fibre.

Another example of a dense interface is found in the work of Akers and Garrett[111] who studied the microstructure and failure mode of the fibre–matrix interfacial region in asbestos fibre reinforced air-cured cement. They likewise reported a mutual interlocking of asbestos fibres with the cement hydrate, as had been observed with WFRC. But, more importantly, energy dispersive X-ray measurements showed no evidence of a calcium hydroxide enriched zone adjacent to the interface, as had been reported for other fibre cement materials. They proposed that fiberizing asbestos fibres increased the fibre surface available for water absorption, which helped to distribute the water-filled voids in the interfacial region. Thus the nature of the fibre–cement interface need not be the same as is described for materials such as glass, but appears to depend on the characteristics of the reinforcing fibre.

1.5.7 *Durability of WFRC composites*

Considerable doubt has been cast on the ability of natural fibres to resist deterioration in cement matrices,[112,113] yet no evidence has been put forward to support these claims. When poor mechanical performance of the composite has been offered as confirmation of fibre failure, due consideration should be given to the potential of the fibre to 'poison' the matrix surrounding the fibre, resulting in poor mechanical properties (see section 1.3.2). An extensive review by Gram[118] reflects this uncertainty. On the other hand, there have been claims which suggest that natural fibre reinforced cement products are durable after 30 years' service.[84,114]

Sharman and Vautier[115–117] have done some excellent work on the durability testing of autoclaved WFRC products at the Building Research Association of New Zealand. They discuss the possible ageing mechanisms of corrosion, carbonation, moisture stressing and microbiological attack. Tests have been used to accelerate these factors on the mechanical properties of WFRC sheet.

A test developed by the Building Research Establishment[119] to assess the durability of glass reinforced cement required specimens to be soaked in water at 50°C. Harper[17] used this test to show that 'Masterclad' could survive a three-week soak with no damage, whereas Sharman and Vautier[117] soaked their autoclaved WFRC samples for a year with no major effect. Air-cured WFRC specimens, on the other hand, experienced some reduction in flexural strength after 84 days.[120]

A French test, V313,[121] designed to evaluate moisture-resistant chipboard, has also been used. This test comprises 72 h in water at 20°C, then 24 h at −12°C and finally 72 h in an air oven at 70°C. Again, autoclaved WFRC products have passed this test[17, 117] suggesting moisture stressing of the wood fibres has little effect on the strength of autoclaved WFRC products, or on the fibre–cement bond.

Gram[118] has developed an accelerated ageing test by subjecting specimens to moistening and cooling, by spraying them with water, and then heating to 105°C. The heat is applied for 5.5 h before water is sprayed in again. His study of air-cured WFRC composites[122] showed that the samples become brittle with ageing and suggested the cellulose fibre was 'mineralized' by precipitation of lime compounds into the fibre lumen. Replacing about 30% of the cement with finely-grained, reactive silica fume reduced the pH value of the pore water to such an extent that no 'mineralization' of the fibre took place.

Tests involving artificial carbonation of asbestos products showed that about 35–40% $CaCO_3$ is formed, with resultant increases in tensile and flexural strengths and decreases in moisture movement and fracture toughness.[123] Tests on autoclaved WFRC samples[116] again indicated about 35% $CaCO_3$ was formed, and although tensile strength, internal bond and moisture movement increased, there was no apparent decrease in fracture toughness.

The effects of natural weathering on the mechanical properties of cellulose fibre reinforced cement products were reported to be satisfactory after several years.[18,124] Parameswaran and Broker[114] have investigated the micromorphology of a cement-bounded insulation board and a wood-cement moulding which had been in use for 25 years in a building in Switzerland. The samples showed no structural changes in their wood tissues even after such a time span.

Moisture movement was reported as being a problem in cellulose cement roofing sheets based on paper pulp, used in Scandinavia.[122] In Australia, ceramic-tile-faced asbestos cement sheet was formerly commonly used in wet internal areas such as bathrooms. It has been noted that WFRC sheet has a higher moisture movement than asbestos cement,[125] and as a result early problems arose concerning its use in such applications. When used in accordance with the manufacturers' current guidelines, the problem does not arise.

Fungal cellar tests on WFRC materials[18,116] and other natural organic fibres in cement have shown little microbiological activity. Parameswaran and Broker[114] found that a sample of a wood–cement composite, after burial under soil for more than 30 years, did show colonization and attack of the outermost surface by fungal hyphae; the wood cells were degraded, possessing decay patterns resembling soft rot and white rot. The inner part of the sample was, however, completely free from invasion.

Natural weathering studies are being conducted[117] in conjunction with the accelerated durability tests.[116] After one year, flexural strength, tensile strength and moisture movement of WFRC sheet had increased significantly. These increases were seen as the result of carbonation under relatively high atmospheric relative humidity levels. It was suggested that the effect of weathering on cellulose fibre reinforced cement composites will depend on the pretreatment of the fibres prior to their inclusion in the composite, and also on the curing mechanism used for the matrix. It was concluded that autoclaved WFRC products based on kraft pulp fibre should be durable in service.

1.6 Applications, specifications and test properties

In this section we will limit the discussion to those products in commercial production in Australia, New Zealand and Malaysia. The Australian brand names will be used for local convenience, but, as is obvious from previous data, basically we will be dealing with products made from refined wood fibre (*P. radiata*) in a matrix of ordinary portland cement and ground sand which has been autoclaved at temperatures above 150°C. Other additives are present in small amounts which for reasons of confidentiality will not be discussed. Within the product range, formulation and fabrication will vary somewhat.

1.6.1 *Sheet products*

The sheet products manufactured by James Hardie & Co Pty Ltd vary in performances, application and finish. Internal wall and ceiling linings (Villaboard II/Versilux II/Fiesta Ceiling sheets) are the products which possibly experience the least demand in performance. Unfortunately, the manufacturer has not provided data concerning the mechanical performance of these products. Some data are available from external sources[125,128] (see Table 1.7). All of these products are easy to cut and fix. They are unaffected by sunlight and steam and will not split or rot. The hard-wearing surfaces will readily accept PVA and acrylic paints.

The fibre cement sheets will not burn and have the following 'Early Fire Hazard Indices' tested to Australian Standard AS 1530 Part 3–1982: Ignition, 0; Flame spread, 0; Heat evolved, 0; Smoke developed, 0.

Hardie's fibre cement sheeting is also used as underlay for vinyl and cork at a thickness of 5 mm, or for ceramic tiles at a thickness of 6 mm. The product is rodent-proof, has been tested by the CSIRO and found to be totally unaffected by termites, and meets the requirements of AS 1884–1985 (Floor Coverings–Resilient Sheet and Tiles–Laying and Maintenance Practices). In wet areas, Compressed Sheet II products are recommended – see later.

Independent tests by the Cyclone Testing Station in Queensland have confirmed that Hardibrace fibre cement sheets can be used as a bracing sheet in areas subjected to high wind and cyclonic conditions. Timber stud walls can be stiffened against longitudinal racking or distortion, in accordance with AS 1684–1979 ('SAA Timber Framing Code'). Sheets are pre-printed with fastening positions for ease of installation (Figure 1.22).

Hardie's Roofing Shingles II provide a distinctive alternative to other roofing materials. They are virtually maintenance-free and are easily fixed and recreate the appearance and character of a traditional slate or shingle roof.

Figure 1.22 Timber-framed house under construction using 'Hardibrace' as structural wall bracing for high-wind and cyclone areas.

Table 1.7 Mechanical and physical properties of commercial WFRC materials

Product	Flex. strength (MPa)		Modulus of elasticity (GPa)		Frac. toughness (kJ/m^2)		Density g/cm^3	Water Absorption %	References
	RH	Wet	RH	Wet	RH	Wet			
Villaboard II (6 mm)	19.2 (L)* 13.5 (T)** 16.3 (av)	12.9 (L) 8.6 (T) 10.7 (av)	9.2 (L) 8.5 (T) 8.9 (av)	6.1 (L) 5.2 (T) 5.7 (av)	6.7 (L) 3.0 (T) 4.9 (av)	9.0 (L) 5.6 (T) 7.3 (av)	1.31	32.1	127
Hardiflex II	24 (L) 12 (T) 18 (av)	—	—	—	—	—	1.4	30–31	124
Compressed Sheet II (12 mm)	29.4 (L) 22.1 (T) 25.7 (av)	21.0 (L) 15.9 (T) 18.4 (av)	16.6 (L) 15.3 (T) 15.9 (av)	12.7 (L) 12.3 (T) 12.5 (av)	4.2 (L) 2.4 (T) 3.3 (av)	19.5 (L) 8.1 (T) 13.8 (av)	1.62	18.6	127

Test, BS 6432:1984 (4-point bend), used to determine flexural strength, density and water absorption.
Modulus of elasticity measured by ASTM D790–71. Fracture toughness measured from area under load deflection curve (4-point bend) divided by specimen cross-sectional area.

*L: longitudinal direction of manufacture
**T: transverse direction

They can be pitched as low as 17° on roofs, and are also suitable as a fascia or wall cladding. The shingles are integrally coloured as part of the manufacturing process, with colour pigments dispersed throughout the materials. The shingles are not recommended for use on structures having a design wind velocity exceeding 33 m/s. (All non-exposed areas in non-cyclone areas are likely to have design wind velocities less than 33 m/s.)

Products which are exposed to more demanding applications, such as external walls, porches, gable ends and fences, have somewhat better mechanical performance (Table 1.7).

Hardiflex II is the original asbestos-free fibre cement product produced in Australia.[15] Being tough and flexible, it is an ideal choice for many general building purposes, both internally and externally (Fig. 1.23).

In New Zealand the WFRC products, manufactured by identical processes

Figure 1.23 A residence clad in 'Hardiplank II' and featuring Hardie's 'Roofing Shingles II'.

to the Australian range of products, have received an Appraisal Certificate (No. 93–1983) from the Building Research Association of New Zealand, which allows scope for use in similar applications such as lining and cladding materials described in NZS 3604 or other standards.

James Hardie & Co Pty Ltd has used high-pressure techniques to develop a new range of materials. These materials have an increased density and improved mechanical performance. The high-performance materials in flat sheet products are the compressed sheet II products (Table 1.7), which in the home area are used for wet area applications. Mosaic, ceramic and quarry tiles in bathrooms, kitchens, laundries and toilets should be laid on Compressed Sheet II. The high-density product offers a flat trafficable surface that is ideal for external decking applications such as verandahs, pool surrounds and upper-storey balconies.

Architects are using the Compressed Sheet II products in both traditional and innovative applications where economy, fire resistance and performance are considered to be absolutely essential. The architecturally acclaimed Australian Institute of Sport Swimming Hall (Figure 1.24) is one example that has an external facade of Hardie's Compressed Sheet II which consists of 9 mm thick panels fixed to a light steel framework.

The cladding material is being used in increasing quantities for the construction of fascias and spandrels and steel-framed urban commercial

Figure 1.24 Modern buildings are now being extensively clad with 'Compressed Sheet II'.

Figure 1.25 WFRC materials used for constructing fascias and spandrels.

buildings. In many commercial and industrial buildings, spandrels and facades are not only important visual elements; they are used to protect the building from excessive solar heat gain and enhance the quality of the interior environment. Generally, spandrels are horizontal cladding elements such as columns in the case of low-rise commercial or industrial buildings, or they can be used in multi-storeyed buildings. In these cases spandrels normally extend from the window head of the floor below to the window sill above (Figure 1.25).

In structural use, Hardie's Compressed Formwork Sheet II offers many advantages in the construction of bridge decking. Construction time is reduced, as prefabricated sheets are simply positioned into a rebate and no formwork removal or cleaning is required. As no scaffolding is needed, there is no interference with traffic flow (Figure 1.26).

Although the performance of Compressed Sheet II has been reported in Table 1.7, for purposes of design data, the properties are as follows: flexural strength, minimum value for saturated sheet (in direction indicated on sheet), 18 MPa; modulus of elasticity in static cross bending, 10.3 GPa (average value for saturated sheet); compressive strength, 150 MPa (load applied to face of sheet); shear strength (load applied at 90° to laminations), 39 MPa.

Figure 1.26 Hardie's Compressed II Formwork sheet being used in bridge construction.

The sheet sizes are 18, 24, 30 and 40 mm thick, 900 and 1200 mm wide, and lengths will be supplied to order. The design criteria for thickness of formwork should be selected from the manufacturer's chart (Figure 1.27). The chart is based on the requirements of AS 1509–1974 ('SAA Formwork Code').

1.6.2 *Pipes*

The process for manufacturing compressed building products results in a higher ratio of wet to dry strength than occurs for non-compressed products. This suggested to Hardie that a pipe-making process similar to the compressed sheet process in raw materials and product properties might be used to make pressure pipes. They have in fact developed, tested and are currently manufacturing a low-pressure pipe suitable for drain and irrigation

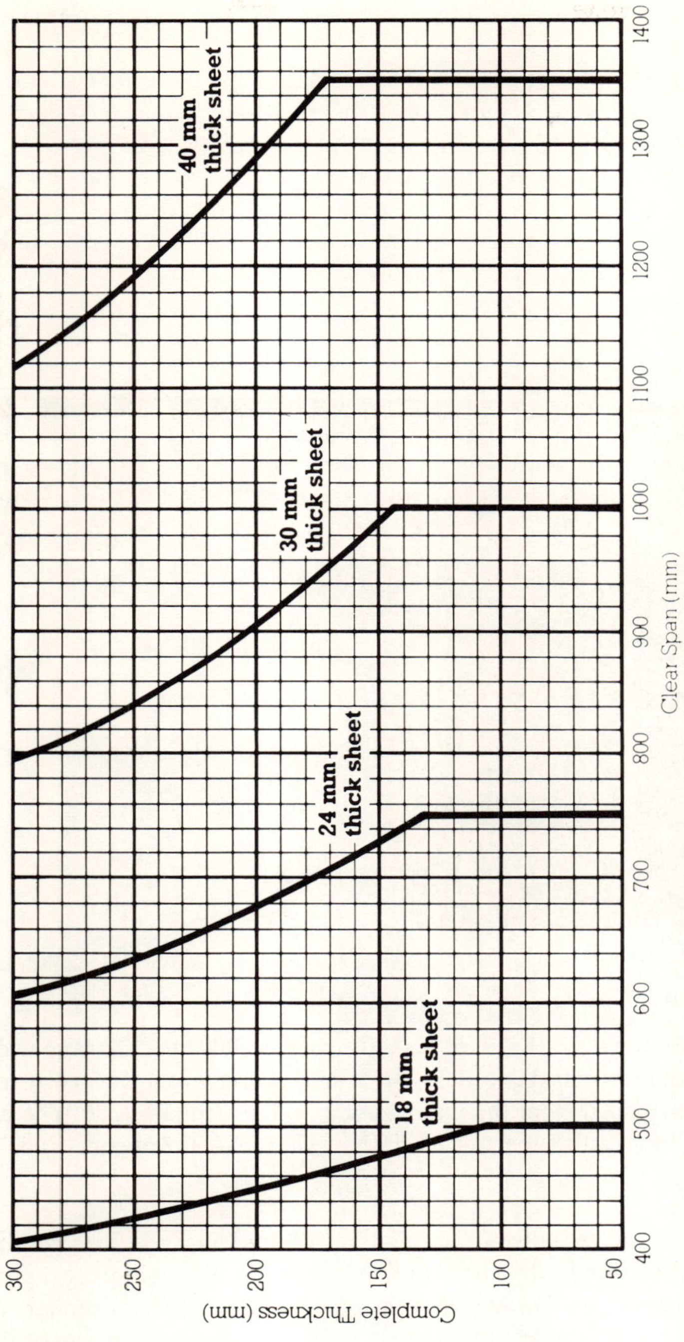

Figure 1.27 Thickness selection chart for Compressed Sheet II Permanent Formwork.

piping. The Mazza process is the basic method of manufacture, along with the autoclave curing process.

The WFRC drainage pipe is used to replace asbestos cement drainpipe and is designed to follow the same class structure as steel reinforced concrete pipe as set out in AS 1342–1973 ('Precast Concrete Drainage Pipes').

Although the drainpipe industry showed some concern about the 'bio-degradation' of cellulose products, studies by the company indicate no apparent difference between the resistance of asbestos cement and WFR cement to attack by insects, organisms and acid.

The pipes have been tested for long-term survival under load by conducting crush tests on saturated WFRC pipe in accordance with AS 1712–1976 ('Asbestos Cement Sewer Pipes, Joints and Fittings').

The abrasion resistance of WFRC pipes has been studied for the Concrete Pipe Assocation of Australia.[126] Concrete, asbestos cement and WFRC pipes were compared in a rotating pipe test. The average loss of wall thickness was 6, 41 and 73% respectively. The appropriateness of the test could be challenged,

Figure 1.28 Injection moulding machine can produce Telecom cable joining pits up to 2 m long.

in that the loss of 41% wall thickness in asbestos cement pipes (after 900 h of testing) would suggest that, if the test was meaningful, considerable failure in asbestos piping should have been recorded, considering that it has been used in Australia and New Zealand for decades. Such is not the case, and the product has been accepted by various Government departments for its prescribed application.

1.6.3 *Moulded products*

The range of roofing accessories and special moulded fittings which traditionally has been used with 'Super Six' roofing sheets is now made with non-asbestos fibres.

During 1985, the newly developed pressed corrugated sheet was introduced in Western Australia. This was made available with the commissioning of a A$4–5 M press. This product was the last of the building products range to be manufactured with asbestos fibres, and so completed the major change from asbestos to wood fibre as the sole source of reinforcement.

For many years Telecom Australia used asbestos fibre reinforced cement as the construction material for cable pits. As a result of the adverse publicity associated with asbestos, Telecom investigated the possibility of using WFRC as the material of construction.[127] The old process involved hand-moulding 'green' sheets into the required shape. More recently, injection moulding has been used. During 1984–85, Hardie's installed a much larger Gremigni injection moulding machine at the Camellia factory in New South Wales (Figure 1.28).

1.7 Concluding remarks

Cement and mortar are cheap matrix materials that can be made with unsophisticated equipment from indigenous minerals. Likewise, natural fibres suitable as reinforcement for such matrices are readily available in most developing countries of the world.

The work described in this and the other chapters of this book shows the beginning of a research area. Adequate field experience has been acquired to enable safe design and construction, but for developing countries to make adequate use of these findings, less sophisticated manufacturing processes are needed, and pretreatments for the natural fibre must be found to avoid 'poisoning' of the matrix by the non-cellulosic materials.

If fibre cement is to continue to compete with aluminium and steel cladding in the developed countries, then decorative coatings need to be developed which are cheap, durable and architecturally acceptable. Inorganic coating formulations are needed to eliminate UV-degradation and the high cost of current organic coatings.

WFRC composites are now readily available, but if they are to etch a permanent position in the market for some time to come, more work needs to be done on the theory of reinforcement for this system; test methods must be developed for both mechanical and physical properties, and means for testing the long term durability of the products established.

References

1. Anon. (1987) Asbestos use steady despite bans in West. *Chem. Ind.* 16 Feb., 101.
2. Harding, J. (1987) Superior pollution – Asbestos in the Great Lakes. *New Ecologist* **2**, Mar./Apr., 47–50.
3. Anon. (1979) Aborigines and asbestos. *New Scientist*, 29 March, 1020.
4. Anon. (1981) Double standards: Asbestos in India. *New Scientist*, 26 Feb., 522–523.
5. Anon. (1982) Turkish cancer deaths – It was asbestos after all. *New Scientist*, 6 May, 353.
6. Crabtree, J.D. (1986) Past, present and future developments in industrial fibre cement technology. Keynote Speech, *2nd Int. Rilem Symp.*, Sheffield, 15 July.
7. US Patent No. 291, 114, 1 Jan. 1884.
8. US Patent No. 635, 221, 17 Oct. 1899.
9. US Patent No. 658, 590, 25 Sept. 1900.
10. The Concrete Society (1973) Fibre reinforced cement composites. *Technical Report* 51.067, 77 pp.
11. James Hardie Industries Ltd (1984) *The name behind the names* 71 pp.
12. James Hardie and Co. Pty Ltd. *Report on paper pulp sheets*, Internal Report, November 1947.
13. Seach, B.G. (1987) Private communication.
14. Greenwood, R. (1983) James Hardie and asbestos – a tough bond to break. *Business Review Weekly*, 12 Nov., 10–13.
15. Anon. (1981) New – a wood-fibre cement building board. *CSIRO Industrial Research News* **146**, 1–2.
16. Coutts, R.S.P. and Ridikas, V. (1982) Refined wood fibre-cement products. *Appita* **35**, 395–400.
17. Australian Patent No. 515 151, issued March 19th, 1981, James Hardie and Co. Pty Ltd.
18. Harper, S. (1982) Developing asbestos-free calcium silicate building boards. *Composites*, April, 123–128.
19. Pedersen, N. (1980) Commercial development of alternatives to asbestos sheet products based on short fibres. *Proc. Symp. on Fibrous Concrete*, London, 16 April 1980, 189–93.
20. James Hardie Industries (1984) Annual Report.
21. Stamm, A. J. (1964) *Wood and Cellulose Science.* Ronald Press, New York, 549 pp.
22. Kollmann, F.F.P. and Cote, W.A. Jr. (1968) *Principles of Wood Science and Technology.* Vol. 1, *Solid Wood*, Springer Verlag, New York, Chaps. 1 and 2.
23. Britt, K.W. (1970) *Handbook of Pulp and Paper Technology.* 2nd edn., Van Nostrand Reinhold, New York, Chaps. 1–3, 5.
24. Clark, J.d'A. (1987) *Pulp Technology and Treatment for Paper.* Miller Freeman, San Francisco, Chaps. 6–8, 12–14.
25. Casey, J.P. (1980) *Pulp and Paper Chemistry and Chemical Technology.* 3rd edn., Vol. 1, John Wiley, New York, Chap. 4.
26. Moslemi, A.A. and Lim, Y.T. (1984) Compatibility of southern hardwoods with Portland cement. *Forest Prod. J.* **34**, 22–26.
27. Sandermann, W. and Kohler, R. (1964) Industrial woods mixed with mineral products. VI. A rapid test to determine whether an industrial wood is suitable for mixing with cement. *Holzforschung* **18**, 53–59.
28. Singh, S.M. (1979) Investigations into the causes of poor strength of Portland cement bonded lignocellulosic materials, *J. Ind. Acad. Wood Sci.* **10** (1), 15–19.
29. Thomas, N.L. and Birchall, J.D. (1983) The retarding action of sugars on cement hydration, *Cement Concr. Res.* **13**, 830–842.
30. Weatherwax, R.C. and Tarkow, H. (1964) Effect of wood on setting of Portland cement, *Forest Prod. J.* **14** (12), 567–570.
31. Michell, A.J. (1980) Composites of commercial wood pulp fibres and cement. *Appita* **33**, 461–463.

32. Australian Patent No. 17, 342/34 (1934).
33. Hillis, W.E. and Brown, A.G. (1984) *Eucalypts for Wood Production*. CSIRO/Academic Press, Sydney, Chap. 12.
34. Kurdin, J.A. (1983) The pulp of the twenty-first century. *Tappi* **66** (6), 9–10.
35. Bamber, R.K. and Burley, J. (1983) *The wood properties of Radiata pine*. 1st edn., Commonwealth Agricultural Bureaux, Slough, Chap. 3.
36. Higgins, H.G. (19386) Private communication.
37. Arno, J.N., Frankle, W.E. and Sheridan, J.L. (1974) Zeta potential and its application to filler retention. *Tappi* **57** (12), 97–100.
38. Britt, K.W. and Unbehend, J.E. (1974) Electrophoresis in paper stock suspensions as measured by mass transport analysis. *Tappi* **57** (12), 81–84.
39. Herrington, T.M. (1986) The surface potential of cellulose. *Paper Technol. and Ind.*, Dec. 1985/Jan. 1986, 383–387.
40. McKenzie, A.W. (1978) The structure and properties of paper. Part XX. The tensile properties of paper and papermaking fibres. *Appita* **32**, 207–212.
41. Kroon-Batenburg, L.M.J., Kroon, J. and Northolt, M.G. (1986) Chain modulus and intramolecular hydrogen bonding in native and regenerated cellulose fibres. *Polymer Comm.* **27**, 290–292.
42. Page, D.H., El-Hosseiny, F., Winkler, K. and Lancaster, A.P.S. (1977) Elastic modulus of single wood pulp fibres. *Tappi* **60** (4), 114–117.
43. Jayne, B.A. (1959) Mechanical properties of wood fibers. *Tappi* **42** (6), 461–467.
44. Spiegelberg, H.L. (1966) The effect of hemicelluloses on the mechanical properties of individual pulp fibers. *Tappi* **49** (9), 388–390.
45. Page, D.H. (1970) The physics and chemistry of wood pulp fibres. *Tappi STAP* No. 8, p. 201.
46. Page, D.H., El-Hosseiny, F. and Winkler K. (1971) Behaviour of single wood fibres under axial tensile strain. *Nature* (*London*) **229**, 252–253.
47. Kim. C.Y., Page, D.H., El-Hosseiny, F. and Lancaster, P.J. (1975) Mechanical properties of single wood pulp fibres. III. Effect of drying stress on strength. *Appl. Polym. Sci.* **19** (6), 1549–1562.
48. Tam Doo, P.A. and Kerekes, R.J. (1982) The flexibility of wet pulp fibres. *Pulp and Paper Canada* **83** (2), 46–50.
49. Kerekes, R.J. and Tam Doo, P.A. (1985) Wet fibre flexibility of some major softwood species pulped by various processes. *J. Pulp and Paper Sci.* **11** (2), J60–J61.
50. Gordon, J.E. and Jeronimidis, G. (1980) Composites with high work of fracture. *Phil. Trans. R. Soc. London Ser. A*, **294**, 545–550.
51. Nevell, T.P. and Zeronian, S.H. (1984) *Cellulose Chemistry and its Applications*. Ellis Horwood, Chichester, 506–530.
52. Hannant, D.J. (1978) *Fibre Cements and Fibre Concretes*, John Wiley, Chichester, 1–50.
53. Piggott, M.R. (1980) *Load Bearing Fibre Composites*. Pergamon, Toronto, 277 pp.
54. Taylor, G.D. (1983) *Materials of Construction*. 2nd edn., Construction Press, London, Chap. 6.
55. Hodgson, A.A. (1985) *Alternatives to Asbestos and Asbestos Products*. Anjalena Publications Ltd. UK, 129–158.
56. Daniel, J.I. and Anderson, E.D. (1986) Acrylic fibre reinforced cement composites. Proc. RILEM Symp. 1986, RILEM Technical Committee 49-TFR, Paper 2.8, 8 pp.
57. Swamy, R.N. and Hussein, M.W. (1986) Effect of curing conditions on the tensile behaviour of fibre cement composites. Proc. RILEM Symp. 1986, RILEM Technical Committee 49-TFR, Paper 4.9, 10 pp.
58. Hillerborg, A. (1980) Analysis of fracture by means of the fictitious crack model, particularly for fibre reinforced concrete. *Int. J. Cement Comp.* **2** (4), 177–184.
59. Kanninen, M.K., Rybicki, E.F. and Brinson, H.F. (1977) A critical look at current applications of fracture mechanics to the failure of fibre-reinforced composites. *Composites* **8** (1), 17–22.
60. Andonian, R., Mai, Y.M. and Cotterell, B. (1979) Strength and fracture properties of cellulose fibre reinforced cement composites. *Int. J. Cement Comp.* **1**, 151–158.
61. Laws, V. (1983) On the mixture rule for strength of fibre reinforced cements. *J. Mater. Sci. Lett.* **2**, 527–531.

62. Swift, D.G. and Smith, R.B.L. (1979) The flexural strength of cement-based composites using low modulus (sisal) fibres. *Composites* **10**, 145–148.
63. Davies, G.W., Campbell, M.D. and Coutts, R.S.P. (1981) An SEM study of wood fibre reinforced cement composites. *Holzforschung* **35**, 201–204.
64. Coutts, R.S.P. and Kightly, P. (1982) Microstructure of autoclaved refined wood-fire cement mortars. *J. Mater. Sci.* **17**, 1801–1806.
65. Morrissey, F.E., Coutts, R.S.P. and Grossman, P.U.A. (1985) Bond between cellulose fibres and cement. *Int. J. Cem. Comp. and Lightweight Concr.* **7** (2), 73–80.
66. Coutts, R.S.P. and Warden, P.G. (1985) Air-cured, wood pulp, fibre cement composites. *J. Mater. Sci. Lett.* **4**, 117–119.
67. Coutts, R.S.P. (1987) Eucalyptus wood fibre reinforced cement. *J. Mater. Sci. Lett.* **6**, 955–957.
68. Coutts, R.S.P. (1984) Autoclaved beaten wood fibre-reinforced cement composites. *Composites* **15**, 139–144.
69. Coutts, R.S.P. (1983) Flax fibres as a reinforcement in cement mortars. *Int. J. Cement Comp. and Lightweight Concr.* **5**, 257–262.
70. Gordon, J.E. (1976) *The New Science of Strong Materials – or Why You Don't Fall Through the Floor.* 2nd edn., Penguin Books, 287 pp.
71. Mai, Y.W., Hakeem, M.I. and Cotterell, B. (1983) Effects of water and bleaching on the mechanical properties of cellulose fibre cements. *J. Mater. Sci.* **18**, 2156–2162.
72. Coutts, R.S.P. and Kightly, P. (1984) Bonding in wood fibre-cement composites. *J. Mater. Sci.* **19**, 3355–3359.
73. Fördös, Z. and Tram, B. (1986) Natural fibres as reinforcement in cement-based composites. Proc. RILEM Symp. 1986, RILEM Technical Committee 49-TFR, Paper 2.9, 6 pp.
74. Hibbert, A.P. and Hannant, D.J. (1982) Toughness of fibre cement composites. *Composites* **13**, 105–111.
75. Mindess, S. and Bentur, A. (1982) Technical Notes: The fracture of wood fibre reinforced cement. *Int. J. Cement Comp. and Lightweight Concr.* **4**, 245–249.
76. Mai, Y.M. and Hakeem, M.I. (1984) Slow crack growth in bleached cellulose fibre cements. *J. Mater. Sci. Lett.* **3**, 127–130.
77. Mai, Y.M. and Hakeem, M.I. (1984) Slow crack growth in cellulose fibre cements. *J. Mater. Sci.* **19**, 501–508.
78. Hughes, D.C. and Hannant, D.J. (1985) Reinforcement of Griffith flaws in cellulose reinforced cement composites. *J. Mater. Sci. Lett.* **4**, 101–102.
79. Zurer, P.S. (1985) Asbestos – The fiber that's panicking America. *Chem. Eng.*, 4 March, 28–41.
80. Michaels, L. and Chissick, S.S. (1979) *Asbestos*. John Wiley, New York, 67–114.
81. Green, A.K. and Pye, A.M. (1978) Asbestos–characteristics, applications and alternatives. *Fulmer Special Report* No. 5 (ISSN 0427–7457), 28 pp.
82. Australian Patent No. 512 457 (filed 15th August 1977), CSIRO Australia.
83. Coutts, R.S.P. (1979) Wood fibre reinforced cement composites. *Research Review 1979*, Division of Chemical Technology, CSIRO, 1–16.
84. Campbell, M.D. and Coutts, R.S.P. (1980) Wood fibre-reinforced cement composites. *J. Mater. Sci.* **15**, 1962–1970.
85. Campbell, M.D., Coutts, R.S.P., Michell, A.J. and Willis, D. (1980) Composites of cellulosic fibres with polyolefins or cement – a short review. *Ind. Eng. Chem. Prod., Res. Dev.* **19**, 596–601.
86. Coutts, R.S.P. (1986) High yield wood pulp as reinforcement for cement products. *Appita* **39**, 31–35.
87. Simatupang, M.H. and Lange, H. (1987) Lignocellulosic and plastic fibres for manufacturing of fibre cement boards. *Int. J. Cement Comp. and Lightweight Concr.* **9** (2), 109–112.
88. Coutts, R.S.P. (1986) Sticks and Stones...!! *Forest Products Newslett.* **2**, 1–4.
89. US Patent No. 4, 263, 049, 1981. Rockwell International A/S, Denmark.
90. Coutts, R.S.P. and Michell, A.J. (1983) Wood pulp fiber-cement composites. *J. Appl. Polym. Sci., Appl. Polym. Sym.* **37**, 829–844.
91. Hjorth, L. (1982) Microsilica in concrete. *Nordic Concrete Research*, Pub. No. 1, 18 pp.
92. Coutts, R.S.P. (1987) Air-cured woodpulp fibre-cement mortars. *Composites* **18**, 325–328.
93. Mai, Y.M., Hakeem, M. and Cotterell, B. (1982) Imparting fracture resistance to cement

mortar by intermittent interlaminar bonding. *Cement Concr. Res.* **12**, 661–663.
94. Lea, F.M. (1976) *The Chemistry of Cement and Concrete*. 3rd edn. Edward Arnold, London.
95. Coutts, R.S.P. and Campbell, M.D. (1979) Coupling agents in wood fibre-reinforced cement composites, *Composites* **10**, 228–232.
96. Coutts, R.S.P. (1981) The chemistry of titanium coupling agents. *Aust. J. Chem.* **34**, 653–657.
97. Patent Cooperative Treaty, W084/03275, publ. 30 August, 1984, Courtaulds.
98. Coutts, R.S.P. (1983) Wood fibres in inorganic matrices. *Chem. in Australia* **50**, 143–148.
99. Coutts, R.S.P. and Warden, P.G. (1987) Air cured, Abaca reinforced cement composites. *Int. J. Cement Comp. and Lightweight Concr.* **9**, 69–73.
100. Bentur, A., Mindess, S. and Diamond, S. (1985). Pull-out processes in steel fibre reinforced cement. *Int. Cement Comp. Lightweight Concr.* **7**(1), 29–37.
101. Akers, S.A.S. and Garrett, G.G. (1983) Observations and predictions of fracture in asbestos-cement composites. *J. Mater. Sci.* **18**, 2209–2214.
102. Ahn, W.Y. and Moslemi, A.A. (1980) SEM examination of wood–Portland cement bonds. *Wood Sci.* **13**, 77–82.
103. Coutts, R.S.P. and Ward, J.V. (1987) Microstructure of wood fibre plaster composites. *J. Mater. Sci. Lett.* **6**, 562–564.
104. Coutts, R.S.P. (1987) Fibre-matrix interface in air-cured, wood pulp, fibre cement composites. *J. Mater. Sci. Lett.* **6**, 140–142.
105. Coutts, R.S.P. and Ward, J.V. (1987) SEM study of high yield wood pulps in cement matrices. In preparation.
106. Barnes, B.D., Diamond, S. and Dolch, W.L. (1978) The contact zone between Portland cement paste and glass 'aggregate' surfaces. *Cem. Concr. Res.* **8**, 233–243.
107. Pinchin, D.J. and Tabor, D. (1978). Interfacial phenomena in steel fibre reinforced cement I: Structure and strength of the interfacial region. *Cem. Concr. Res.* **8** (1), 15–24.
108. Bentur, A., Diamond, S. and Mindess, S. (1985) The microstructure of the steel fibre-cement interface. *J. Mater. Sci.* **20**, 3610–3620.
109. Stucke, M.S. and Majumdar, A.J. (1976) Microstructure of glass fibre-reinforced cement composites. *J. Mater. Sci.* **11**, 1019–1030.
110. Singh, B. and Majumdar, A.J. (1981) Properties of GRC containing inorganic fillers. *Int. J. Cement Comp. Lightweight Concr.* **3** (2), 93–102.
111. Akers, S.A.S. and Garrett, G.G. (1983) Fibre-matrix interface effects in asbestos-cement composites. *J. Mater. Sci.* **18**, 2200–2208.
112. Cook, D.J. (1980) Natural fibre reinforced cement and concrete – recent developments. *Proc. Symp. Advancement in cement-matrix composites*, Boston, 251–258.
113. Mansur, M.A. and Aziz, M.A. (1982) A study of jute fibre reinforced cement composites. *Int. J. Cement Compos. and Lightweight Concr.* **4** (2), 75–82.
114. Parameswaran, N. and Broker, F.W. (1979) Micromorphological investigations of wood-cement composites after long-term use. *Holzforschung* **33** (4), 97–102.
115. Sharman, W.R. (1983) Durability of fibre cement sheet claddings. *N. Z. Concr. Constr.* **27**, 3–7.
116. Sharman, W.R. and Vautier, B.P. (1986). Durability studies of wood fibre reinforced cement sheet. Proc. RILEM Symp. 1986, RILEM Technical Committee 49-TFR, Paper 7.2, 9 pp.
117. Sharman, W.R. and Vautier, B.P. (1986) Accelerated durability testing of autoclaved woodfibre-reinforced cement-sheet composites. *Durability of Building Materials* **3**, 255–275.
118. Gram. H-E. (1983) Durability of natural fibres in concrete. Swedish Cement and Concrete Inst. fo., 1.83, Stockholm, 255 pp.
119. Building Research Establishment (1979) Properties of GRC: ten year results. Information Paper 1 P 36/79, BRE, Garston, 4 pp.
120. Wells, R.A. (1982) Future developments in fibre reinforced cement, mortar and concrete. *Composites* **13** (2), 169–172.
121. Association Française de Normalisation (1972) Accelerated aging test by the V 313 method. Paris, NF B51–263.
122. Gram, H.-E. (1986) Durability studies of natural organic fibres in concrete, mortar or cement. Proc. RILEM Symp. 1986, RILEM Technical Committee 49-TFR, Paper 7.1, 5 pp.
123. Jones, F.E. (1946) Weathering tests on asbestos cement roofing materials. *DSIR Building Res. Tech. Paper* **29**, London, 38 pp.
124. Sinha, U.N., Dutta, S.N., Chaliha, B.P. and Iyengar, M.S. (1975) Possibilities of replacing

asbestos in asbestos cement sheets by cellulose pulp. *Ind. Concr. J.*, August 1975, 228–232.
125. Cole, W.F. and Pepplinkhouse, H.J. (1985) A wall-tiling evaluation project 1. Properties of tiles and cement sheets. *J. Aust. Ceramic Soc.* **21** (2), 45–54.
126. Concrete Pipe Association of Australia (1987) *Abrasion Resistance of Concrete Pipes*, issued June 1987.
127. Telecom Australia (1984) Replacement of asbestos fibre by cellulose in cable pits. Review of activities 1983–1984.
128. Coutts, R.S.P. and Warden, P.G. (1987) Unpublished results.

2 Wood fibre reinforced concrete

A. SARJA

Abstract

Wood fibre concrete is the definition of a wide range of modified materials. The material is produced as lightweight wood wool for insulation and acoustic purposes, as wood chip cement for masonry blocks, as wood particle cement boards and as loadbearing and non-loadbearing elements made from structural wood fibre concrete. The common basic principle of all types is the coating of wood fibres with hydraulic cement paste, which guarantees durability through so-called mineralization of the wood fibres. The material properties can be varied across a wide range. All types of wood fibre concrete products can be fabricated by mechanized production processes. Advantages in the manufacture of insulation and structural elements are light weight, good workability and ductility of the hardened material. Wood fibre concrete is suitable for use in developing and industrialized countries. Its properties are also well suited to renovation construction. Apart from wood fibre, several other vegetable fibres can also be applied in the same production technology.

2.1 Definition and classification

Wood fibre concrete is defined here as a composite material containing at least hydraulic cement, water and wood fibres of macro size, the majority of fibres exceeding 5 mm in length. The content of wood fibres is high, usually more than 20% in volume. In addition to the basic constituents, wood fibre concrete may also contain sand or other aggregate and substitutes such as pozzolanas, fly ash or blastfurnace slag and admixtures such as accelerators, plasticizers and impregnation solutions.

Depending on the constituents, mix design and manufacturing method, wood fibre concrete may have very different properties and applications. The various types can be classified as follows:

(i) Wood fibre concretes consisting of cement paste and wood fibres such as
  1 Wood wool
  2 Lightweight wood chip composites (wood chip cement)
  3 Wood–cement composites (wood particle cement)

(ii) Wood–mineral aggregate–cement composites (structural wood fibre concrete).

This chapter deals with the main characteristics of the various wood fibre concretes, which are summarized in Table 2.1.

From the manufacturing point of view, all types of wood fibre concrete are highly suited to industrialized production processes. Highly mechanized and automated processes are therefore available on the market. More or less manual versions of these processes are also possible for some wood fibre concretes.

The various types of wood fibre concrete are illustrated in Figures 2.1 to 2.4. The history of *wood wool* and *wood chip* cement dates back to the 1930s, when the industrial production of wood wool panels and wood chip blocks first began. Since then these products have been used extensively in countries such as Germany, Switzerland, Britain, the USA and Northern Europe. Dense *wood particle cement boards* were developed in the early 1970s in Germany and Switzerland. The production of wood cement chipboard has since spread to many countries, including the USA, Canada, Northern Europe and several developing countries. The technology of *structural wood fibre concrete* was developed at the Technical Research Centre of Finland between 1984 and 1986[71–74]. The first factory to produce wall elements from this concrete was established in Singapore in 1987[75].

The evolution of wood fibre concrete technology has made it possible to provide a wide range of products for construction[71–75]. The material is

Table 2.1 Main characteristics of various wood fibre concretes

Type	Basic constituents	Density (kg/m³)	Tightness	Principal applications
None-dense concretes				
Wood wool	Long wood chips (length > 80 mm), cement, water, admixtures	350–600	None	Thermal insulation, acoustic panels, moulding panels
Wood chip cement	Short wood chips (length < 20 mm), cement, water, admixtures	400–600	Pure	Blocks for masonry wall to be filled with ordinary concrete. Thermal insulation
Tight concretes				
Wood particle cement	Short wood chips (length < 20 mm), cement, water, admixtures	1000–1200	Considerable	Boards for partition walls, ceilings, floors and facades
Structural wood fibre concrete	Short wood chips (length < 30 mm) cement, aggregate, water, admixtures, substitutes	1200–1800	Good to considerable	Structural wall and slab components, facade elements

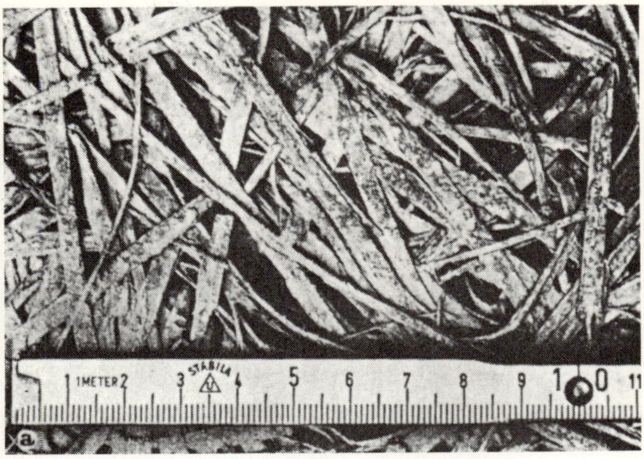

Figure 2.1 Material structure of wood wool.

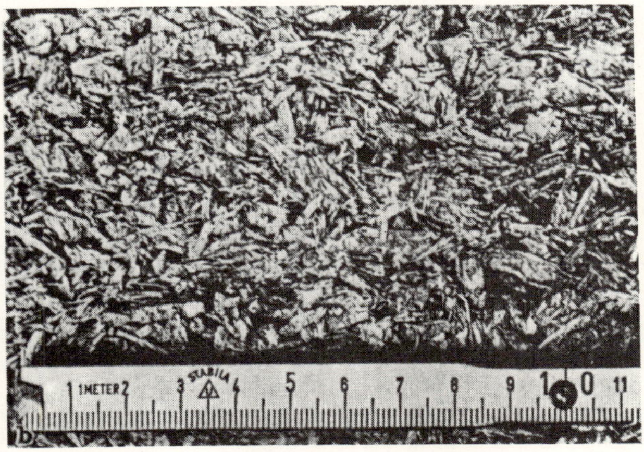

Figure 2.2 Material structure of wood chip cement.

Figure 2.3 Wood particle cement board.

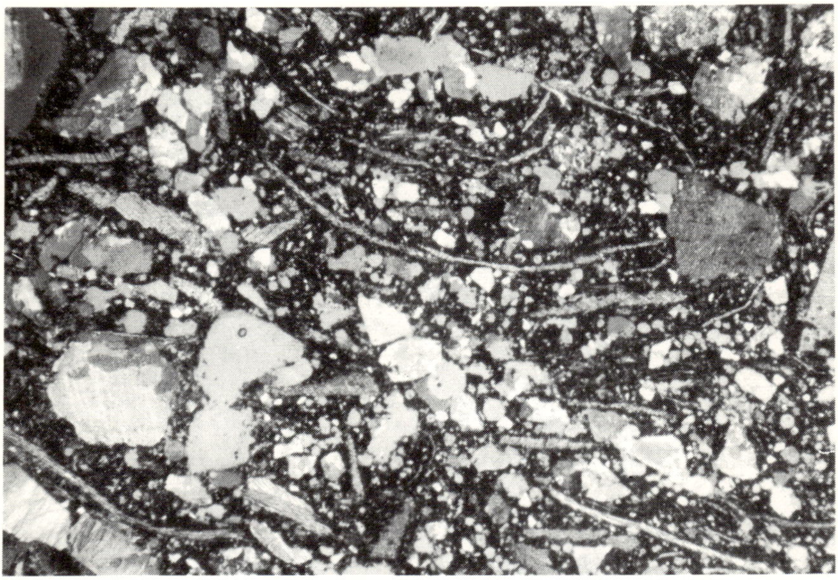

Figure 2.4 Material structure of structural wood fibre concrete.

suitable for use in both developing and industrialized countries. Its properties are also highly suited to renovation construction. The various wood fibre concretes each have their own role and in this way fulfil all technological requirements. In practical applications it is important that the user familiarizes himself with the properties and adaptability of the different types in order to make the correct choice of material and product according to their intended purpose.

In the above definition, the hydraulic cement is portland cement, various blended cements or special cements. In some cases, other types of inorganic binder, such as calcined gypsum, magnesite or magnesium oxysulphate are used, e.g. in chipboards. The following discusses mainly the applications of portland or blended cements. The same principles mostly apply also to concretes with other mineral binders. Some very important differences between binders should be remembered, such as the lack of water resistance in the case of gypsum.

2.2 Functional principles of wood fibre concrete

2.2.1 *Mineralization*

The mixing of hydraulic cement paste with wood fibres causes coating of the fibres with cement and the immediate filling of pores and voids in the chips with minerals. The diffusion of minerals into the wood fibres continues over a

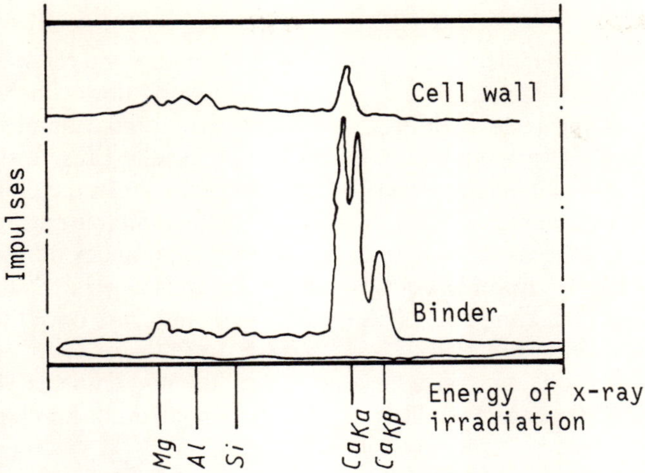

Figure 2.5 X-ray analysis spectra of binder and wood cell walls in a cement-bound spruce board.

longer period of time and is known as the mineralization process. The process alters the properties of the wood fibres, making them resistant to fire, rot and fungi[9,39]. The mineralization process is somewhat similar to that which occurs under natural conditions, where natural constituents between cellular fibres are changed into silicates over thousands of years.

Examination under the electron microscope and microanalyses show mineralization to have occurred in the cell walls of wood bound with cement. Figure 2.5 shows the microanalysis spectra of both a binder and the cell walls of the wood which it binds. Magnesium, silicon, aluminium and calcium, which are typical elements in the binder, are detectable within the cell wall. This can only result from the diffusion of these elements from the binder into the wood fibre[39]. The diffusion is a continuous process and the degree of wood mineralization depends on the thickness of the wood fibre and on how long the effect lasts[9].

The diffusion caused by mineralization does not alter the macro-structure of the wood fibres – examination under the electron microscope shows the cells to have remained intact. However, impregnation is associated with changes in the fine structure of the cell wall. These changes include reduced crystallinity of the wood fibre, as shown by X-ray diffraction analysis. In the case of wood fibres, the reduction is probably caused by a change from pulp I to pulp II due to the binder components. The binder components reach the cell wall by diffusion along the capillary pores in the wall. As a change in pulp fibres is not possible until at least part of the hemicellulose particles covering them have disappeared, it is probable that this is also the case in wood fibre concrete. This is supported by the fact that water soluble hemicellulose and its conversion products have been observed in wood fibre concretes. They have also been found to hinder hydration[41].

2.2.2 *Durability of wood fibres*

The precise significance of mineralization, i.e. impregnation of wood cell walls with binder components, for the physical properties of wood fibre concrete products remains as yet unclear. Both practical experience and laboratory studies have established that humidity and air, which usually cause disintegration of wood, cannot influence wood fibre concrete. Obviously, therefore, mineralized wood does not provide bacteria and fungi with a suitable nutrient medium. In particular, the alkaline environment (pH $>$ 11) caused by diffused calcium hydroxide effectively hinders the growth of fungi and bacteria[9,39]. It is possible that other binder components also have a toxic influence once they have penetrated the cell wall[39].

Unlike ordinary wood, mineralized wood fibres are not damaged by exposure to light, and they are not influenced by lack of light, air or humidity if hardening of the concrete is already complete[9].

Mineralization makes wood fibres more brittle, which has been observed to cause brittleness in cellular fibre materials with a small volume of fibres[66]. Where there are few fibres these are unable to plasticize owing to their brittleness, and cracks continue to grow unhindered. In the case of wood fibre concretes, the gradual increase in mineralization of the fibres does not cause brittleness problems, since the large number of fibres effectively prevents the progressive growth of cracks.

2.3 Material properties

The material properties of the various wood fibre concretes vary tremendously. For this reason it is important to treat each type separately.

2.3.1 *Mechanical properties*

The basic mechanical properties of the various wood fibre concretes are summarized in Table 2.2

The most variable type is structural wood fibre concrete; its density can vary between 1200 and 2000 kg/m^3 and in some cases even between 800 and 2000 kg/m^3. Sections 2.3.1.1–2.3.2.3 discuss the properties of structural wood fibre concrete[72,75]. The same principles can also be partly applied to wood particle cement and wood chip cement, taking into account the differences in density. Correlations between compressive and tensile strengths and density are shown in Figure 2.6.

These correlations can be numerically estimated from the following formulae:

$$f_c = 4.5 \cdot \left(\frac{\rho}{1000}\right) \quad f_{ct} = \frac{\rho}{500}$$

Table 2.2 Basic mechanical properties of wood fibre concretes

	Type of wood fibre concrete			
Property	Wood wool	Wood chip cement	Wood particle cement	Structural wood fibre concrete
Density (kg/m^3)	350–600	400–600	1000–1200	1200–2000
Bending tensile strength (N/mm^2)	0.4–1.7	0.7–1.0	10–15	2–6
Compressive prism strength (N/mm^2)	(1)	1.2–2.0	(2)	6–20
Modulus of elasticity (N/mm^2)	(1)	300	3000	3000–20 000
Workability after hardening	Easy sawing; nails have no bond	Easy sawing and nailing; nails have no bond	Nailing, sawing and drilling with light tools; nails have limited bond	Nailing sawing and drilling with light or heavier tools; nails have limited or good bond

(1) Undefined. Compression strain at a compressive strength of 0.3 N/mm^2 varies between 10% and 25%.
(2) Undefined.

where
f_c is the compressive cube strength (N/mm^2)
f_{ct} is the bending tensile strength (N/mm^2)
ρ is the density of concrete dried at RH 40% (kg/m^3).

The modulus of elasticity as a function of density is shown in Figure 2.7.

The numerical approximation for the modulus of elasticity can be derived as follows:

$$E_c = 1500 \cdot \left(\frac{\rho}{1000} \right)^4$$

where
E_c is the modulus of elasticity (N/mm^2)
ρ is the density of concrete dried at RH 40% (kg/m^3).

2.3.1.1 *Effects of moisture on mechanical properties.* The moisture content influences the strength and modulus of elasticity in the same way as in ordinary concrete. The tests[71] compare the properties of wood fibre concretes stored at

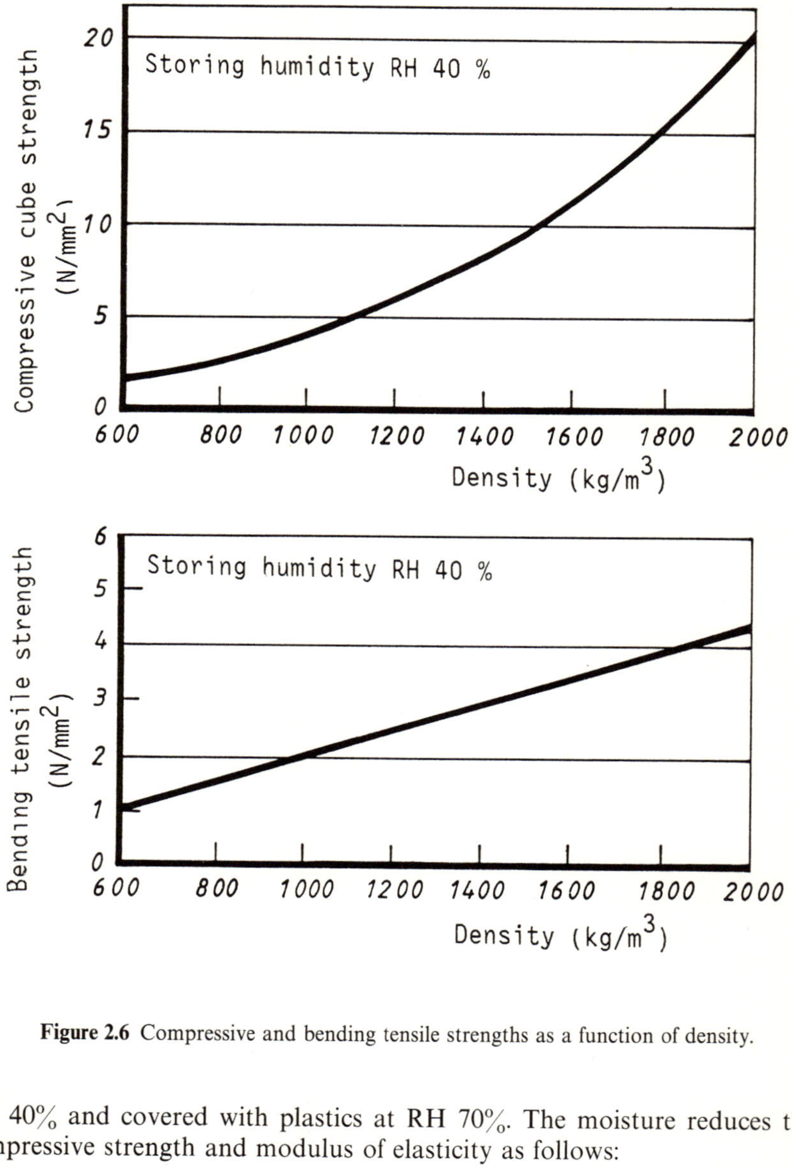

Figure 2.6 Compressive and bending tensile strengths as a function of density.

RH 40% and covered with plastics at RH 70%. The moisture reduces the compressive strength and modulus of elasticity as follows:

$$f_{cRH} = f_c \cdot \left(1 - 4 \cdot \frac{RH - 40}{\rho}\right)$$

$$E_{cRH} = E_c \cdot \left(1 - \frac{RH - 40}{\rho}\right)$$

where

f_{cRH} is the compressive strength at a given humidity RH (%)

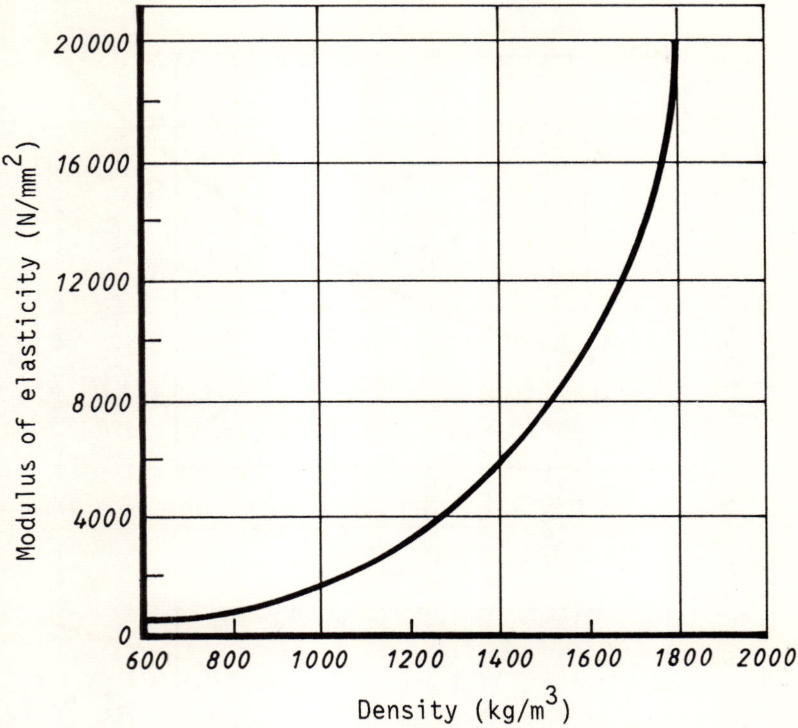

Figure 2.7 Modulus of elasticity as a function of density.

f_c is the basic compressive strength
E_{cRH} is the modulus of elasticity at a given humidity RH (%)
E_c is the basic modulus of elasticity
ρ is the density of concrete at RH 40% (kg/m^3)
RH is the relative humidity.

For example, at a density of 1200 kg/m^3, the difference in compressive strength at RH 40% and 80% is 13% and that in the modulus of elasticity 5%.

2.3.1.2 *Creep and shrinkage.* The creep coefficient of wood fibre concrete increases with decreasing density. Based on test results, the basic creep coefficient can be approximated as follows:

$$\phi = 3 \cdot \frac{2400}{\rho}$$

where

ϕ is the creep coefficient
ρ is the density of concrete dried at RH 40% (kg/m^3).

Shrinkage includes the effects of drying and chemical shrinkage. In natural fibre concrete, the natural fibre controls the moisture behaviour of the cement

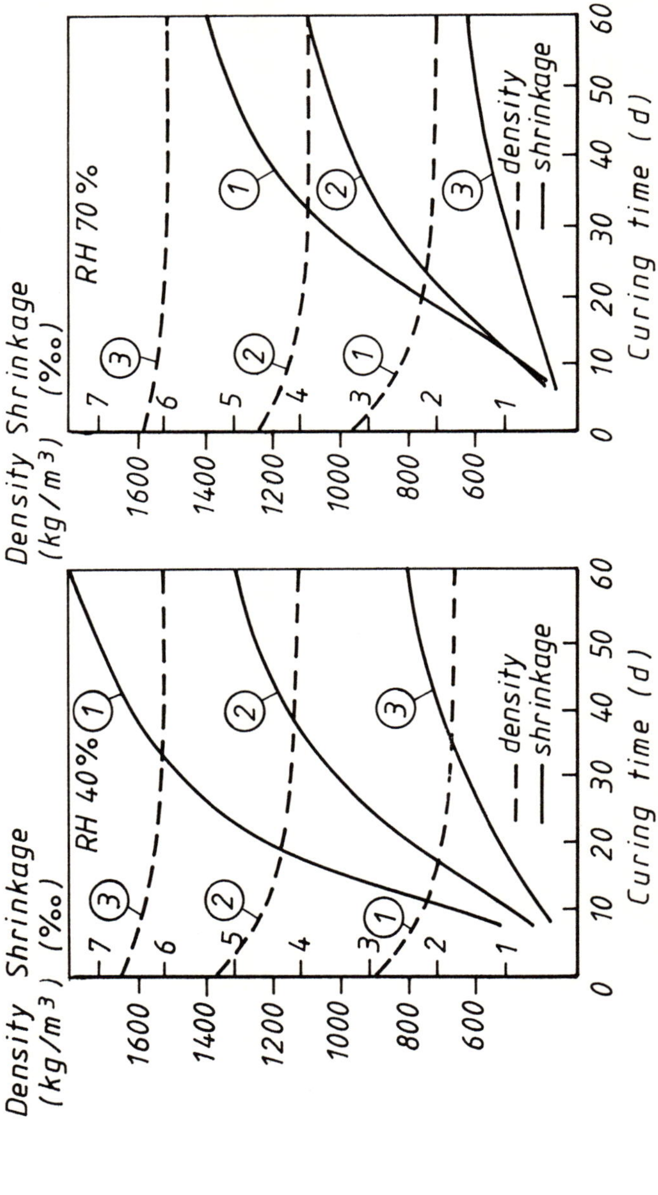

Figure 2.8 Shrinkage of wood fibre concretes in relation to density and relative humidity.

paste. Changes in the moisture content of the cement paste in turn affect the strain throughout the material. The shrinkage behaviour of natural fibres is therefore primarily controlled by the cement paste, while the moisture of the material is itself controlled by the natural fibres.

Since natural fibres give capillary porosity to the material, it naturally follows that the moisture content of this type of concrete will change rapidly with varying external moisture conditions. This also means rapid shrinkage of the material.

Figure 2.8 shows the relationship between shrinkage and density in wood fibre concretes at RH 40% and 70%.

The final shrinkage after casting can be estimated as follows:

$$\varepsilon_{sh} = 12\left(\frac{2800-\rho}{2400}\right)^2\left(\frac{200-\mathrm{RH}}{200}\right)\times 10^{-3}$$

Table 2.3 Basic physical properties of wood fibre concretes

	Type of wood fibre concrete			
Property	Wood wool	Wood chip cement	Wood particle cement	Structural wood fibre concrete
Thermal conductivity	0.10–0.15	0.10–0.15	0.25–0.35	0.37–0.80
Moisture conductivity (kg/Pa × 10^{-12})	Very high	50–25	20–15	15–10
Thermal straining coefficient (1/K × 10^{-5})	1.7–1.6	1.7–1.6	0.20–0.25	1.0–1.2
Swelling (%)	(1)	0.3–0.5	2–5	0.20–0.25
Health	No injurious constituents	No injurious constituents	No injurious constituents	No injurious constituents
Sound insulation	None	Moderate	Good	Good to high
Sound absorption	High (over 60%)	High (roughly 60%)	Good (roughly 40%)	Good (30–40%)
Flammability (2)	Poorly flammable	Poorly flammable	Non-flammable	Non-flammable
Fire resistance	Class 1, BS 476: Part 7, 1968	DIN B1	DIN B1	DIN A2

(1) Undefined DIN B1 = flame-resistant
(2) DIN 4102 DIN A2 = incombustible

where

ε_{sh} is the shrinkage strain $\times 10^{-3}$

ρ is the density of concrete dried at RH 40%

RH is the relative humidity (%).

The shrinkage naturally varies according to the water/cement ratio or the concrete mix and must be checked for every mix. In prefabrication manufacture the shrinkage of products is much less than the above figures because the greater part of shrinkage takes place almost immediately after casting, during storage in the factory.

2.3.2 *Physical properties*

The physical properties of the various wood fibre concretes are summarized in Table 2.3.

The physical properties of structural wood fibre concrete can also be related to the density[72] as follows.

2.3.2.1 *Moisture swelling and water absorption.*

The moisture swelling of wood fibre concrete has been tested by increasing the relative humidity around the hardened concrete from 40% to 90%. The test specimens showed no swelling strain during the test. In another test the specimens were cured at RH 65% and stored for 24 h in water. At densities of 700 and 1500 kg/m^3 the swelling strain was 3×10^{-3} and 2×10^{-3} respectively.[71] For the sake of comparison it should be noted that in this test the permitted swelling strain at the usual thickness of wood chip board is 8%.

The total water absorption was studied according to RILEM Recommendation No. 11.1, 'Absorption of water by immersion'. The following approximation was derived:

$$w = \left(\frac{800}{\rho}\right)50$$

where

w is the total water absorption compared with the dry weight of concrete oven-dried at 105°C (%)

ρ is the density of oven-dried concrete (kg/m^3)

2.3.2.2 *Thermal conductivity.*

The thermal conductivity can be evaluated as follows:

$$\lambda = (0.4\rho \times 10^{-3}) - 0.1$$

where

λ is the thermal conductivity (W/Km)

ρ is the density of concrete (kg/m^3)

2.3.2.3 *Sound insulation and sound absorption.*

Owing to its heterogeneity and

low modulus of elasticity, natural fibre concrete has good sound insulation properties. The numerical value of sound insulation is dependent on the structure as a whole and must therefore be evaluated on a case-by-case basis. A good example of excellent sound insulation is a Fibrecon hollow core wall panel 70 mm thick which, when rendered gives a sound insulation value exceeding 40 dB[71].

The sound absorption ratio is roughly

$$10\left(1+\frac{3000}{\rho}\right)\%,$$

where ρ is the density of concrete (kg/m^3).

2.3.2.4 *Thermal straining coefficient.* The thermal straining coefficient can be evaluated as follows:

$$\lambda = 0.08\left(1+\frac{600}{\rho}\right)\times 10^{-5}$$

where
λ is the thermal straining coefficient (1/K)
ρ is the density of concrete (kg/m^3).

2.3.2.5 *Moisture conductivity.* The moisture conductivity can be calculated as follows:

$$\lambda = \frac{1500}{\rho}\times 10^{-11}$$

where
λ is the moisture conductivity (kg/m s Pa)
ρ is the density of concrete (kg/m^3).

2.3.3 *Durability properties*

The durability properties of the various wood fibre concretes are summarized in Table 2.4.

2.4 Production processes

2.4.1 *Wood wool products*

Wood wool is composed of long chips and cement. The length of the chips is over 80 mm, their width 3–6 mm and thickness 0.2–0.5 mm.[9] The cement to wood ratio in parts per volume varies from 1:2 to 1:3.5, with more chips per unit volume in thicker panels than in thinner ones.[43]

Table 2.4 Durability properties of wood fibre concretes

	Type of wood fibre concrete			
Property	Wood wool	Wood chip cement	Wood particle cement	Structural wood fibre concrete
Moisture resistance	resistant	resistant	resistant (1)	resistant
Frost resistance	(2) (3)	(2) (3)	resistant (3)	resistant (3)
Rot and fungus resistance	resistant	resistant	resistant	resistant
Termite	(4)	(4)	(5)	resistant (6)

(1) Resistant when the binder mineral is portland cement, blended cement or activated blastfurnace slag
(2) Nonresistant if there is no coating; resistant when coated with plaster
(3) Resistant when the basic cement mortar meets the resistance properties and the outer surface is coated with paint, rendering or plaster
(4) No exact results exist; evaluated on the basis of tests with lightweight structural wood fibre concrete to be fairly resistant
(5) No exact results exist; evaluated on the basis of tests with structural wood fibre concrete to be moderately resistant
(6) Termite resistance tested and found to be as good or better than that of ordinary concrete[71].

2.4.2 *Wood chip cement products*

Typical wood chip products include blocks and bricks.

Wood chip cement is mixed in a specially manufactured power mixer at a water–cement ratio[43] of 0.8 to 1.0. The mix is spread on to the form and compacted at a compressive stress not exceeding 0.15 MPa. The panels are then superimposed, together with the forms, and kept under load for one day. They are then dismantled from the forms and transferred to a humid, suitably warm storage to harden for two weeks. The manufacturing process can be accelerated by using finely ground cement or magnesite[9] as a binder or by hardening the panels under steam at low pressure after dismantling. Finally, the panels are dried before being delivered to the customer.

The manufacture of wood chip cement products is illustrated in Figure 2.9.

2.4.3 *Wood particle cement boards*

The manufacturing method is rather similar to that of conventional wood particle board. The particles are dried and mixed with magnesium and its salt solution, and the mix is spread out and compressed onto a board. On the finishing line the panel is cooled, sanded, trimmed and packed.

Figure 2.10 shows a typical manufacturing line for cement particle boards. The mix is spread on to mobile forms. After casting the panels are sawn and superimposed along with their forms. The pile is then compressed hydraulically and locked in such a way that the compressive force is maintained while

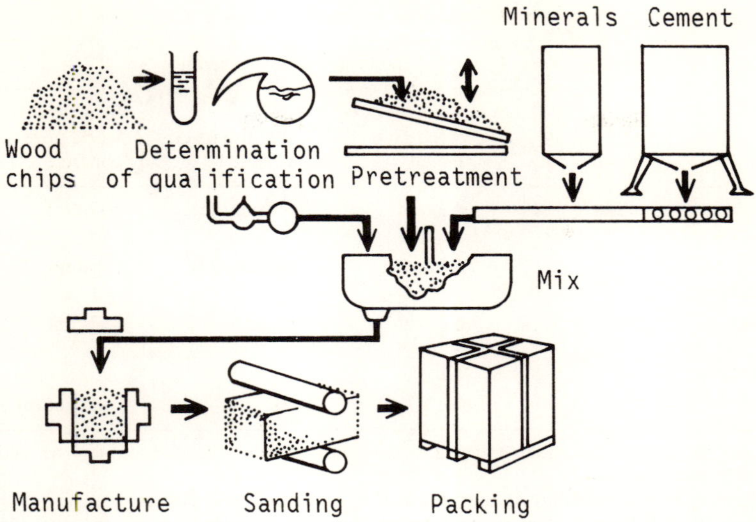

Figure 2.9 Manufacture of wood chip cement products[3].

the pile is transferred to pre-storage. The compressive stress varies from 2.0 to 2.5 MPa.[42] While in pre-storage the panels attain suitable strength for dismantling of the forms. The storage time depends on the temperature, which is generally around 60°C, and varies from six to eight hours.[52]

Although the panels themselves are not actually heated, heat develops mainly as a result of cement hydration, and the temperature is maintained by external heating. Once the forms are dismantled, the panels are transferred to final storage, where they are dried to an equilibrium reflecting climatic conditions. Finally, the panels are trimmed and the surfaces sanded as required.

2.4.4 *Structural wood fibre concrete products*

Figure 2.11 shows the material flow and processing of structural wood fibre concrete products. Wood chips are transported by road from the silos and then pneumatically to the flake mill which renders them more uniform. The chips are then blown pneumatically into the storage silo before being transferred by screw conveyor via a batching device into the mixer. Before the actual preparation of the concrete mix the wood fibres are treated with chemicals to make them more compatible with the cement.

Cement is brought to the station in pressurized containers and is pneumatically transferred to the storage silo. Where a cement is transported in sacks a receiving silo and sack shredder are required. Aggregates (mixtures of sand and filler) are brought to the receiving hopper at the mixing plant by means of a front-loader and lifted into the storage hoppers by cup elevator.

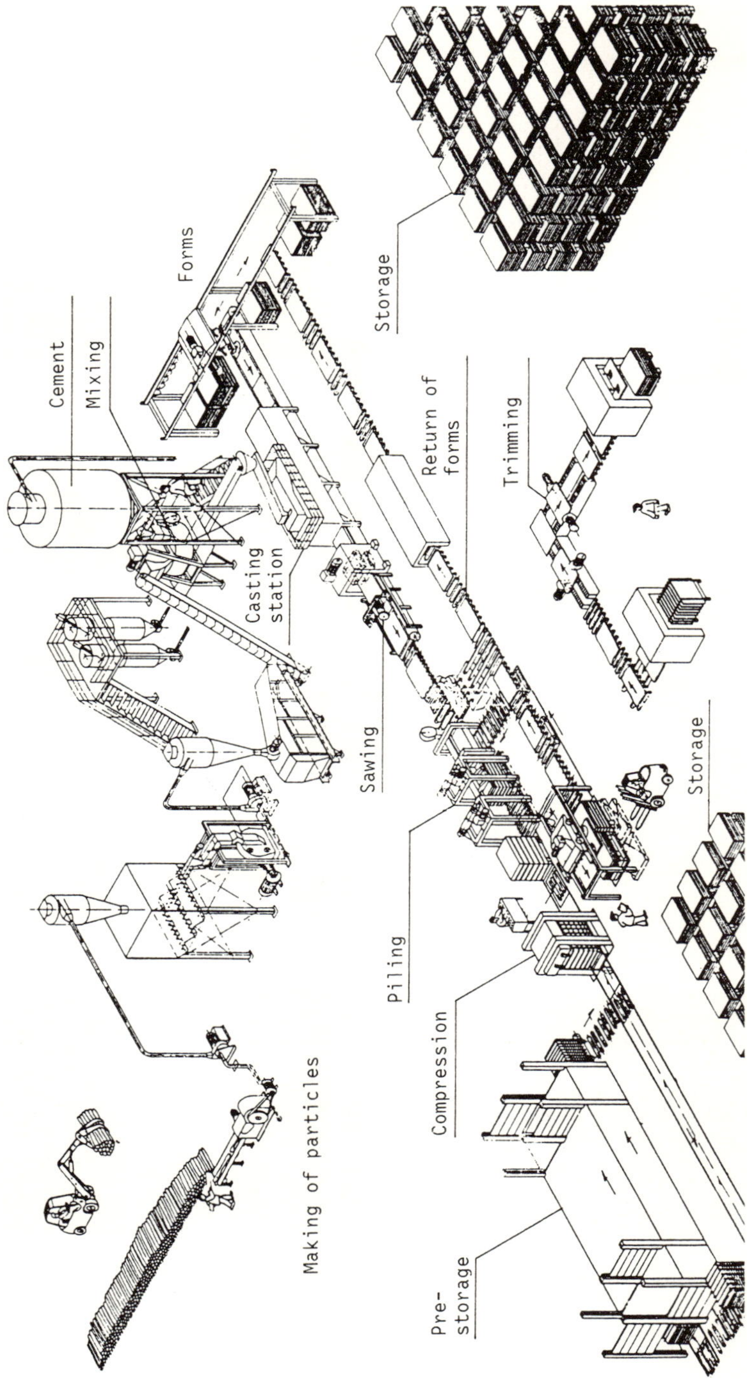

Figure 2.10 Typical manufacturing line for particle cement boards[55].

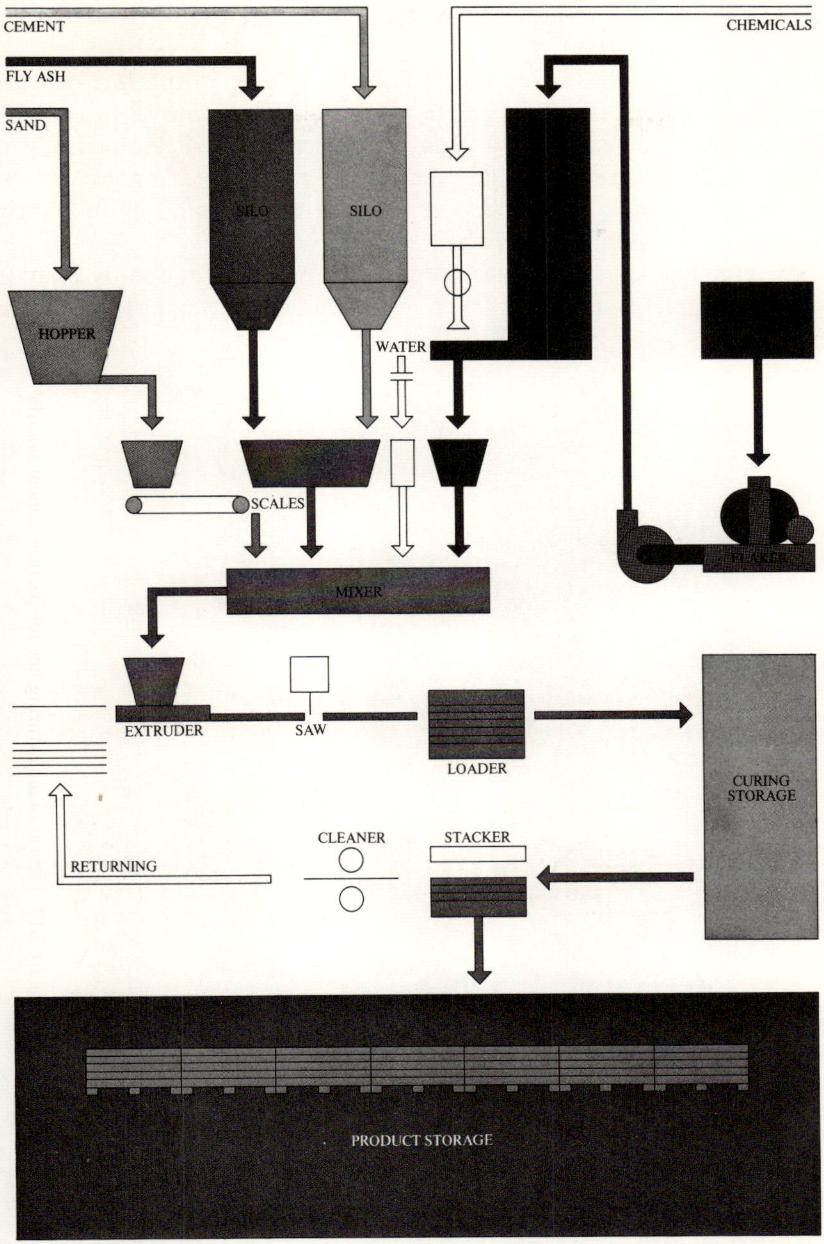

Figure 2.11 Factory production of structural wood fibre concrete[75].

The concrete mix is fabricated in batches in a special mixer. Continuous mixing is also possible.

After mixing, the material is brought to the receiving hopper of an extruder-type casting machine which formulates the mix into elements of the desired design. Following the extruding process, the element is sawed automatically to its final length. The dimensioned elements are then stacked in pallets of ten onto a special loader. The pallets are transported automatically to curing storage for prehardening over 24 h.

Less automated and mechanized production processes and mobile plants have also been planned.[75] Test houses have been built by mixing the concrete

Figure 2.12 Pumping of structural wood fibre concrete into walls built from polystyrene blocks[71,72].

at an ordinary ready-mix plant and adding the structural wood fibre concrete to the structure via an ordinary concrete pump.[71,72] Such a production process is illustrated in Figure 2.12.

2.5 Products and their applications in construction

The possibilities afforded by wood fibre concrete products in house building are endless. These products are suitable for almost any light, non-loadbearing structure, structural wood fibre concrete also being suitable for loadbearing structures. They are most suitable, however, where those properties that exceed those of other materials, such as fire technical properties, acoustic properties, resistance to humidity, durability and lightness, can more effectively be exploited. In addition to ordinary building these products can also be used in the construction of small houses, renovation and noise shelters.

2.5.1 *Wood wool*

Wood wool panels are widely used as acoustic surface panels, thermal insulation in walls, floors and roofs and as moulding panels. A typical wood wool panel structure is shown in Figure 2.13. Wood wool panels are light, relatively strong products intended for different protective and coating purposes. Standards for their properties, testing and use have been set at least in Britain, the Federal Republic of Germany and Austria (BS 1105[59], ÖNORM B3465[58], DIN 1101[14]). Table 2.5 shows the properties of wood wool panels according to the Austrian standards.[58]

Figure 2.13 A typical wood wool panel structure.

Table 2.5 Properties of wood wool panels[58]

Panel length, mm (+ 5, − 10 mm)[1]			2000			
Panel width, mm (± 3 mm)[1]			500			
Panel thickness, mm (+ 3, − 2 mm)[1]	15	25	35	50	75	100
Weight, kg/m^2	8.5	11.5	14.5	19.5	28.0	36.0
Density, kg/m^3	570	460	415	390	375	360
Flexural tensile strength minimum mean value, MN/m^2	1.7	1.0	0.7	0.5	0.4	0.4
Compression strain (%) at 0.3 MN/m^2 load – maximum mean value	10	15	18	22	25	25

[1]Tolerances

2.5.2 *Wood chip cement*

Various insulation boards and moulding elements for casting on site are manufactured in Central Europe from wood fibre concrete. They are made of chip cement and their manufacturing process is similar to that of form blocks. Even blast blocks for ventilation ducts[20] can be made from wood fibre concrete, as can blocks, floor covering slabs and bound stones[43]. The international market also offers form elements for walls, made from wood fibre concrete[45]. These are rectangular, hollow and room-sized. As with form block walls, the inside is filled with concrete. This form of construction is known as case concrete construction.

Masonry wood chip cement panels are used in the German Democratic Republic and are rectangular panels with grooved edges (Figure 2.14). In addition to the standard sizes indicated in the figure, the product assortment includes shortened panels with a minimum length of 160 mm. The panels are manufactured from inorganic binder and generally from cement and wood chips. Their density averages 800 kg/m^3 and compressive strength 1.5 MN/m^2. Form blocks are used in wall construction and are illustrated in Figure 2.15.

In Central Europe, wood fibre concrete is used as insulation in the core of sandwiched external wall elements (Figure 2.16). The outer envelope, which is roughly 30 mm thick, consists of ordinary concrete. The outer surface can be smooth or grooved, and even a crushed surface is possible. The wood fibre concrete – which in this case is wood chip cement – is spread over the fresh concrete, ensuring good adherence between layers. The inner envelope consists of mortar.

Close to the outer envelope in the wood chip cement layer are vertical ventilation ducts, connected to open horizontal joints. Their purpose is to reduce structural moisture and remove condensed water.

Wall elements fall into two types, normal-sized and large. During manufacture it is possible to place extra insulation within the elements thereby decreasing their thermal transmission coefficient.

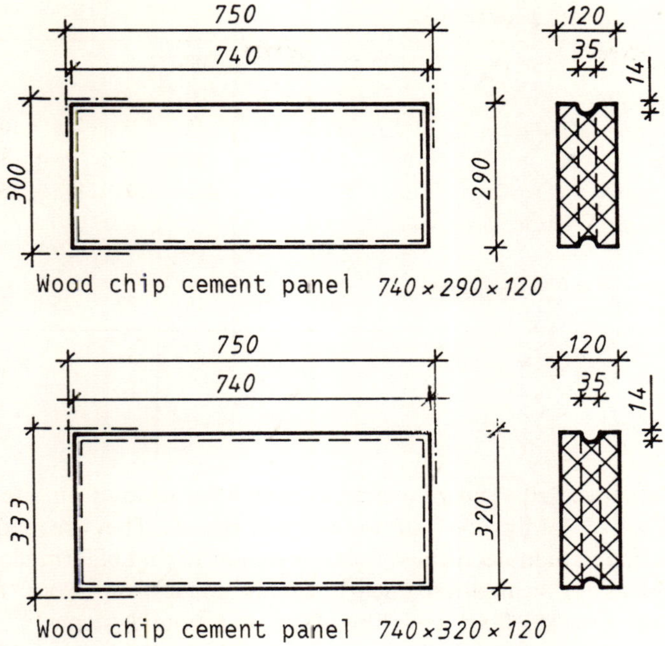

Figure 2.14 Masonry wood chip cement panels[7].

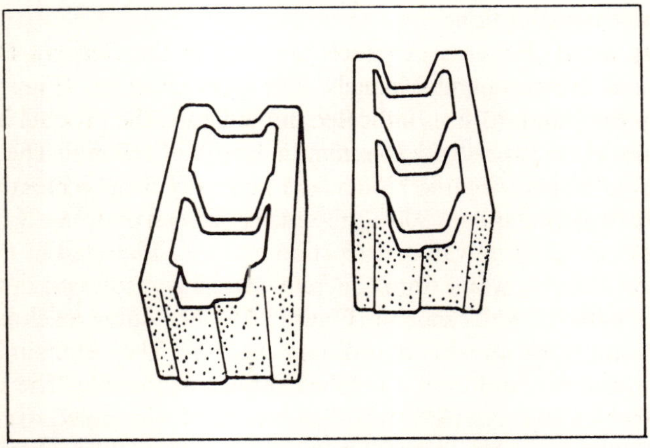

Figure 2.15 Form blocks made from wood chip cement[9].

2.5.3 *Wood particle cement boards*

Wood particle cement boards can be used for indoor walls, fire walls, ceilings and floors of houses, flats, institutions, offices, shops, restaurants, schools, hospitals, industrial buildings, exhibition halls, most public buildings, agricultural buildings, warehouses, etc. Further applications include fire doors,

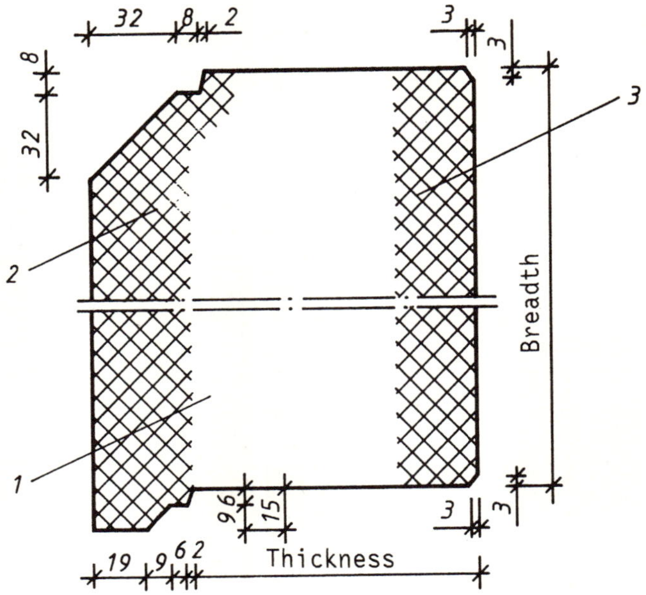

1 - wood chip cement

2 - concrete layer (fine aggregate)

3 - mortar layer

Figure 2.16 Wood chip cement in the core of a sandwiched external wall element[3].

prefabricated houses, renovation of old buildings, double flooring of computer rooms, control rooms, offices etc., and also furniture in which the supply of board materials has been limited for fire protection purposes – previously only asbestos cement and vermiculite boards were available.

2.5.4 *Structural wood fibre concrete*

The first practical application of this concrete so far has been a wall panel system for non-bearing internal and external walls of multi-storey buildings and for bearing walls and slabs of small houses[75]. The products have also been applied to structural systems – for instance the wall systems of multistorey buildings and low-cost small houses. Finnish companies have developed special applications for low-cost construction in warm or hot climates[75].

These components can be used to build load-bearing and non-bearing walls, ceilings, roofs and floors and load-bearing slabs. In principle, low-rise one- and two-storey buildings can be built entirely from the products shown in Figure 2.17[75]. The automated production of structural wood fibre concrete panels is illustrated in Figure 2.18[75] and a typical construction system in Figure 2.19.

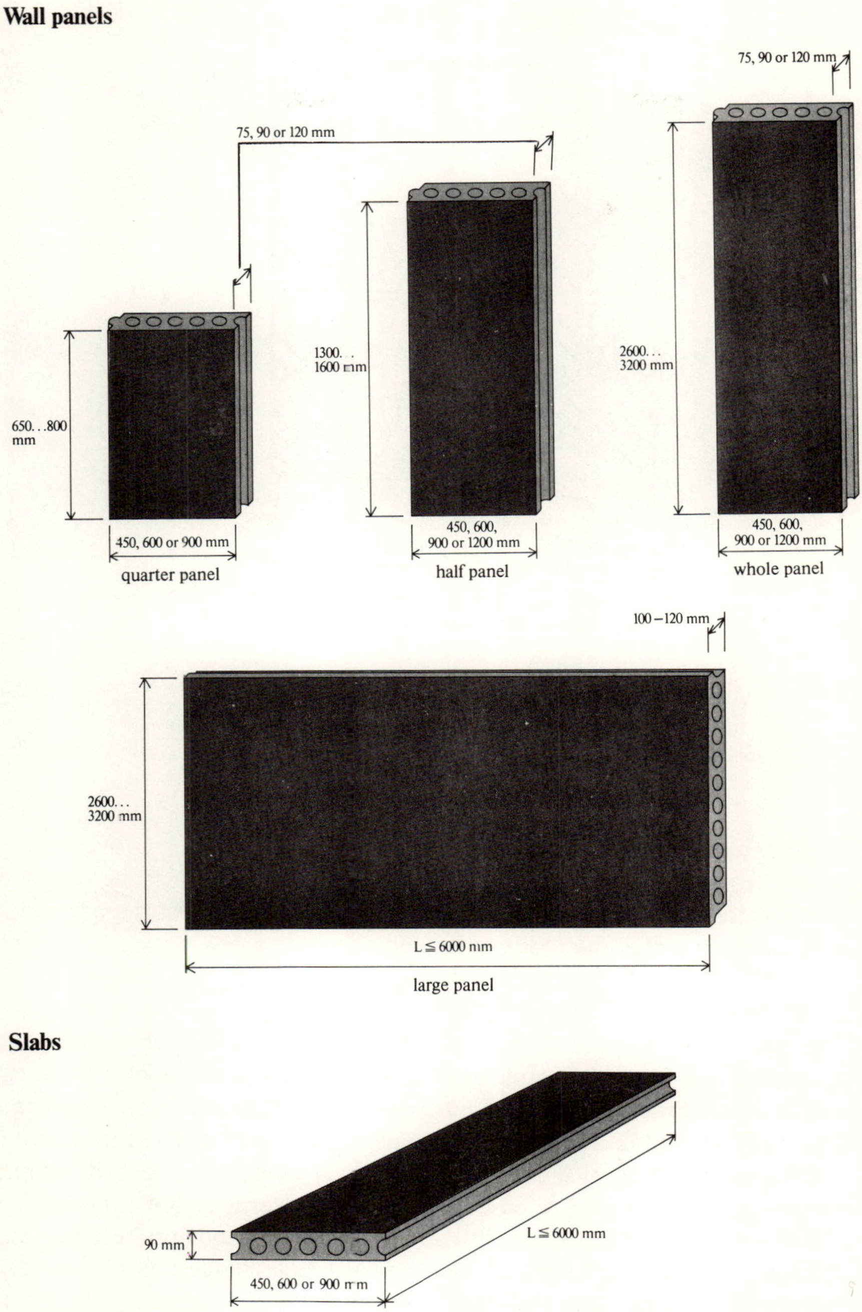

Figure 2.17 Types of structural wood fibre concrete panels and their basic dimensions[75].

Figure 2.18 Automated production plant of structural wood fibre concrete panels[75].

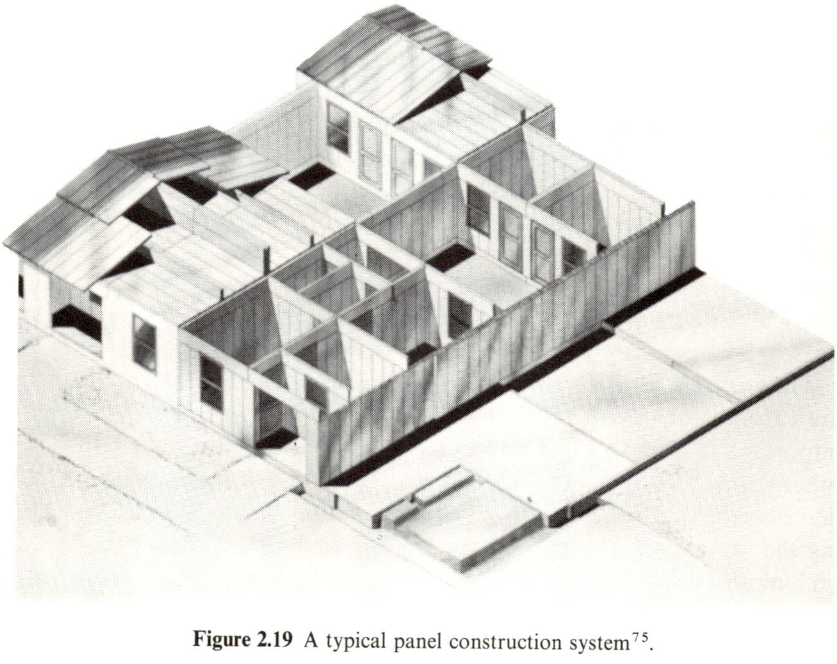

Figure 2.19 A typical panel construction system[75].

Walls. The panel walls are constructed vertically with the aid of supporting rods. The panels are attached by injecting cement into the vertical seams.

A row of horizontal panels is installed above the vertical panels at the upper end of the wall. The panels are attached to each other and to the wall by the injection of cement. This row of panels forms the beam structure above the windows and doors and stabilizes the wall structure.

Ceiling (unloaded). Panels with a slightly modified side profile are used for the ceiling. They are installed side by side and supported at their ends by the wall. The seams are reinforced with a 6–12 mm reinforcing bar, and concrete is injected under pressure into the seam.

Upper floor (loaded). The loaded floors are built in the same manner as the ceiling. The thickness of the element should be chosen in accordance with the load and span. The upper surface is smoothed with a filler.

Roof. The roof can also be constructed from panels. The structure resembles the upper floor but should have an inclination of at least 1:3. Water resistance of the seams is ensured by elevating the cement in the seam.

In demanding climates the roof must be covered with a suitable roofing material.

2.6 Other vegetable fibres

Until now, natural fibre concrete technology has primarily focused on wood fibres from sources such as conifer-type and rubber trees. Extensive studies on other natural fibres, however, such as bamboo, flax, rice straw, jute, coir, rice husk and coconut, show that most of these are in fact suitable for a wide range of products[31,60–62,66,70,72,80–82].

When beginning to apply new fibres – be they new types of wood or other vegetable fibres – it is extremely important to become thoroughly acquainted with not only the production process but also the mechanical, physical and durability properties of the product. This can be done by means of a feasibility study under consultation of the suppliers of the manufacturing process, who possess the necessary knowledge.

The mechanical, physical and durability properties described in this chapter are a useful preliminary design tool. However, even for wood fibre concrete it is important to check all parameters separately for each product. When using other natural fibres, many of the properties may differ extensively from those described here. Advice should therefore be sought from laboratories possessing the necessary knowledge in the appropriate feasibility study or phase of preliminary design.

2.7 Concluding remarks

The evolution of wood fibre concrete technology has made it possible to provide a wide range of products and production processes for construction. The material has major assets in both developing and industrialized countries.

In developing countries the use of local materials can be improved. In addition to wood fibres, many other natural fibres are applicable on the same principles. The properties of wood fibre concrete vary extensively, enabling the adjustment of properties according to structural and production needs. Depending on local economic and social conditions, the choice can be made between a highly automated production process and simple manual methods.

In industrialized countries the good properties of wood fibre concrete serve a particular purpose, for example in acoustics, fire resistance and durability. Its high workability is also helpful in the manufacture of products in factory and on site assembly. The automation of production in the factory produces economic results. Special areas of application are the construction of new buildings and renovation.

Much research has been carried out over the last few decades on the chemical, physical, mechanical and durability properties of wood fibre concrete. Today there is also a vast amount of practical experience. Certain wood fibre concretes have already been in use for more than fifty years. It is on this that practical knowledge concerning the new types of wood fibre concrete is based.

In the design and manufacture of wood fibre concrete materials, it is important that the user familiarizes himself with the properties and suitability of both the raw materials and final product in order to make the correct choice of material and product according to their intended purpose.

References

1. Absorption of water by immersion. RILEM recommendation No 11.1/-CPC.
2. Aussenwandelemente, DAP-1. Lieferprogram. Villmergen 1983, Durisol AG, 6 pp.
3. Aussenwandelemente, DAP-NH-2. Konstruktionsblätter. Villmergen 1982, Durisol AG, 8 pp.
4. Aussenwandelemente, DAP-NV-2. Konstruktionsblätter. Villmergen 1982, Durisol AG, 8 pp.
5. Bröker, F.W. Schwindverhalten zementgebundener Holzwerkstoffe. BFA 1973:93, 133–141.
6. Bröker, F.W. and Simatupang, M.H. (1974) Ursachen der Dimensionänderung zementgebundener Holzwerkstoffe. *Holz als Roh- und Werkstoff* **32**(4), 150–155.
7. Bursian, G. *et al.* (1973). *Holzbeton. Produktion, Anwendung und Erfahrungen.* Schriftenreihen der Bauforschung, Reihe Baustoffe, Heft 20, 23 pp., Berlin.
8. Culetto, K. (1970) Elementbauweise mit Durisol. *Schweizerische Schreinerzeitung* **81**(4), 69–74.
9. Cziesielski, E. (1975) Beton mit Fasern aus Holz. *Holz als Roh- und Werkstoff* **33**(8), 303–307.
10. Dachelemente, DDP-1. Lieferprogramm. Villmergen 1981, Durisol AG, 6 pp.
11. Dachelemente, DDPG-2. Konstruktionsblätter. Villmergen 1982, Durisol AG, 8 pp.
12. Dachelemente, DDPN-2 Konstruktionsblätter. Villmergen 1982, Durisol AG, 8 pp.
13. Dietrichs, H.H. (1964) Das Verhalten von Kohlenhydraten bei der Holzverkernung. *Holzforschung* **18**(1/2), 14–24.

14. DIN 1101. Holzwolle-Leichtbauplatten. Masse, Anforderungen, Prüfung. Deutsches Institut für Normung, 1980, 6 pp.
15. DIN 4102. Teil 1. Brandverhalten von Baustoffen und Bauteilen. Baustoffe. Begriffe, Anforderungen und Prüfungen. Deutsches Institut für Normung, 1981, 15 pp.
16. DIN 53 122. Teil 1. Prüfung von Kunststoff-Folien, Elastomerfolien, Papier, Pappe und anderen Fläckengebinden. Bestimmung der Wasserdampfdurchlässigkeit. Gravimetrisches Verfahren. Deutsches Institut für Normung, 1974, 4 pp.
17. Duripanel, DUP-1. Lieferprogramm. Villmergen, Durisol AG, 4 pp.
18. Duripanel, DUP-2 Anwendung. Villmergen 1982, Durisol AG, 17 pp.
19. Fassaden mit Schalungssteinen und Plattenverkleidung. *Baumarkt* **74** (1985)32, 829–830.
20. Herzig, E. (1966) Durisol-Verfahren, Durisol-Produkte und deren Anwendungen. In: *Chemie und chemische Technologie mineralgebundener Holzwerkstoffe*, eds. W. Sanderman and G. Schmitz, Institut für Holzchemie und chemische Technologie des Holzes der Bundesforschungsanstalt für Forst- und Holzwirtschaft, Hamburg, 106–116.
21. Hickin, N.E. (1971) *Termites – a World Problem.* Hutchinson/Rentokil, Essex, 59–65.
22. Hourdis, DH-1. Lieferprogramm. Villmergen 1981, Durisol AG, 4 pp.
23. Hourdis, DH-2. Verlegeanleitung. Villmergen 1982. Durisol AG, 8 pp.
24. Hourdis, DKH-1. Lieferprogramm. Villmergen 1981, Durisol AG, 4 pp.
25. Hourdis, DKH-2. Verlegeanleitung. Villmergen 1982, Durisol AG, 8 pp.
26. Huffaker, E.M. (1962) Use of planer mill residues in wood-fiber concrete. *Forest Prod. J.* **12** (7), 298–301.
27. Isolierplatten, DP-1. Lieferprogramm. Villmergen 1981, Durisol AG, 4 pp.
28. Isolierplatten, DP-2. Verlegeanleitung. Villmergen 1982, Durisol AG, 8 pp.
29. Isopanel, die aussergewöhnliche Bauplatte für universelle Beanspruchung. Fulgurit (1980) 15 pp.
30. Johnson, R.H. and Graft, S.H. (1930) The properties of cement-sawdust mortars, plain and with various admixtures. Oregon State College, Bulletin series **3**, Corvallis, 26 pp.
31. *Joint RILEM/NCCL/CIB symposium, on use of vegetable plants and their fibres as building materials*, Baghdad, 7–9 October 1986, 452 pp.
32. Kaiser, E. (1949) *Holz-, Schaum- und Gasbeton. Exproble Verfahren.* Verlag Jacob Mendelsohn, Nürnberg, 39 pp.
33. Kasai, Y. *et al.* (1982) Production and properties of cemented crushed wood cement board made of used timber. *Trans. Japan Concr. Inst.* **4**, 17–24.
34. Konventionelle Aussführung in Holzspan-Schalungssteinen. *Bauhandwerk* **33**(1981)6, 491–493.
35. Mauersteine, DM-1. Lieferprogramm. Villmergen 1983, Durisol AG, 8 pp.
36. Mauersteine, DM-2. Verlegeanleitung. Villmergen 1982, Durisol AG, 27 pp.
37. Nicht, J. *et al.* (1971) Verarbeitung von Sägespänen und Rinde zu plattenförmigen Werkstoffen (III). *Holzindustrie* **24** (11), 326–329.
38. Paramasivam, P. and Loke, Y. (1980) Study of sawdust concrete. *Int. J. Lightweight Concr.* **2**(1), 57–61.
39. Parameswaran, N. *et al.* (1977) Zur Mikrotechnologie mineralgebundener Holzwerkstoffe. *Holzforschung* **31**(6), 173–178.
40. US Patent 2,623,828. Cementitious material and method of making. (L.P. Dove). (30.12.1952). Appl. 83,527, 25.11.1949.
41. US Patent 4,406,703. Composite materials made from plant fibers bonded with portland cement and method of producing same. (B.M. Guthrie & R.B. Torley), (27.9.1983). Appl. 213,040, 4.12.1980.
42. Problemlösung Holz-Zement-Bauplatten. *Holz-Zentralblatt* **102** (1976)82, 118–119.
43. Probst, E. (1982) *Handbuch der Betonsteinindustrie mit Fachwörterbuch.* 7th edn., Carl Marhold, Berlin, 668 pp.
44. Sandermann, W. (1966) Chemie und Technologie mineralgebundener Holzwerkstoffe. *Holz-Zentralblatt* **92**(13), 238–239.
45. Sandermann, W. (1969) Untersuchungen über das System Holz-Zement. *BFA* **73**, 13–22.
46. Sandermann, W. *et al.* (1960) Studien über mineralgebundene Holzwerkstoffe. *Holzforschung* **14** (3), 70–77.
47. Sandermann, W. and Brendel, M. (1956) Die zementvergiftende Wirkung von Holz-

inhaltsstoffen und ihre Abhängigkeit von der chemischen Konstitution. *Holz als Roh- und Werkstoff* **14**, 307–313.

48. Sandermann, W. and Kohler, R. (1964) Über eine kurze Eignungsprüfung von Hölzern für zementgebundene Werkstoffe. *Holzforschung* **18** (1/2) 53–59.
49. SFS 3516. Wood particle boards. Methods of test, 1984, 9 pp.
50. SFS 4152. Particle boards. V-313-boards (type III). Properties. Finnish Standards Association, 1984, 3 pp.
51. SFS 4153. Particle boards. V-313-boards. Methods of test. Finnish Standards Association, 1984, 6 pp.
52. Trennwand, W.T.W. Bison System. Bison element Bähre & Greten GmbH & Co KG, D 3454 Bewern, Flüttenweg.
53. Weatherwax, R.C. and Tarkow, H. (1967) Decayed wood as an inhibitor. *Forest Prod. J.* **17** (7), 30–32.
54. Weatherwax, R.C. and Tarkow, H. (1964) Effect of wood on setting of portland cement. *Forest Prod. J.* **14** (12), 567–570.
55. *Wood-Cement Board.* H.V. Verkor, Lauwe, Belgium, 7 pp.
56. Önorm B 3208. Mantelsteine. Österreichisches Normungsinstitut 1976, 6 pp.
57. Önorm B 3352. Mantelbetonwände. Österreichisches Normungsinstitut 1973, 8 pp.
58. Önorm B 3465. Holzwolle-Leichtbauplatten. Österreichisches Normungsinstitut 1965, 4 pp.
59. BS 1105. Specification for wood cement slabs up to 125 mm thick. British Standards Institution London, 1981, 8 pp.
60. Arnaouti, C. and Illston, J.M. Tests on cement mortars reinforced with natural fibres. The Hatfield Polytechnic School of Engineering, Division of Civil Engineering, 1979.
61. Balaguru, P.N. and Shah, S.P. (1985) Alternative reinforcing materials for less developed countries. *Int. J. for Devel. Technol.* **3**, 87–105.
62. Coutts, R.S.P. (1983) Flax fibres as a reinforcement in cement mortars. *Int. J. Cement Comp. and Lightweight Concr.* **5** (4) 257–262.
63. Ehrentreich, W. (1976) Problemlösung Holz-Zement-Bauplatten. In Wunstorf bei Hannover laeuft die Produktion auf vollen Touren. *Holz-Zement-Blatt* **102** (82), 1118–1119.
64. Elmendorf, A. (1951) New fiber boards made with noncritical binders. Forest Products Research. Society.
65. Friberg, W.R. Wood fibre diatomic concrete. Lightweight building material. University of Idaho, College of Agriculture, Extension bulletin **179.**
66. Gram, H.-E. Durability studies of natural organic fibres in concrete, mortar or cement. *RILEM Symposium FRC 86, Third Int. Symp. on Developments in Fibre Reinforced Cement and Concrete*, Vol. 2.
67. Merz, T. (1978) Die dichte Betonschale. Anwendung von Holzwolle-Leichtbauplatten. *Bauplatt* **89** (10) 3–7.
68. Neuentwicklung: Durisol-Schalungsstein mit Lüftungskanälen. *Dachdecker Meister* **29** (1976)3, 20–24.
69. Prestemon, D.R. (1976) Preliminary evaluation of a wood-cement composite. *Forest Prod. J.* **26**, 2.
70. Ramaswamy, H.S., Ahuja, B.M. and Krishnamoorthy, S. (1983) Behaviour of concrete reinforced with jute, coir and bamboo fibres. *Int. J. Cement Comp. and Lightweight Concr.* **5**(1), 3–13.
71. Sarja, A. Paroll, H., Hakkarainen, T., Koskinen, P. and Pyy, H. Structural wood fibre concrete (in Finnish). Manuscript 1986. 54 p. + appendices.
72. Sarja, A. Structural concrete with wooden or other vegetable fibres. *Joint RILEM/NCCL/CIB Symp.*, Baghdad, October 1986, 117–126.
73. Sarja, A. (1985) New structural fibre concrete. *Finnish Trade Review* **7**, p. 28.
74. Sarja, A. Method for the transformation of certain properties. International patent application PCT/FI85/00005.
75. Structural natural fibre concrete technology. Product description. Advanced Construction Technology, Helsinki 1987.
76. The structural use of chipboard for flooring. A state-of the-art report prepared jointly by the Institution and the Timber Research & Development Association. *Struct. Engineer.* **62** (1984)1, 35–38.

77. Wood fibre building boards and adhesives. *Wood* **20** (1955)1, 30–32.
78. *Wood Fibre Concrete*. Concrete Products. (1987)**3**, 1 p.
79. Zollinger, R.M. (1956) Das Schalen mit Holzfaserplatten. *Bauhandw*. **8** (2), 66–88.
80. Spence, R.J.S. and Cook, D.J. (1983) *Building Materials in Developing Countries*. New York, 338 pp.
81. Appropriate building materials for low cost housing. African region, Vol. I & II. *Proc. Symp.*, Nairobi, November 7–14, 1983, Spon, London, 1983.
82. Building materials for low-income housing. Asian and Pacific region. *Proc. Symp.*, Bangkok, Thailand January 20–26, 1987, Spon, London, 1987, 357 pp.

3 Bamboo reinforcement for cement and concrete

L. ROBLES-AUSTRIACO and R.P. PAMA

Abstract

Bamboo is the fastest-growing and highest-yielding renewable natural resource. It is used as it naturally occurs for housing, utility structures and for many other articles. As reinforcement for cement mortar and concrete, bamboo is gaining acceptance in developing countries. Its wider use will be greatly influenced by the understanding of its mechanical properties and physical behaviour. The properties of bamboo not only vary with individual species, but also depend upon many factors, such as species, age, moisture content, extent of seasoning, time of cutting, maturity, soil and climatic conditions.

The principal problems associated with the use of bamboo as a reinforcement are: volume change, moisture susceptibility, consequent loss of bond and decay. Solutions recommended by researchers and engineers to this problems are presented. In addition, topics included are research and development on bamboo and its properties, bamboo as reinforcement for slabs, beams, walls, roofs, columns, water containers, well and reservoir linings, silos, septic tanks, slabs on grade and soil-cement. Also discussed are composites reinforced with bamboo fibres, pulps and strips. Also highlighted are recommended practices and practical applications.

3.1 Introduction

Bamboo, a perennial grass belonging to the class Monocotyledoneae, exists abundantly in many developing countries. More than 10 million tonnes of bamboos are produced annually in the world. Almost all these come from Asia. It is estimated that in China alone there is an annual growth of 3.5 tonnes[1].

Bamboo grows in tropical, subtropical and temperate zones. It is the fastest-growing and highest-yielding renewable natural resource. The genera and species are widely distributed, but nowhere is their usefulness as great as in Asia. Bamboo is used as it naturally occurs for housing, utility structures and for many articles of everyday use. It is also used for fuel, food and feed. However, bamboo research has not received the necessary recognition in any country except Japan and perhaps China, and in fact has gained importance only during the last decade. Now research is in progress on methods of seed storage and identification of bamboo by characters of the culm sheaths on the new shoots and rhizomes. Experiments on vegetative propagation techniques are in progress in Bangladesh, and the possibility of harvesting bamboos annually is being investigated by the University of Agricultural Sciences, Dharwar, India[1].

Considerable work is in progress at the Forest Research Institute, College, Laguna, Philippines, to grow artificially shaped bamboos for use in housing. Bamboos (*Bambusa vulgaris*) have been made to grow into square, rectangular and triangular shapes for low-cost housing, handicrafts, decorative items and cottages. Other on-going research on bamboos in the Philippines deals with anatomy, preservation and harvesting method[1–4]. Japan has also successfully grown artificially shaped bamboos. Production is simple. A wooden mould, in the desired shape, is installed on the young bamboo shoot. After about six months, when the bamboo has taken on the desired shape, the mould is removed[1].

More research on seasoning and treatment of bamboo by easily procurable and cheap methods has been initiated in different countries. Also of interest is research on methods of establishing bamboo plantations (Figures 3.1, 3.2), their management technique, economic evaluation and chemical treatment[5–7].

Figure 3.1 *Thyrsostachys siamensis* plantation. Courtesy of the Royal Forest Department, Ministry of Agriculture and Cooperatives, Thailand.

Figure 3.2 Seedlings of *Dendrocalamus asper*. Courtesy of the Royal Forest Department, Ministry of Agriculture and Cooperatives, Thailand.

The potential of bamboo as a reinforcement for cement matrix was recognized only after the pioneering investigation of Chu, using bamboo to reinforce concrete, in 1914 at the Massachusetts Institute of Technology[8]. In China, the use of bamboo for reinforcing concrete has been practised since 1918. A reinforced concrete roof over a school at Canton was constructed in 1918, in which bamboo was used. The roof spans 3 m and is about 110 mm thick. The structure was reported to be in good condition after 40 years[9]. In 1919, bamboo reinforcement was used in concrete piles in railways[10].

Research scientists in Germany[11], Italy[12] and the United States[13–19] have investigated, to varying extents, the problems, advantages and disadvantages of using bamboo as reinforcing material. Early attempts have been made by Datta[11] and De Simone[12] to find a use for bamboo in reinforced concrete structures. Glenn[13] conducted an extensive study on bamboo reinforced concrete, involving tests on beams and construction of small structures. The use of bamboo as reinforcing material in light, emergency concrete construction was investigated by the US Army Engineer Waterways Experiment Station[14–18], and Youssef[20] conducted research on *Arundinaria gigantea*, a species of bamboo found in Egypt.

Bamboo, which is cheap and available in abundance in most parts of Asia, has attracted the attention of many research scientists in India[21–31], Indonesia[32], Japan[33], the Philippines[34–36], Malaysia[37–40] and

Thailand[41–51]. In India, research programmes on composite construction for testing the feasibility of bamboo as reinforcement have been undertaken. Purushotham[25] completed a large number of studies on the properties of Indian bamboo. Similar studies were also conducted by the Bamboo Research Committee for the Manila Bureau of Public Highways, Philippines.[35].

The report published in 1972 by the United Nations[52] on the use of bamboo in building construction served as a stimulus for further interest in bamboo as a reinforcement material. Since then considerable effort has been expended in various parts of the world in determining its mechanical properties and in predicting its behaviour as accurately as possible.

3.2 Bamboo and its properties

3.2.1 *Physical properties*

Bamboo can be classified into two main groups according to its growth pattern, namely sympodial and monopodial (Figure 3.3). About 75 genera and 1250 species occur in different countries of the world[1]. Each species has widely differing characteristics affecting its usefulness as a building material. The woody pointed stems of bamboo, commonly called culms, grow closely together in clumps (Figures 3.4–3.6). The growth of bamboo culm is rapid, about 70 mm per day, and can be as much as 350 mm to 400 mm per day[13,14]. The culms complete their growth within a period of four to six months. Once

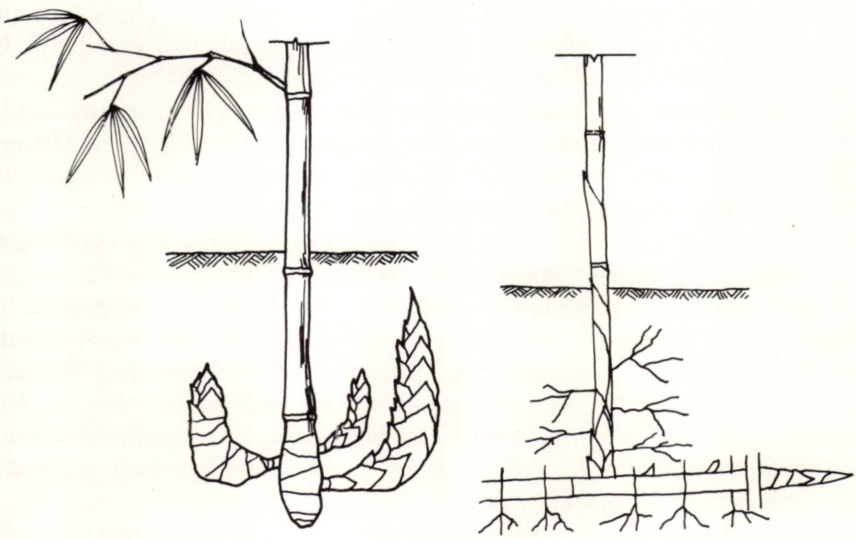

(a) Sympodial (b) Monopodial

Figure 3.3 Growth patterns of bamboo.

Figure 3.4 *Thyrsostachys oliveri* Gamble, Botanical Garden, Chiang Mai, Thailand. Courtesy of the Royal Forest Department, Ministry of Agriculture and Cooperatives, Thailand.

the maximum height is attained, lignification of the culm takes place during the subsequent two or three years. The culm reaches maturity after the fifth or sixth year, or even later, depending on the species[52].

Bamboo culms are generally cylindrical and smooth, with diameters ranging from 29 mm to 300 mm. The culms are divided into nodes and internodes (Figure 3.7). They are usually hollow, and have transverse dividing walls at the nodes. The culm is covered outside and inside by hard waxy cuticles which offer considerable resistance to the absorption of water, particularly when dry. The cavity of each internode is very variable in diameter and in some species this may be solid.

The tissue of the bamboo culm is built up by parenchyma cells and vascular bundles consisting of vessels, thick-walled fibres and sieve tubes. The movement of the water in the culm takes place through the vessels, while the fibres are responsible for the strength of the bamboo. Fibres constitute 60% to

Figure 3.5 *Gigantochloa* sp. Botanical Garden, Chiang Mai, Thailand. Courtesy of the Royal Forest Department, Ministry of Agriculture and Cooperatives, Thailand.

70% by weight of the wood of the bamboo, and fibre content is greater in the periphery than inside. Fibre distribution is highest in the internodes situated at one-quarter to one-half the height of the culm from the bottom. This region also contains the largest and most mature fibres. The average length of bamboo fibre is about 100 times its diameter[13,52].

3.2.2 *Mechanical properties*

The random nature of bamboo makes it necessary to either determine or verify all previously reported mechanical properties of different species. The properties not only vary with individual species, but also depend upon many factors, such as species, age, moisture content, extent of seasoning, time of cutting, maturity, soil and climatic conditions. One other immediate problem with tests on bamboo is that no agreed standards exist. Thus, comparisons of results are extremely difficult.

The principal problems associated with the use of bamboo as a reinforcement are volume changes, moisture susceptibility, consequent loss of bond and decay. The advantages cited in the use of bamboo as reinforcement are reasonably high tensile strength, good availability, its replenishable nature and negligible cost.

3.2.2.1 *Strength.* Bamboo attains its greatest strength after three years, when

Figure 3.6 *Dendrocalamus sericeus*, Botanical Garden, Chiang Mai, Thailand. Courtesy of the Royal Forest Department, Ministry of Agriculture and Cooperatives, Thailand.

it assumes a brownish colour. A high value of the modulus of elasticity is always associated with a correspondingly high tensile strength. On average, the node of a bamboo is weaker in tension than the internode. The tensile strength of seasoned bamboo is generally greater than that of the unseasoned bamboo. Youssef observed that green bamboo, with 39% moisture, possessed only about 60% of the tensile strength and between 30% and 35% of the compressive and flexural strengths compared with seasoned bamboo[20]. Table 3.1 gives the ultimate tensile strength and modulus of elasticity in direct tension of various species of bamboo. The tensile strength is found to vary from 38 N/mm^2 to 505 N/mm^2. This variation is understandable, considering the influence of different factors and the different types of test specimens used (in some cases full bamboo rounds, in others bamboo splints). In most cases, the node was the weaker section of a culm, due to a large number of soft

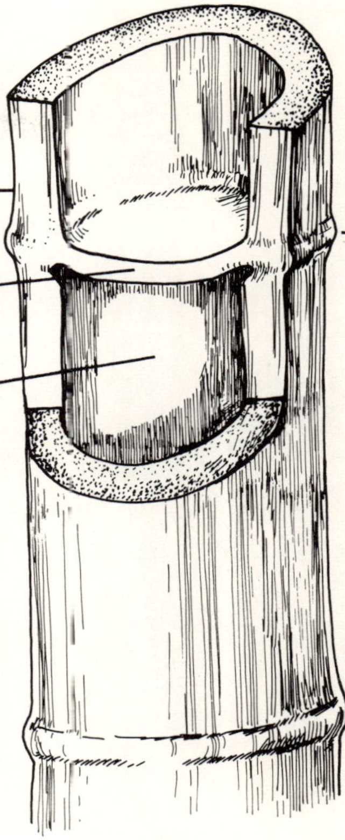

Figure 3.7 Bamboo culm.

proliferative cells in the growth zone, and to liquids being transported from stem to leaf stalk through a number of vessels crossing the nodal area. Specimens with the greatest distance between nodes usually had the greatest tensile strength. Pama *et al.*[50] concluded that the stress–strain relation of bamboo was essentially linear and exhibited a brittle type failure in tension (Figure 3.8).

Cox and Geymayer[17] reported tests for determining the tensile and bond strength, coefficient of thermal expansion, change of value due to variation in moisture content and load deformation characteristics of bamboo under sustained load. They observed that the coefficient of expansion of bamboo culm in the longitudinal direction is about one-third of that of concrete, and radially, it is four times that of concrete. The short-time static strength of bamboo culm was found to be higher than that in the case of sustained load. They also found that the Poisson's ratio of bamboo ranged between 0.250 and 0.409, with an average value of 0.317. Poisson's ratio for *Bambusa blumeana* reported by Janssen[53] was 0.46 for outer skin, 0.32 for cross-section and 0.34 for inner side, well within Cox and Geymayer's findings.

Table 3.1 Ultimate tensile strength and modulus of elasticity of bamboo

Location	Species	Ultimate tensile strength (N/mm^2)		Modulus of elasticity (N/mm^2)		Reference
		Min.	Max.	Min.	Max.	
Egypt	*Arundinaria gigantea*					
	Green	115.0*	161.0**	11 257.0*	12 385.0**	(20)
	Seasoned	188.0*	256.0**	18 612.0*	22 220.0	
Ghana	*Oxytenanthera abyssinica*	52.0	400.0	—	—	(59)
India	Batamana	106.4^{+}	—	14 420.7^{+}	—	(29)
	Chiwli	202.8^{+}	—	24 623.1^{+}	—	
	Perai	151.6^{+}	—	14 911.2^{+}	—	
	Farasmana	115.3^{+}	—	17 658.0^{+}	—	
	Dendrocalamus strictus	102.5	—	13 300.0	—	(24)
	D. strictus	89.9	—	16 600.0	—	
	Bambusa balcooa	96.5	—	16 548.0^{+}	—	(23)
	D. strictus					
Indonesia	*D. giganteus*	180.1	192.9	12 013.7	16 882.8	(32)
	D. asper	200.1	217.8	11 975.4	14 674.5	
	Gigantochloa robusta	173.3	202.7	9061.2	10 714.2	
	B. vulgaris var. *striata*	117.3	136.5	5950.0	8662.0	
	G. apus	198.7	297.2	5666.2	16 927.4	
	G. atter	259.5	384.1	12 794.7	19 278.9	
	G. pseudoarundinacea	230.0	320.3	11 550.0	14 883.3	
	B. arundinacea	112.3	168.7	5053.3	9204.9	
Japan	Madake of the *Shiudosasa* genus	38.3	371.6	—	—	(33)
Malaysia	*B. vulgaris* var. *striata*	—	295.4	—	—	(37–40)
	D. asper	—	440.6	—	—	
	B. blumena	—	372.8	—	—	
	D. giganteus	—	319.4	—	—	
Philippines	*B. spinosa*	138.0	288.0	13 859.0	19 340.0	(35)
Singapore	Unknown	190.0	—	29 500.0	—	(58)
Thailand	*Thyrosostachys oliveri* Gamble	164.8*	273.7*	—	—	(42)
		257.9**	505.0**	19 788.7	30 338.0	
		142.7*	206.9*	—	—	(44)
		275.8*	310.3**	24 132.5	48 265.0	
		128.9*	—	—	—	
		166.9**	—	14 479.0^{+}	—	(43)
		82.8*	109.3**	9650.0*		(48)
USA	*A. tecta, A. phyllostachys*	124.0^{+}	344.0^{+}	13 770.0	30 983.0	(13)
	A. tecta	76.5	172.7	14 400.0	27 700.0	(18)
	A. tecta	56.0	73.0	10 500.0	15 300.0	(15)
	Unknown	48.3	—	—	—	(19)
West Germany	Unknown	194.8^{+}	—	—	17 582.0^{+}	(11)

$^{+}$average *nodal **inter-nodal

Results of tests conducted to determine the compressive strength and modulus of elasticity of bamboo have been reported by Youssef[20], Espinosa[34], Purugganan[35], Cook[48], Glenn[13], and Fang[54]. Results of compressive strength tests have also been reported by Widjaya[32], Datta[11] and Abang Ali[37–40]. A summary of the results, shown in Table 3.2, indicates that the compressive strength varies from 23.2 N/mm^2 to 117.7 N/mm^2 while the compressive

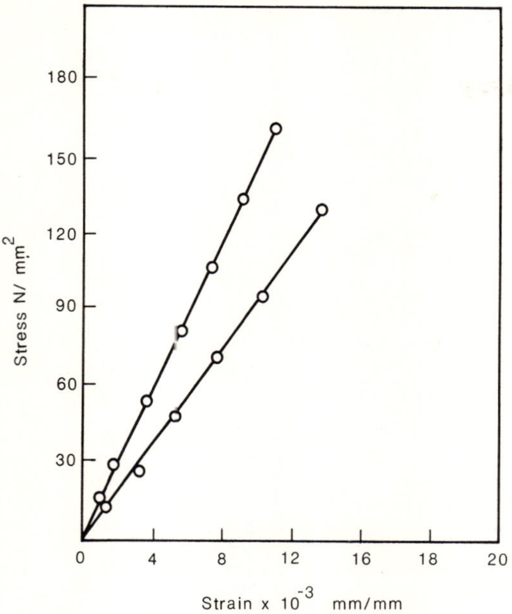

Figure 3.8 Stress–strain curve of bamboo[43].

modulus of elasticity varies from 4600 N/mm^2 to 19 350 N/mm^2. In comparing compressive with tensile properties, it may be noted that bamboo possesses a significantly lower strength in compression than that in tension.

The strength of bamboo generally increases until maturity. With few exceptions, after the age of five years, the increase in strength is less significant than the effect of moisture content on compressive strength. It has been found to be similar to that of wood – the strength nearly doubles on drying from the green condition to the air-dry state[52].

The culm has a tubular structure, stiffened at intervals by the nodes, which prevents its buckling and collapse. The location of the nodes is significant in the bending strength of bamboo. In static bending tests, specimens with a node at the loading point showed a higher strength but lower stiffness than those having the load point between nodes. In bamboo splints, the compressive strength of the outer layers from the culm is higher than for the inside layers. Bamboo splints with and without nodes, when tested in static bending gave higher values when the face nearest the periphery was in compression[52].

Shear strength is important in bamboo because it is the weakest feature. Meyer and Ekelund studied the bamboo to demonstrate that bamboo is as strong as wood in tension and deflection but much weaker in shear. In their opinion, the shear strength of bamboo is only 8% of its compressive strength[53]. When failure occurs during bending, it is not because of a shortage of tensile

Table 3.2 Compressive strength and modulus of elasticity of bamboo

Location	Species	Compressive strength (N/mm²) Min.	Max.	Modulus of elasticity (N/mm²) Min.	Max.	Reference
Egypt	*Arundinaria gigantea*					
	Green	29.0*	31.7**	4600.0*	5800.0**	(20)
	Seasoned	87.3*	91.6**	6700.0*	9600.0**	
Indonesia	*Dendrocalamus giganteus*	59.1	63.4	—	—	(32)
	D. asper	55.5	62.7	—	—	
	Gigantochloa robusta	50.0	52.3	—	—	
	Bambusa vulgaris var. *striata*	40.9	47.5	—	—	
	G. apus	37.3	69.0	—	—	
	G. atter	68.6	84.0	—	—	
	G. pseudoarundinacea	61.8	117.7	—	—	
	B. arundinacea	28.4	34.1	—	—	
Malaysia	*B. vulgaris* var. *striata*	50.5	—	—	—	(37–40)
	D. asper	64.7	—	—	—	
	B. blumena	56.2	—	—	—	
	D. giganteus					
	I	44.7	—	—	—	
	II	56.9	—	—	—	
Philippines	*B. spinosa*	35.0	73.0	—	—	(34)
	B. spinosa	45.0	59.0	9350.0	14617.0	(35)
Thailand	*Thyrosostachys oliveri* Gamble					
	Seasoned	55.3	—	19350.0	—	(48)
USA	*Phyllostachys* sp.	31.0	75.0	5325.0	17527.0	(13)
	A. tecta	23.2	27.3	7950.0	9500.0	(54)
	Unknown	—	—	5722.9		(19)
West Germany		—	72.6	—	—	(11)

*nodal **inter-nodal

strength in the fibres, but because of incohesiveness between the fibres, in which shear plays a role.

An internode appears to be better in shear than a node. In an internode all fibres run precisely along each other, but in a node they are interrupted by many vessels that cross them to reach the diaphragm inside the node.

Atrops[55] found that the shear strength increases with decreasing wall thickness from bottom to top along the culm. He argued that the percentage of sclerenchyma increases with decreasing wall thickness. Grosser and Liese[56] report similar findings.

3.2.2.2 *Swelling-shrinkage-bond problems.* Low bond with concrete is one of the principal problems associated with the use of bamboo as reinforcement. Understanding bond mechanism is essential for effective crack control. Apart from the fact that there is low natural adhesion between the bamboo and concrete, the major cause of inferior bond is the initial swelling of the bamboo when it absorbs water from wet concrete followed by shrinkage, thus separating bamboo from the concrete.

Bamboo culms have been observed to undergo diameter changes of about 5% and length changes of up to about 0.05% with variations in the moisture content. In addition, the coefficient of thermal expansion of bamboo is as low as one-third that of concrete longitudinally and as high as 10 times that of concrete diametrically. These differences will contribute to cracking of concrete cover and to a loss of bond. Cox and Geymayer suggested presoaking of seasoned bamboo prior to use, and mechanical interlocking between bamboo and matrix[17,18].

In India, studies have been carried out on waterproofing bamboo for reinforcing concrete[21]. It has been shown that a coat of resin dissolved in alcohol, with a subsequent application of white lead, can cut down the absorption to a stage where the corresponding change in volume as a result of absorption is almost negligible. Absorption and consequent variation in volume are reduced in China by bituminous coatings or impregnation with 2% solution of aluminium sulphate[21].

Fang and Mehta[54] found sulphur treatment to reduce water absorption potential significantly during the concrete curing process. Bamboos were first sand-blasted to remove the smooth surface, soaked in liquid sulphur for approximately one hour, then coated with a layer of sand and allowed to dry completely before use. Table 3.3 gives bond strength values obtained from various species of bamboo. Considerable variance is found in the bond strength of bamboo reinforced cement matrix due to several factors such as treatment and condition of bamboo, size of protrusions, spacing of nodes, age, curing conditions and other properties of concrete. Seasoned bamboos possess higher bond strength compared to unseasoned bamboos due to lower water absorption and to reduced swelling and shrinkage of the bamboo.

Tests carried out by Mentzinger and Plourde[57] on untreated, varnish-treated and hydrocarbon sealer-treated bamboo indicated substantial improvement in bond strength between bamboo and concrete. It was found that treatment of the bamboo with varnish and hydrocarbon sealers did not significantly affect the tensile strength of the samples, a favourable result. The two methods of treatment had different effects on the modulus of elasticity of bamboo, varnish increasing its value, and the sealer decreasing its value, possibly eliminating the use of the sealer as a waterproofing agent because of this adverse effect.

On the other hand, the results of the bonding tests performed placed the sealer in a most favourable position. The sealer-treated bond strengths averaged 0.72 N/mm^2, compared to the untreated bond strength average, 0.24 N/mm^2. Varnish too was proven to be an effective waterproofing treatment, averaging 0.30 N/mm^2 without being coated with an abrasive, and 0.38 N/mm^2 when coated with an abrasive. The bonding strength of the sealer sample was found to decrease when coated with the abrasive[57].

Table 3.3 Ultimate bond strength of bamboo

Location	Species	Treatment	Embedment length (mm)	Testing age (days)	Average concrete strength (N/mm²)	Water–cement ratio	Bond strength (N/mm²)	Reference
Egypt	*Arundinaria gigantea*	Untreated	150	28	15.21	0.6	0.56	(20)
			—	—	22.56	0.4	0.68	
		Two coats varnish	150	28	15.70	0.6	1.11	
					22.07	0.4	1.27	
Ghana	*Oxytenanthera abyssinica*	Green, untreated	280	28	31.20	—	0.33	(59)
		Seasoned, two coats bituminous paint	280	28	30.40	—	2.6	
India	*Bambusa balcooa, Dendrocalamus strictus*	Untreated	—	—	—	—	0.39	(23)
		Fine sand sprinkled over wet paint	—	—	—	—	0.62	
	Unknown	Previously soaked in water, one coat of resin alcohol and white paint	—	28	—	—	1.03–1.24	
		Polyester-sand	—	—	—	—	>0.56; >0.81	
		Epoxy-sand	—	—	—	—	1.13; >1.15	
	Perai	Untreated	—	—	—	—	0.29	(27)
		Ciba epoxy	—	—	—	—	1.13	
	Unknown	Dry, untreated	—	—	—	—	0.24	
		Dry, varnished	—	7	9.3–31.9	—	0.30	(57)
Philippines	*B. spinosa*	Seasoned culm, coal tar treated	—	—	—	—	0.81	(35)
		Seasoned culms asphalt emulsion treated	—	—	—	—	0.62	
		Seasoned culms untreated	—	14	—	—	0.66	
	B. spinosa	Untreated	—	14–28	—	—	0.4–1.53	(36)

Singapore	Unknown	Untreated	—	—	—	—	1.95⁺	(58)
		with water sealer	—	—	—	—	2.60⁺	
Thailand	*Thyrosostachys oliveri* Gamble	Untreated, splints with node	101.6	7	26.20	0.5	1.24	(43)
		One day immersion in water	101.6	7	26.20	0.5	0.55	
		Zinc chloride	101.6	7	26.20	0.5	1.14	
		Ferric chloride	101.6	7	26.20	0.5	0.52	
		Calcium chloride	101.6	7	26.20	0.5	0.62	
		Newtex Neoprene Adhesive 101 with surface treatment	101.6	7	26.20	0.5	1.38	
	T. oliveri Gamble	Seasoned, untreated	100.0	28	37–48	—	0.24–1.47	(48)
USA	*A. gigantea*, *A. tecta*	Dry, untreated	300.0	21	—	0.6	0.69†	(13)
		Green, untreated	—	—	—	—	0.50†	
		Dry, coat of asphalt emulsion	—	—	—	—	1.16†	
	A. tecta	Seasoned splints, untreated	—	—	—	—	0.34–0.38	(18)
		Wound with steel wire	—	—	—	—	0.52	

†average

The ultimate strength tests performed on the saturated strips very ably proved the damaging effect of water in softening the bamboo and decreasing its tensile strength. The small decrease in tensile strength of the varnish-treated sample again proved the material's usefulness as a waterproofing treatment.

Durrani[43] observed that the bond stress decreased as the period of immersion in water or 2% of ferric chloride, zinc chloride and calcium chloride solution increased.

Youssef found that the bond strength increases with increase in cement content and with reduction in water–cement ratio. It is well accepted that an effective moisture-proof treatment is necessary for bamboo reinforcement. Several different treatments have been adopted with varying degrees of success[20].

In Singapore[58], an experimental investigation was conducted on cement mortar reinforced with woven bamboo mesh in a manner similar to ferrocement. The specimens for the composite were divided into three series, A, B and C, each consisting of a group of specimens. The bamboo meshes used in series B and C were coated with a thin layer of water sealer under the trade name 'Throseal' in an attempt to improve the bond between bamboo and mortar. In series C the casting pressure was varied for a constant volume fraction of bamboo.

A comparison of the specimens in the A- and B-series showed that the use of water sealer improved both the cracking and ultimate strengths. The modulus of elasticity of the composite in tension also increased significantly. This may be attributed to the improvement in bond. Application of casting pressure in addition to water sealer further improved the strength and elastic modulus. For obvious reasons, the ultimate strength of the composite increased with increasing quantity of bamboo strips.

According to Perry *et al.*[59] the ultimate bond strength of bamboo generally increased with each of the following:

(i) The seasoning period of the bamboo
(ii) The presence of protrusions at the nodes of the splints, which became embedded in the concrete test specimens
(iii) The function introduced by filing the smooth surfaces of the splints
(iv) The number of coats of waterproofing paint applied to the splints
(v) Sand treatment to the coating of waterproofing paint
(vi) Compressive strength of the concrete.

3.3 Bamboo as reinforcement

Worldwide inflation, scarcity of steel and the critical housing shortage in developing countries have accelerated the search for alternative materials as suitable reinforcement for cement and concrete. Bamboo, an abundant renewable resource with adequate strength, has been used as reinforcement for cement mortar and concrete. Structures constructed with bamboo reinforcement are specially popular in developing countries because of low cost. Bamboo as reinforcement for structural components, water resources structures, utility structures, soil-cement and cement composites is discussed.

3.3.1 *Bamboo reinforced structural components*

The housing need in developing countries is becoming increasingly important due to an expanding population, rapid industrial development and migration. However, it is difficult to provide low-cost housing for the rural poor because of limited funds and the increasing cost of traditional building materials. It is important to emphasize and encourage the use of cheaper and abundantly available indigenous material. One such material is bamboo as reinforcement for structural components such as slabs, beams, walls and roofs.

3.3.1.1 *Slabs.* The use of bamboo as reinforcement for one-way and two-way slabs has been reported by a number of researchers. Cox and Geymayer[17,18] observed the formation of yield patterns very similar to that of steel-reinforced slabs. They suggested a tentative and simple design method using a minimum of 1.5–2.5% of split and presoaked bamboo reinforcement to give sufficient ductility to the slab. This reinforcement was neglected in the strength calculation and the slab design is as an unreinforced concrete section. Kowalski[60] recommended control of deflections in one-way solid and voided slabs. He also suggested a conservative approach to span–depth ratio. Kankam *et al.*[61] performed laboratory tests on ten simply supported one-way bamboo reinforced concrete slabs subjected to concentrated line loads and eleven bamboo reinforced concrete two-way slabs with simply supported boundaries subjected to concentrated central loading. In the one-way slab, three different modes of failure were observed: first, concrete in compression, which is symptomatic of over-reinforcement; secondly, both shear and concrete in compression; and thirdly, bamboo in tension.

It was observed that the onset of cracking is an important serviceability limit-state in bamboo reinforced concrete in which deflections increase rapidly after cracking. A maximum permissible design moment of M_{cr}, calculated from the concrete section only, gave a short-term factor of safety of about 2 against cracking and about 5 against collapse. The cracking moment, M_{cr} of concrete, based on the modulus of rupture,[61] is given by

$$M_{cr} = f_{tc}bh^2/6 \qquad (1)$$

where b = width and h = depth of the section.

The slabs used bamboo culms of the *Oxytenanthera abyssinica* of at least three years of age. The splints were waterproofed with two coats of bituminous paint at intervals of 24 h and dusted with sand immediately after the second coat, then allowed to dry for 24 h. The percentage of longitudinal bamboo reinforcement varied from 2.8 to 4.5 and from 1.0 to 1.4 along the transverse direction. The mix proportions were 0.58:1.0:2.2:3.7 (water:cement:fine

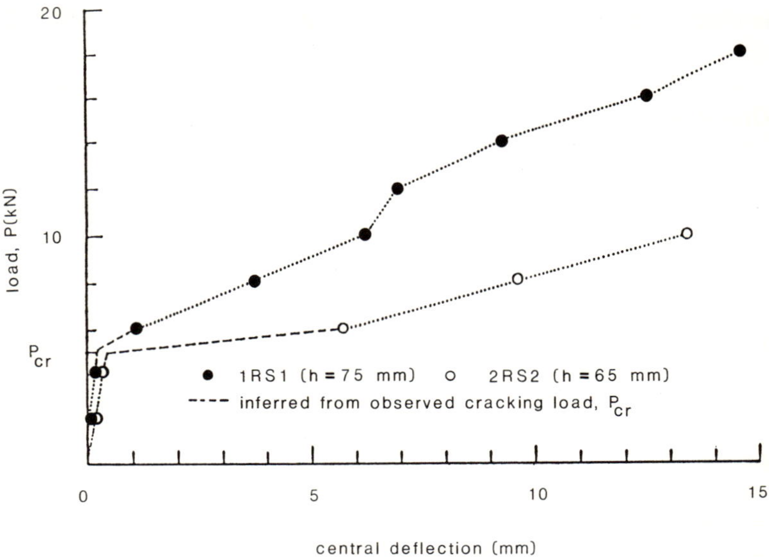

Figure 3.9 Typical load–deflection curves[64].

aggregate:coarse aggregate) by weight. These types of slabs did not exhibit any bond problem between the concrete and the bamboo reinforcement from repeated wetting and drying or from low-load cycling tests.

The M_{ult}/M'_{ult} ratio of the speciments was found to vary from 1.0 to 3.2. The latter value indicates that the slabs are governed by the internodal strength and therefore over-reinforced. The danger in this type of slab is sudden failure. Figure 3.9 shows the typical load-deflection curves of slabs. The gradient of the curves is relatively large until cracks appear at the bottom of the slab. The deflection curves flatten conspicuously as the cracks widen and multiply until failure occurs.

For two-way slabs, the details of the specimen were similar to the one-way slab. Concrete strength, span–depth ratio, bamboo percentage and the treatment of the bamboo were varied. Kankam *et al.*[62] used the yield line theory developed by Johansen[63] assuming isotropic reinforcement in the bottom face only. Fan mechanism and overall flexural mechanism of failure were assumed, to derive the following equations:

(i) *Fan mechanism* (R = radius of fan, a = radius of loading platen)

$$P + \frac{q}{3}(R^2 + Ra + a^2) = \frac{2m\pi(1+i)}{(1 - a/R)} \tag{2}$$

If q is negligible in comparison with P then

$$P = \frac{2m\pi(1+i)}{(1 - a/R)} \tag{3}$$

If the effect of the loading platen is neglected

$$P = 2m\pi(1 + i) \tag{4}$$

(ii) *Overall flexural mechanism*

$$P = \frac{8m}{(L - C)}[2\sqrt{(2L - C\sqrt{2})(L - C)} + C(\sqrt{2} + 1) - 2L] \tag{5}$$

if

$$\delta = \frac{2a}{L} = \frac{c}{L}$$

then

$$\gamma = \frac{1}{2}(2 - \delta\sqrt{2}) - \sqrt{(2 - \delta\sqrt{2})(1 - \delta)}$$

where

P = concentrated load
q = self-weight of the slab/unit area
R = radius of the fan
a = radius of the loading platen
m = sagging moment/unit length of the yield line
C = diameter of the loading platen = $2a$
L = span of the square slab
γL = shift of a yieldline from a corner of the slab.

In the slabs tested, the bamboo percentages varied from 2.86 to 4.0. Punching load was always greater than the load predicted by yield line theory. There was no shear failure.

On average, the experimental cracking loads were 15% of the theoretical punching loads based on the CP 110 method, neglecting the effect of bamboo. If the effect of bamboo was considered, the CP 110 method generally overestimated the punching loads, except for those slabs with treated bamboo. The treatment improved bond strength in the slab, resulting in a higher punching load.

It has been observed that bamboo does not make a significant contribution to the flexural stiffness of bamboo reinforced concrete sections. Kankam *et al.*[62] recommend a ratio of slab to overall depth not exceeding 23 to avoid excessive deflections and cracking under service loads.

Ali[42] tested bamboo reinforced cement mortar slab elements in direct tension, compression, flexure and torsion. Bamboo reinforcement was in the form of mesh and the only variable was the volume fraction of bamboo reinforcement, which was varied from 1% to 3% of the total volume of the specimen. For all series, the thickness of the specimen was kept constant at 38.1 mm, and *Thyrosostachys oliveri* Gamble, a variety of bamboo abundant in Thailand, was used. The mortar had a cement–sand ratio of 1:1.75 by weight, and a water–cement ratio of 0.35.

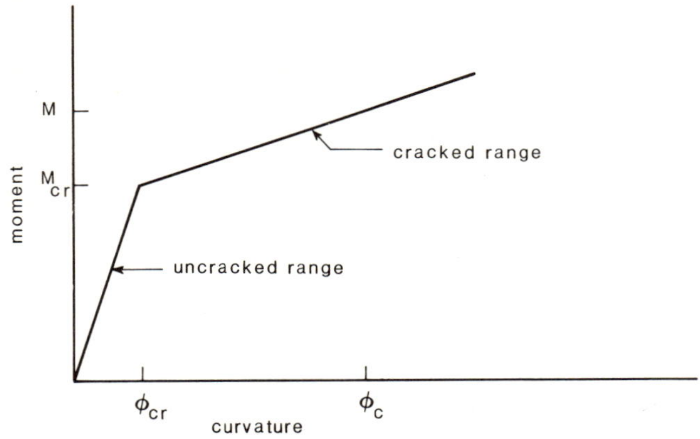

Figure 3.10 Moment–curvature curve of bamboo reinforced slab[49].

In bending, the composites behaved as a two-layered elastic material in the cracked range, with different modulus of elasticity in compression and tension. The presence of bamboo reinforcement considerably increases the ductility of the slab element when the volume fraction of bamboo used is greater than the critical volume fraction required for ductility. The force deformation in flexure of the slab is shown in Figure 3.10.

In axial compression, local buckling of bamboo reinforcement causes premature spalling of mortar. The ultimate strength in axial compression of bamboo reinforced slab is less than the ultimate strength of the mortar. The low experimental ultimate tensile strength values are due to the failure in bond of the reinforcement.

The theoretical structural behaviour of bamboo reinforced slabs was determined from the law of mixtures of composite material. The following equations were derived.

Before cracking

$$E_1 = V_m E_m + V_b E_b \tag{6}$$

After cracking

$$E_{cr} = \alpha_1 V_m E_m + V_b E_b \tag{7}$$

$$f_t = \frac{3(1+\beta')}{\beta'} \frac{M}{bh^2} \tag{8}$$

$$f_t = 3(1+\beta') \frac{M}{bh^2} \tag{9}$$

at failure

$$f_{tu} = V_b f_{bf} \tag{10}$$

$$M_u = 0.85 f'_c V_m \frac{h_0^2}{2} + \frac{1}{3} f_{bf} V_b (h - h_0)^2 \tag{11}$$

where

E_1, E_{cr} = elastic moduli of the composite, before and after cracking
E_m, E_b = elastic moduli for the matrix and the bamboo respectively
V_m, V_b = volume fractions of the matrix and the bamboo reinforcement respectively
α_1 = factor for reduced contribution of matrix (equal to zero for advanced stages)
f'_c, f_{bf} = strengths of matrix in compression, and of bamboo in tension
f_c, f_t = stress in compression and in tension
f_{tu} = failure strength of composite
$\beta' = E_1/E_{cr}$
M, M_u = moment at any stage and ultimate moment respectively
b, h = width and thickness of bamboo reinforced mortar
h_0 = neutral axis depth at ultimate condition

$$= h\frac{0.5V_b f_{bf}}{0.85V_m f'_c + 0.5V_b f_{bf}}$$

Reasonably good agreement was reported between theory and the experimental results[42,49].

An experimental investigation was conducted in cement mortar reinforced with woven bamboo mesh in Singapore[58]. The main parameters of the study were the volume fraction of bamboo and its surface treatment. The effect of casting pressure was also included.

Local bamboo having a light yellow colour was used. Strips, 5.5 mm wide and 0.88 mm average thickness, and seasoned for at least two months, were woven manually into meshes, keeping the desired opening size constant as far as practicable. These meshes were then cut into shapes and sizes of the test specimens. The bamboo meshes used in series A were untreated, but those in series B and C were coated with thin layer of water sealer to improve the bond between bamboo and mortar. The ultimate tensile strength increased with increasing quantity of bamboo strips, as expected. Comparison of the results in different series shows that use of water sealer improved both the cracking and ultimate tensile strengths. This is attributed to the improvement in bond. Application of casting pressure in addition to water sealer further improved the tensile strength and elastic modulus of the composite.

In bending, the amount of bamboo apparently has no influence on cracking strength. But an increase in the area fraction of bamboo improved the ultimate strength significantly (Figure 3.11). An application of casting pressure in addition to water sealer improved the bending strength. However, it appeared that beyond the optimum pressure between 2.0 and 2.7 N/mm^2, the strength started to decrease.

Based on full composite action between bamboo and mortar, and considering different moduli of elasticity for the composite in tension and

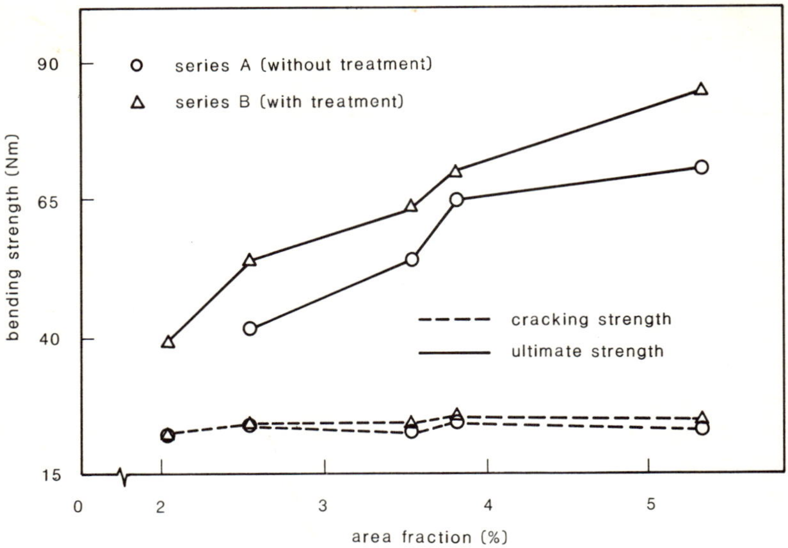

Figure 3.11 Effect of area fraction of bamboo on cracking and ultimate strength in flexure.

compression within the cracking range, the ultimate moment M_u is given by

$$M_u = 0.85 f'_c A_m \frac{h^2 c}{2} - \frac{1}{3} \sigma_{bu} A_b (h - h_c^2)^2 \tag{12}$$

where

M_u = ultimate moment
f'_c = cylinder compressive strength of mortar
A_m = area fraction of mortar
h = overall depth

$$h_c = \frac{0.5\sigma_{bu} A_{bh}}{(0.85 f'_c A_m + 0.5\sigma_{bu} A_b)}$$

= depth of compression zone

The experimental results for treated bamboo meshes approached the predicted values given by expression for M_u. This value is even exceeded when a small amount of casting pressure is applied; however, the results for untreated bamboo are far below the theoretical results.

Toughness index is a measure of the energy required to deflect the flexural specimens a given amount compared to the energy required to bring the specimen to the point of first crack[64]. The use of bamboo mesh considerably increased the flexural toughness of mortar. For plain mortar specimens, the toughness index is 1; however, the average toughness index for the specimens with water sealer was about 1.5 times greater than that for specimens with untreated bamboo meshes[58].

3.3.1.2 *Beams.* The use of bamboo as tensile reinforcement for beams has been extensively studied and found to be very similar to that of steel reinforced members[13,17,18,20,65]. Dimensional change due to moisture variation is the primary concern. Bamboo reinforcement absorbs moisture from fresh concrete, and swells, causing severe cracking of concrete before any load is applied. Subsequently, as moisture content decreases, the bamboo shrinks, causing separation of bamboo from the matrix. Several authors have studied this effect and have recommended solutions to overcome the problem[20,21,43,54,57–59]. The solutions, effective in varying degrees, have been chosen for their simplicity and low cost.

For rectangular beams, Cox and Geymayer[17,18] recommend an optimum reinforcement ratio of 3–4%. Bamboo reinforced rectangular beams are capable of developing about three times the flexural strength of unreinforced beams with identical cross-sections. However, the use of inverted T-beam sections[15,16] rather than rectangular beams results in a moderate increase of the ultimate moment capacity for the same total cross-sectional area. The ACI working stress design and ultimate strength design are used, but with allowable stresses of the bamboo. For field design, they neglect the effect of bamboo reinforcement[17,18].

In India, Masani *et al.*[10] reported that load capacity of bamboo reinforced beams was increased by using bamboo splint dowels as the diagonal tension reinforcement along the sections of the beams where the vertical shear was generally high. The findings are that simple beams up to 4.6 m and concrete T-beams up to 6.2 m can be safely reinforced with bamboo splints.

According to Glenn[13], bamboo reinforcement in concrete does not prevent the failure of concrete by cracking at loads beyond that of plain concrete; however, bamboo reinforcement does considerably increase the load capacity of the member at ultimate failure up to an optimum value, compared to an unreinforced member. The load increased to about 4–5 times greater than that of plain concrete of equal dimensions[13].

Smith and Saucier[15,16] observed that beam failure was due to excessive straining of bamboo reinforcement and by some bond slippage between the reinforcement and the concrete. All the beams cracked initially at or near the point of maximum moment. For maximum long-term deflections, they recommended a minimum depth of 101.6 mm for flexural members. They also recommended transverse reinforcement spaced at 76.2 mm centres and placed so as to have a clear cover of about 6.35 mm. These transverse reinforcements were used to prevent flexural failure in the transverse direction during handling.

The tensile strength of seasoned bamboo indicates its suitability as reinforcement; however, its low moduli of elasticity in tension and flexure will require an area of bamboo reinforcement for flexural members of about ten times that of a corresponding steel reinforced member, according to Youssef[20]. Due to this low modulus of elasticity of bamboo, flexural members will always develop cracking under normal service loads[6].

The US Naval Civil Engineering Laboratory developed a design chart for bamboo reinforced concrete based on steel reinforced concrete data[65]. Cross-sectional dimensions of a bamboo reinforced beam, and number and size of bamboo reinforcing rods can be selected from Figures 3.12 and 3.13.

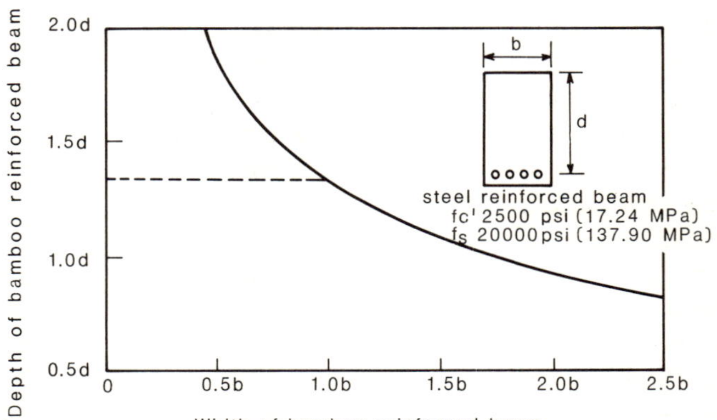

Figure 3.12 Width of bamboo reinforced beam[65].

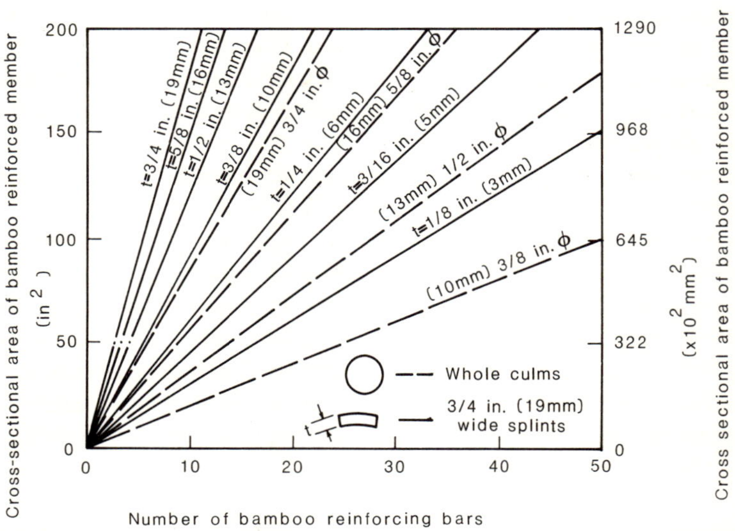

Figure 3.13 Bamboo reinforced beams and reinforcement[65]. Number of bamboo reinforcing bars plotted will provide a member with 3% reinforcement. For other percentage requirements, multiply the number of bars by the percent required +3.

Figure 3.12 provides the cross-sectional dimensions of a bamboo reinforced beam that will have the same bending moment resistance coefficient as a balanced steel reinforced beam, singly reinforced. The curves in Figure 3.13 were drawn for 3% of bamboo reinforcement, the optimum range for flexural members. Designers may develop similar curves that will conform to their national codes.

3.3.1.3 *Walls and roofs.* Bamboo reinforced cement mortar wall was tested in Hasanuddin University[66]. Local bamboo about 2–3 years old from the Tanak Toroja region, and bamboo splint of 25 mm width, the common width used by rural people, were used. The unit cost of 1 m^2 bamboo reinforced cement mortar in 1980 was Rp 2655 (US$4.25) compared to Rp 4410 (US$7.06), the unit cost of conventional red brick wall.

The tests proved that bamboo is adequate reinforcement for cement mortar walls. For the impact test, fracture was recorded at 4.5 m height; for the bending test, failure occurred at 180 kg static load; and the kink test indicated that the panel is stable even under an applied load of 1800 kg[66].

In Indonesia, a public facility for bathing and washing was constructed in the village of Buniwangi and Citarik[67], using bamboo as reinforcement for the wall and roof. The structure was constructed in five days, and in 1982 cost Rp 47 750 (US$76.40).

According to Brink and Rush,[65] bamboo reinforced concrete walls should have a thickness of not less than 127 mm and not less than 1/30 the distance between the supporting or enclosing members, and should be reinforced with at least 19.7 mm diameter culms at 152.4 mm centres in both vertical and horizontal directions. This reinforcement should be provided as one layer of mesh in the middle of the section. Two bamboo culms 12.7 mm or more in diameter should be placed above and at the sides of openings, and two 19.1 mm diameter culms of 1.22 m should be placed diagonally across the corners of openings.

In India, frames of 1 m × 2 m were infilled with bamboo splints and suitably plastered with 1:6 cement-sand mortar; this is commonly referred to as bamboocrete. The wall panel was able to resist 600 kg before cracks appeared[31]. Considering only shear load and neglecting frame action, the ultimate stress of the panel was 274.7 N/m^2. Permissible shear stress of brickwork wall[31] built with 1:1:6 cement, lime and sand mortar is only 147.15 N/m^2. The bamboo reinforced cement roofing was cylindrical in shape. Pretreated bamboo splints were used to form the reinforcement, and two coats of 1:6 cement-sand mortar were applied. The roof was supported by the bamboo reinforced cement walls[31].

3.3.1.4 *Columns.* The behaviour of bamboo reinforced concrete tied columns under concentric and eccentric loads was investigated by Abu Sadeque[44]. The physical properties of all column sections were 127 mm × 127 mm cross-section,

1.45 m length and with a longitudinal reinforcement of full bamboo (*Thyrosostachys oliveri* Gamble) sections with external diameter varying from 9.52 mm to 19.10 mm. The eccentricity was taken as 10% of the lateral dimension of the column. Tie spacing and size were determined on the basis of criteria set out by Bresler and Gilbert,[68] which were modified for bamboo reinforcement as follows:

Criterion I: To develop the maximum strength of the concrete core, ties should have a spacing equal to or less than twice the core dimension of the column.

Criterion II: To prevent the buckling of the longitudinal bamboo at a stress below the ultimate compressive strength of bamboo, the tie spacing should not exceed a distance l, defined by the equation

$$l = d^* \sqrt{\frac{c\pi^2 E_b}{8f_{by}}} \tag{13}$$

where

d^* = the average diameter of the bamboo culm
E_b = the modulus of elasticity
f_{by} = ultimate strength of bamboo culm in compression
$c = 2$

Criterion III: To prevent the longitudinal bamboo from premature buckling due to inadequate stiffness of the ties, the latter must have a spacing defined by the equation

$$l = \left(\frac{3\pi^5 E_b t' b'}{4A' E'_b}\right)^{1/3} d^* \tag{14}$$

where

E'_b = modulus of elasticity of the tie
t' = thickness of the bamboo culm
b' = effective length of tie
A' = area of the tie

Abu Sadeque[44] observed that the controlling design spacing based on the smallest value of the spacings given by these criteria was found to be adequate. The method of estimating ultimate loads of bamboo reinforced column based on the CEB (European Concrete Committee) stress–strain curve of concrete gives a theoretical value of load closer to the experimental values. Also the longitudinal reinforcement should be arranged in a circular core so that the bamboo skin ties can be easily bent[44].

Singh[47] further studied the behaviour of bamboo as reinforcement for short concrete columns subjected to eccentric loads. All columns were 127 mm × 127 mm in section, 889 mm in height and reinforced with 15.2 mm to 25.4 mm diameter bamboos (*Thyrosostachys oliveri* Gamble). The loads were applied at e/t ratios ranging from 0 to 1.2 along the major axis, where e is the eccentricity of load measured from A_b, the area of longitudinal bamboo reinforcement, and t is the thickness of column.

It was observed that for $e/t < 0.15$, the failure is caused by crushing and the whole section remains in compression. At $e/t > 0.5$, failure is caused by crushing of concrete on the compression side followed by the formation of tensile cracks in the concrete on the tension side.

Based on the average properties of all bamboo reinforced concrete columns subjected to uniaxial load, a load–moment interaction curve was drawn (Figure 3.14) for an assumed column, and the experimental results were plotted. The experimental and theoretical load–moment interaction are fairly close to each other.

Cook *et al.*[48] reported that a column with 7.28% of bamboo reinforcement failed at a load marginally lower (2.7%) than the failure load of an unreinforced concrete column having a cube strength of 34.5 N/mm^2. According to them, bamboo reinforcement increases the load-carrying capacity of concrete columns only when part of the column is subjected to tension.

Earlier researchers also reported that bamboo struts can sufficiently improve compressive strength of weaker concrete[11]; however, bamboo reinforcement of about 7.5% did not significantly alter the axial failure load of a concrete column[20].

Brink and Rush[65] recommended that vertical bamboo reinforcement should be approximately 4% of the column cross-section for rectangular

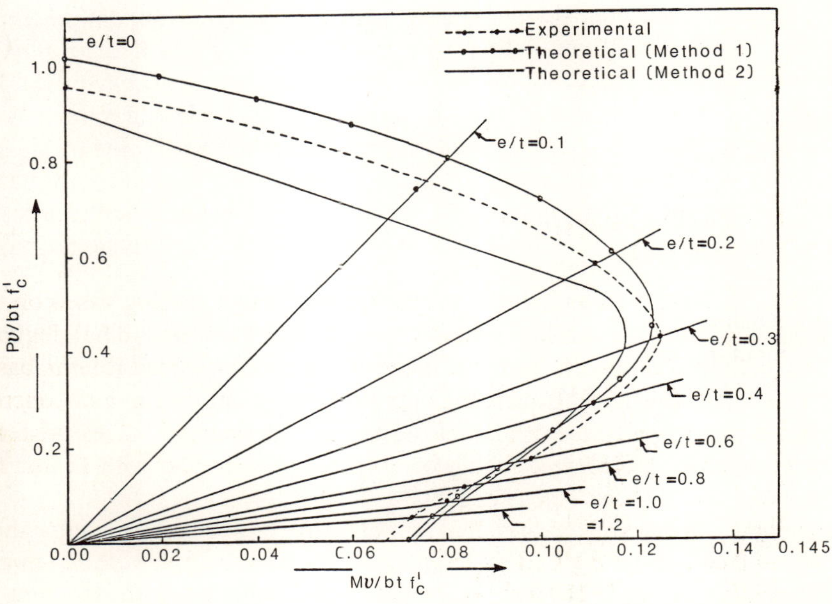

Figure 3.14 Load–moment interaction diagrams for bamboo reinforced concrete column[47].

column, and that bamboo lateral tie reinforcement should be spaced less than 16 times least dimension of the vertical reinforcement, nor further apart than the least dimension of the column. Enough ties should be provided so that every practical bar is held firmly in its designed position and has lateral support equivalent to that provided by a 90-degree corner of a tie. A common rule for determining the size of a tie is that its cross-sectional area is 2% of the area of all vertical reinforcement confined by it. The concrete cross-sectional area of bamboo reinforced rectangular columns should be about 2.25 times the concrete area of steel reinforced rectangular columns, indicating a 50% increase in face dimensions.

3.3.2 *Bamboo reinforced water resources structures*

There is a growing need for water resources structures in developing countries to harness water for national growth. Most developing countries, however, lack the techniques and resources to build water resources system to effectively store, transport, control and transform water. It is in this context that an appropriate material for construction which is simple to construct at reduced cost assumes significance.

3.3.2.1 *Water containers.* Water containers account for a major share of bamboo reinforced cement mortar and bamboo reinforced concrete applications. Dimensions and reinforcement requirements of many tanks are arrived at by approximation on the basis of available data on tanks constructed in the past.

For cylindrical tank design, Watt[69] assumed the tank foundation to be sufficiently rigid and continuous with the shell wall. Sashi Kumar *et al.*[70] provided two alternative approaches to understand the exact behaviour at the base of the tank: they assumed that the shell wall is extremely thin compared to the base plate and idealized the walls as being elastically restrained by means of a moment resisting coil spring, or assumed only partial restraint at the base.

Assuming composite behaviour, Sashi Kumar *et al*[70] suggested the process of a moment resisting coil spring, or assumed only partial restraint at the base.

(i) Given the capacity, establish the dimensions of the tank.
(ii) Assume a suitable thickness and estimate the adequacy of the reinforcement. Equate the stress on the tank wall due to static liquid pressure and the resisting force per unit of wall height.

$$t = \frac{WHR}{\sigma_m} \tag{15}$$

where

t = thickness
W = unit weight of liquid
H = depth of water in the tank

σ_m = allowable tensile stress of the composite which may be taken as equal to the ultimate tensile strength of plain mortar

R = radius of the cylindrical tank

(iii) Determine the Young's modulus of the composite. Use the Poisson ratio of the bamboo (Watt assumed Poisson's ratio as zero[72]).

(iv) Choose an appropriate set of boundary conditions and solve the governing shell equation.

(v) Compute the maximum hoop stress and bending moment, then check against the corresponding permissible values at first crack.

(vi) If the factor of safety obtained is too large or too small, then decrease or increase the thickness and/or spacing of reinforcement accordingly. The choice will depend upon the minimum cover available. Repeat the above process till satisfactory results are obtained.

(vii) The critical stresses in the roofs of tanks are a consequence of temperature changes governed by climatic factors and must be accounted for in the design.

The construction technique involved the preparation of the bamboo mesh reinforcement (Figure 3.15–3.16). The application of mortar or concrete may vary depending upon the resources available. The simplest method is hand trowelling (Figure 3.17). Shotcreting, another method of mortar application, is expensive and requires skilled labour.

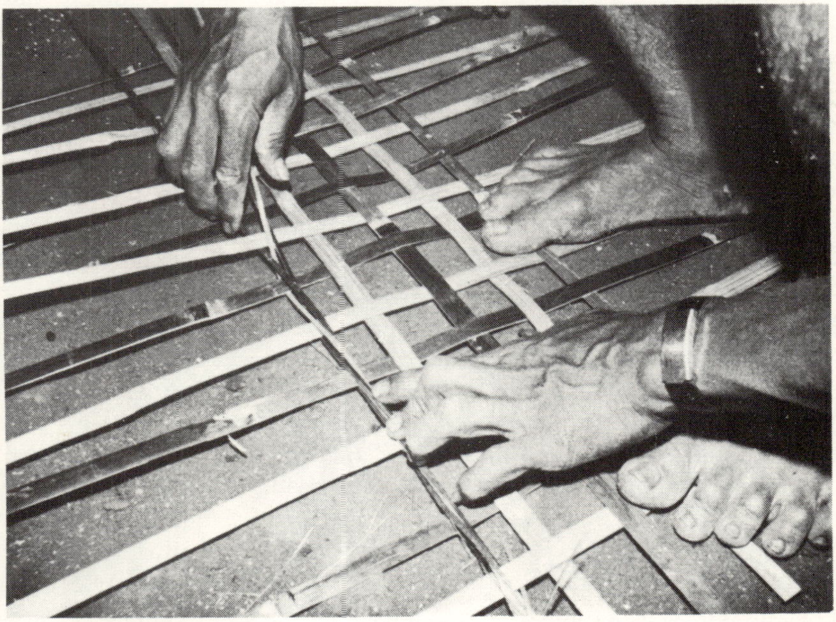

Figure 3.15 Preparing the bamboo mesh.

Figure 3.16 Completed bamboo mesh reinforcement for water tank.

Rainwater collection from roofs has been a common mode for providing drinking water supplies in remote rural areas in developing countries. Bamboo reinforced cement mortar and bamboo reinforced concrete water tanks have been used in many countries in Asia. In Indonesia, the Yayasn Dian Desa recommended a 4.5 m^3 family size tank[71–73]. An instruction manual on how to construct the tanks has been published for the extension workers to follow[74,75]. This type of tank has been sponsored by UNICEF, the World Bank, the Government Bureau for Village Development, the Indonesian Department of Health[76,77], Hasanuddin University, the Development Technology Centre (DTC), and USAID[66,78]. The cost of a bamboo reinforced cement mortar water tank, 1.2 m high, 1.3 m diameter and 30 mm thick, in 1982 is Rp 37 575 (US$60.12).

In Thailand, the Appropriate Technology Action for Development (ATAD) recommended capacities are 3.5 m^3 for a small family and 6 m^3 for a big family[79]. ATAD conducted training for villagers to transfer the technology. It is estimated that 50 000 bamboo reinforced concrete water tanks have been built in Thailand[80]. Some tanks are over ten years old and are still in use[83]. However, some others have failed. The Faculty of Engineering, Khon Kaen University and the Population and Community Development Association, under the financial assistance of the Canadian Government, conducted tests to investigate the long-term behaviour of bamboo and its suitability for tank

Figure 3.17 Plastering a bamboo tank in Indonesia.

reinforcement. Studies on bamboo as reinforcement for water tanks have been undertaken by many researchers in Thailand[81–85] and other countries[86,87].

In the Solomon Islands[87], two sizes of tanks, 500-gallon and 1000-gallon, were designed and constructed with bamboo and mild steel wires as reinforcement. These rectangular tanks were formed from precast panels measuring 1.2 × 1.5 × 1.2 m for the small tank and 1.6 × 1.8 × 1.6 m for the bigger size tank respectively. The moulds for casting the panels are made out of hard wood. The bamboo reinforced cement rainwater tanks are cheaper and easier to construct compared to tanks constructed in ferrocement, concrete and galvanized sheet. The average cost per m^3 of storage in 1983 is US$20.00 for a bamboo cement tank, US$25.50 for a ferrocement tank, US$31.38 for a concrete tank and US$35.86 for a galvanized sheet tank.

In the western region of Pakhara, Nepal, water supply is the main problem for the villages. To lower the cost of water containers, local bamboo baskets have been plastered with mortar. Tanks of 50-litre capacity have been successfully used[88].

3.3.2.2 *Well and reservoir lining.* Shallow surface wells have great potential, not only for family use but also for irrigation. These wells can be constructed rapidly at comparatively low cost. The minimum diameter for hand-dug wells should be kept at 1.0 m if a single person is to work inside and 1.3 m if two people need to work inside.

To ensure optimum utilization of such wells, it is essential to use a well lining to prevent collapse or sloughing, to keep burrowing animals out, and to exclude dirty or polluted surface water; further if the water-bearing aquifer is under artesian pressure, it can prevent locally charging the water. In Indonesia, prefabricated bamboo reinforced cement well casings (Figure 3.18)

Figure 3.18 Prefabricated bamboo reinforced cement well lining in Indonesia.

are used in South Sulawesi[89]. The well lining is of 0.8 m diameter, 1.2 m height and 2.5 mm thickness.

The main source of water for most villagers in the northeast coastal plains of West Java, Indonesia, is the irrigation system. The water is not only polluted but also carries sediments from the hills. Another problem is that once a year for about two months, the canals are empty for maintenance work. To improve the water supply situation, a small community water system (MCKS) was designed according to Ministry of Health requirements. Bamboo reinforced cement mortar was used as lining for a slow sand filter[90] and reservoir[91]. The slow sand filter was used for a system supplying 2 000 people and the storage reservoir is for 500 people.

3.3.3 *Bamboo reinforced utility structures*

3.3.3.1 *Silos.* Traditional storage methods used in most developing countries are unscientific. This results in deterioration in quality of grains besides significant losses from attack by birds, insects, rodents and moulds. Ferro-cement bins, used to improve storage conditions, have been modified using bamboo as reinforcement. In Bangladesh, silos of six to ten tonnes capacity, reinforced with bamboo (Figure 3.19) and two layers of hexagonal mesh on both sides of the bamboo have proved satisfactory[92,93]. At the Asian Institute of Technology (AIT), a 2.80 m^3 conical cement mortar rice bin with bamboo reinforced cement was investigated. Bamboo splints soaked in water for at least three days were used. The reinforcement cover was 10 mm[92]. It is recommended that to maintain the quality of paddy stored, the moisture content should be less than or equal to 12%.

3.3.3.2 *Septic tanks.* A 1.5 m^3 septic tank (Figure 3.20), 0.75 m in diameter and 1.5 m high, was constructed in Babakan Jayanti, Village of Citarik, West Java, Indonesia[78]. The septic tank used bamboo mesh as reinforcement. It was constructed in five days and the cost in 1982 was Rp 65 500 (US$104.80).

3.3.3.3 *Footbridges.* In Kampung Cimmanggu, Village of Buniwayi, Indonesia, a bamboo reinforced cement mortar footbridge[78] was used to replace the traditional coconut trunk bridge (Figure 3.21). The bridge has 7.8 m long span and is 1.5 m wide. The reinforcement consists of a bamboo mesh combined with wire and steel. The bridge was designed for a maximum load of 294.3 N/m^2 and constructed within 12 days. The cost of the bridge including foundation, formwork and scaffolding is Rp 224 850 (US$359.76) in 1982.

3.3.3.4 *Slabs on grade.* Concrete pavements are inherently susceptible to unexpected and often unaccountable cracking, even though a liberal flexural strength can be provided on the basis of a conservative working stress. The primary purpose of reinforcing concrete pavement is to maintain each slab as

Figure 3.19 Bamboo reinforcement for silo in Bangladesh.

an integral unit regardless of the cracking of the concrete, by tying the portion on each side together without appreciable separation at the crack, so that the slab will expand and contract substantially as a unit due to temperature changes. There are no specific limits restricting the opening of the cracks. Based on maximum pavement deformation of 0.51 mm, horizontal crack widening of 0.25 mm due to tension in reinforcement is considered to be a reasonable limit[43,50].

Assuming that full subgrade–pavement contact exists and that frictional resistance is overcome by the tension in the reinforcement, the cross-sectional area of reinforcement required per metre width of slab at the crack is

$$A = \frac{wfL}{2\sigma} \tag{16}$$

Figure 3.20 Bamboo reinforced cement septic tank.

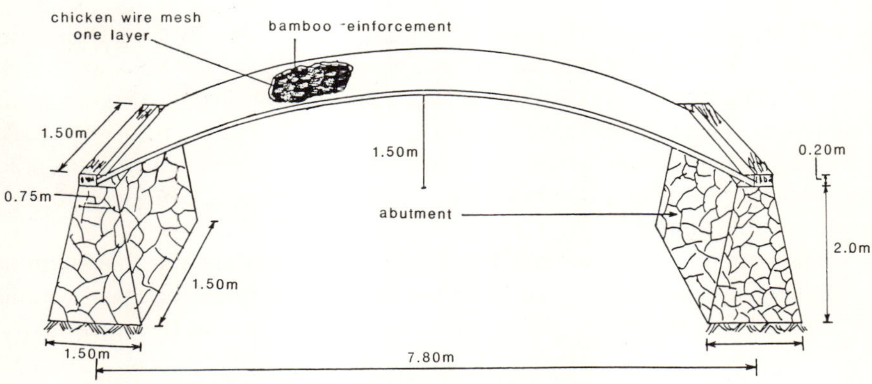

Figure 3.21 Footbridge with bamboo reinforced mesh combined with wire and steel.

where

w = weight of slab in kg/m^2
f = coefficient of friction between slab and subgrade, generally taken as 1.5
L = length of slab in m
σ = allowable reinforcement stress at crack in kg/cm^2.

Durrani[43] concluded that the percentage of bamboo reinforcement needed for slabs ranging from 3.05 m to 12.19 m in length varies from 0.26% to about 1.0%. The function of reinforcement in the pavement is to resist horizontal forces only, so that the most desirable depth for placement is at about 1/3 of the pavement thickness from the top. To ensure a fairly uniform cross-section of the bamboo throughout the length, the distal and basal ends should be alternated in every row and nodes should also be staggered (Figure 3.22).

Brink and Rush[65] recommended that for slabs on grade reinforcement spacing should not be greater than the slab thickness. The size and spacing of the bamboo reinforcement can be determined using Figure 3.23.

3.3.4 *Bamboo reinforced soil-cement*

Bamboo reinforced soil-cement can be used in countries where mud is commonly used for walls. The use of bamboo as reinforcement will considerably improve the structural behaviour of the structural component. Soil with

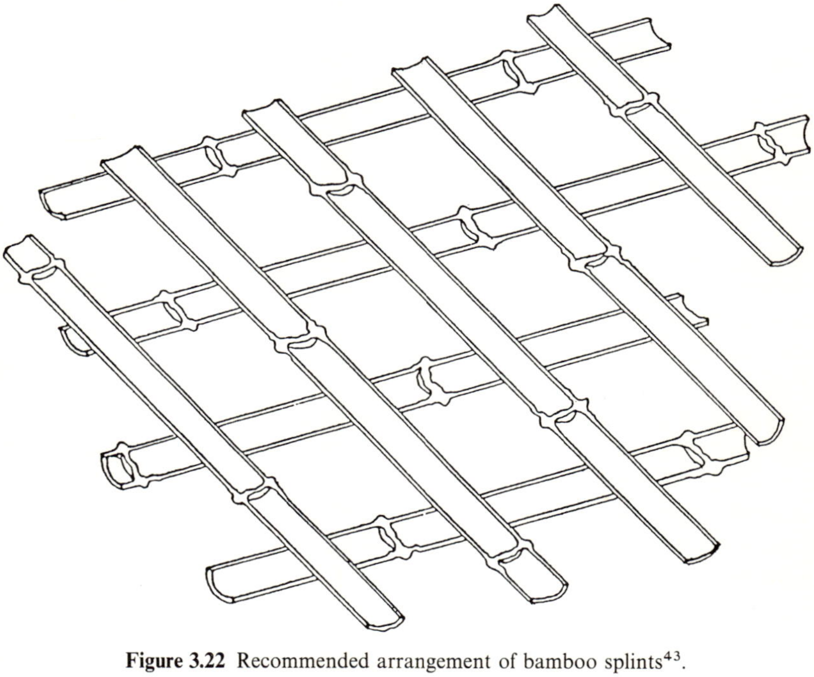

Figure 3.22 Recommended arrangement of bamboo splints[43].

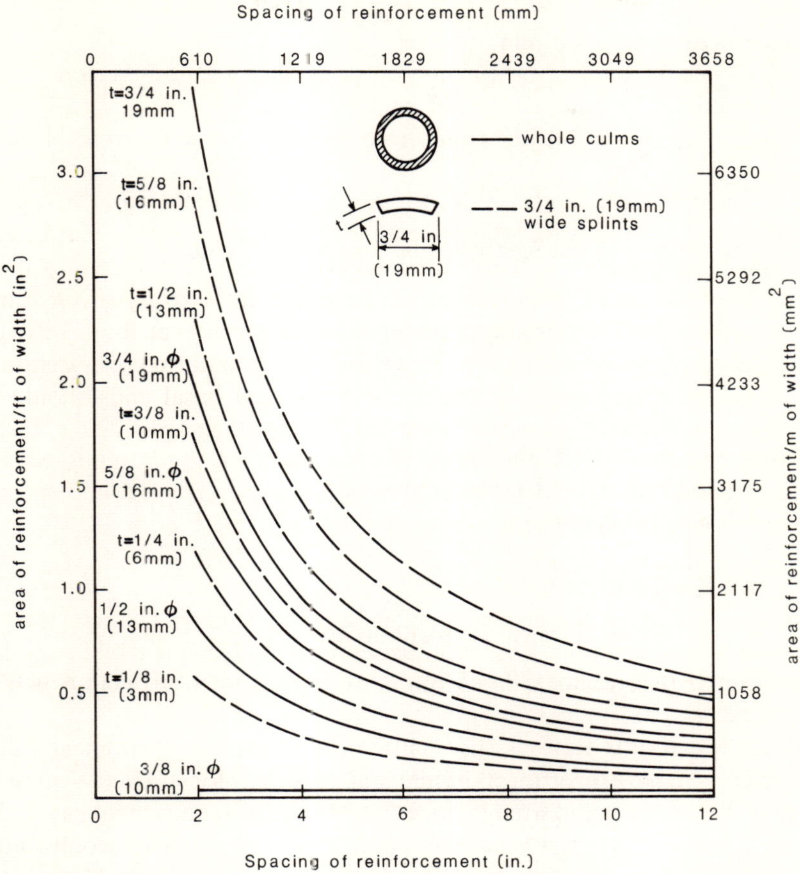

Figure 3.23 Size and spacing of bamboo reinforcement in slabs and walls[65].

plasticity index between 6.5 and 11 with sand content less than 40% was recommended by Mehra *et al.*[23]. The behaviour of bamboo reinforced cement beams have been investigated by researchers in India[22,23,26,99–102] and in Thailand[41].

Hamid[41] observed that soil-cement has a higher modulus of elasticity in compression than in tension. In tension the specimen behaved elastically and gave a brittle fracture, whereas in compression there was evidence of some plastic deformation before failure.

Hamid[41] approached the design of bamboo reinforced soil-cement from the ultimate load concept conforming to ACI Code. The area of bamboo reinforcement was determined from

$$A_B = \frac{M_R}{f_B(d - k_u d)}$$

where

A_B = area of bamboo reinforcement
M_R = resisting moment at the first crack appearance in the beam
f_B = ultimate tensile strength of bamboo at node
d = effective depth of the beam from top fibre to the centroid of the tensile force.
k_u = neutral axis depth factor at ultimate failure condition.

The bond stress was checked with the following equation:

$$\sigma_B = \frac{V}{\Sigma_0 (jd)}$$

where

V = shearing force of the beam
Σ_0 = perimeter of the bamboo reinforcement
j = lever arm factor
$= \left(1 - \frac{k_u}{3}\right)$
σ_B = bond stress between the bamboo and the soil-cement.

If σ_B exceeds the ultimate bond stress, then another trial section will be assumed.

Beam test results showed fairly close agreement with theoretical values predicted from the properties of the materials especially for specimens cured for 28 days. The difference in strength for beams made with soil-cement having 7% and 10% cement content was marginal, and for economy, it was recommended to use 7% cement[41].

In India, it has been observed that to prevent adverse cracking, the tensile reinforcement must be about 1.5% with proper cover and spacing[23,26,94]. It has also been observed that cement content and percentage of reinforcement do not have significant effect on the cracking load. However, cement contents can affect the failure load significantly.

In the beams tested, failure was either in bond or in shear[23,94–96]. Measures to improve bond characteristics have to be adopted. Nainan and Kaplan[95] obtained improved bond strengths by using coir ropes wound around the bamboo reinforcement, driving nails into the reinforcement and sprinkling sand on treated surfaces. Using end anchorage devices[27] or shear connectors[96] was observed to be highly effective in improving bond.

Bamboo reinforced soil-cement has been used as bonded underlay in a 270 m long section in Rohtak, India, and it was reported to have a satisfactory performance[23,97]. Bamboo reinforced soil-cement was also used for lintels in low-cost houses[23,95]. Nainan and Kalam observed that bamboo is not effective as compression reinforcement in soil-cement column[95].

3.3.5 *Composites reinforced with bamboo fibres, pulps and strips*

The two reasons for reinforcing a matrix are to produce a composite which is stiffer than the unreinforced matrix, and to hold the matrix together after it has cracked. The strength of the cracked composite may then be significantly higher than that of the unreinforced matrix; the cracked composite may also extend a great deal more before it breaks, and this may impart superior properties of impact resistance and fracture toughness to the composite to those of the matrix alone.

3.3.5.1 *Bamboo fibres.* Pakotiprapha *et al.*[46,98,99] investigated the mechanical properties of fibre reinforced cement paste composite in compression, tension, bending and torsion, in both elastic and cracked ranges based on the actual bond–slip relationship of the fibre. The fibre selected in the study is bamboo which was obtained by mechanically pressing oven dried short bamboo sticks of specified length. The critical length of the bamboo fibre, l_c, is obtained as

$$l_c = \frac{2a_f\sigma_{fu}}{\rho_f\tau_u}$$

where

τ_u = maximum bond stress
σ_{fu} = ultimate fibre stress
a_f = means of the cross-sectional areas
ρ_f = means of the peripheries of fibres.

With regard to length, an efficiency factor is introduced to take into account the stress distribution at the end portions of the fibres. This efficiency factor is determined with the assumption that any portion of a fibre has an equal probability of occurring on a particular cross-section. For the first mode of failure, the length efficiency factor ψ_b is equal to 0.50; for the second mode of failure, the length efficiency factor ψ_f takes the form

$$\psi_f = 1 - \frac{l_c}{2l}$$

Analytically the material was treated as a composite, the properties of which were derived by the law of mixtures. The post-cracking behaviour was derived, based on a one-dimensional model in which the randomly oriented discrete fibre system is idealized into an equivalent transversely isotropic system with equivalent unidirectional fibres[46,98,99]. The effects of random orientation and finite length of the fibres as well as bond–slip characteristics of the fibre were taken into consideration.

The experimental investigation of the mechanical properties of bamboo fibre reinforced cement paste composite[46,98,99] in compression (Figure 3.24) tension (Figure 3.25) and bending (Figure 3.26) in both elastic and cracked range showed good agreement with the theoretically predicted values. In

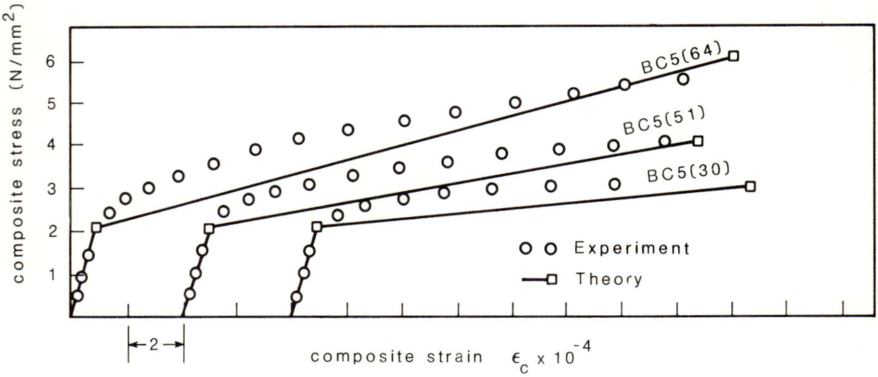

Figure 3.24 Stress–strain curves of the composites in direct tension[99].

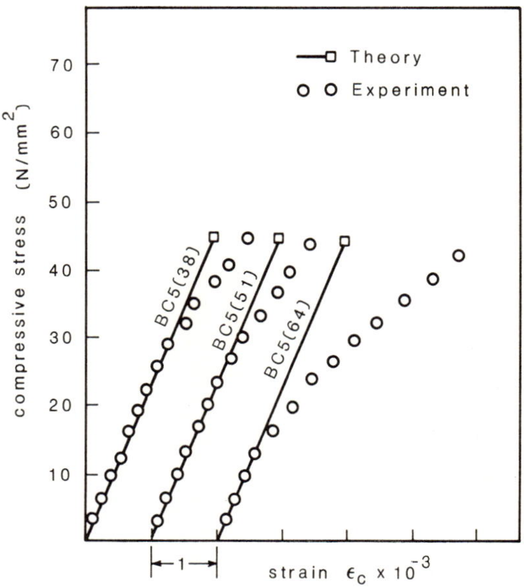

Figure 3.25 Stress–strain curves of the composites in axial compression[99].

direct tension, cracks in the matrix occurred when the composite strain exceeded the cracking strain of the matrix. The post-cracking ductility of the composites is contributed by the bamboo fibres which serve as crack retarders. The amount of bamboo fibres to be used in order to achieve this post-cracking ductility should not exceed the minimum volume fraction, given as

$$V_{fc} = \frac{2r\sigma_{mu}}{\psi_b F_2 l \tau_u} \qquad \text{for } l < l_c$$

$$V_{fc} = \frac{\sigma_{mu}}{\psi_f F_2 \sigma_{fu}} \qquad \text{for } l > l_c$$

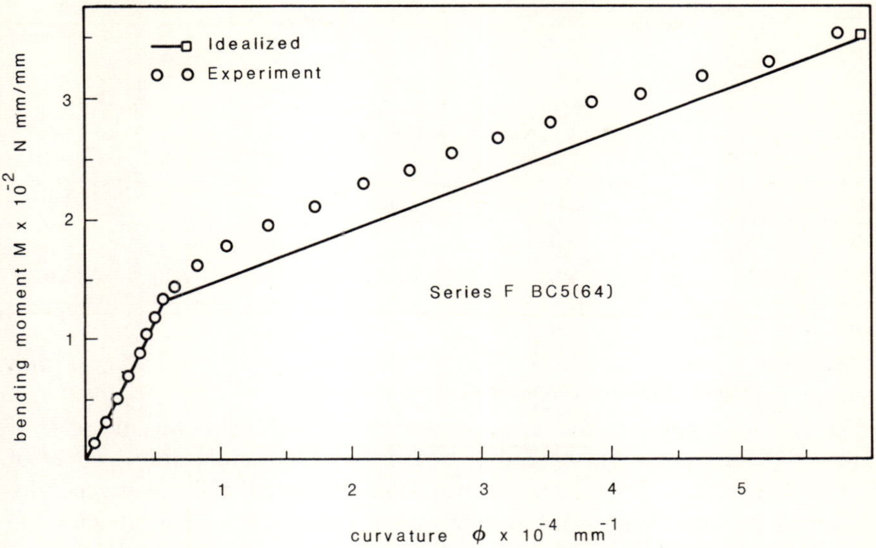

Figure 3.26 Typical moment–curvature relationship of the composite[99].

where

V_{fc} = minimum volume fraction of bamboo fibres required for strengthening to occur under direct tensile stress
r = ratio of cross-sectional area to the periphery of bamboo fibre
σ_{mu} = direct tensile strength or modulus of rupture of cement paste matrix
σ_{fu} = ultimate strength of bamboo
ψ_b, ψ_f = length of efficiency factor when failure is of first mode and second mode respectively
τ_u = maximum bond stress between bamboo fibre and cement paste matrix
F_2 = orientation factor
l = length of bamboo fibre
l_c = critical length of bamboo fibre.

At ultimate condition in compression, local buckling of the fibre causes premature spalling of the matrix which reduces the load-carrying capacity of the composite[46,98,99].

In bending, the composite behaves as a two-layered system in the elastic range. The composite behaves essentially as a bilinear material. The ultimate moment of resistance of the composite can be predicted by assuming two rectangular stress blocks; the cement paste alone resisting the compressive forces and the fibres resisting the tensile forces. This is valid if sufficient volume of fibre is used to achieve the post-cracking ductility of the composite.

Smith *et al.*[100] experimented on bamboo fibre reinforced mortar to make

Table 3.4 Results of tests on bamboo reinforced concrete beams (ref. 102)

Bamboo %	Plain concrete strength: Split (N/mm²)	Plain concrete strength: Flexural (N/mm²)	Split strength of composite (N/mm²)	Flexural load (kN)	Modulus of elasticity (kN/mm²)	Gain in strength, %: Split	Gain in strength, %: Flexural
1	1.97	14.42	2.11	15.13	6.05	7.20	4.90
2	2.18	18.94	2.40	20.00	6.91	10.00	5.60
3	2.34	18.64	2.62	16.00	7.22	12.10	−8.80
4	2.38	16.15	2.54	14.20	8.01	6.70	−12.10

600 × 600 × 10 mm panels for cladding material and 1800 × 1800 × 40 mm planks that could be used as a substitute for wood.

Test results indicated that: average strength of each panel was in excess of that of the plain panels; there appeared to be an optimum fibre content of about 8%; and the greatest strength was achieved when the fibre was generally oriented parallel to the largest side of the panel and coincided with the direction of the bending stress.

The planks were subjected to central point loading tests over a span of 1.7 m. Failure for the reinforced plank was less sudden, and a significant load was carried even after the tensile cracks had been checked by the fibre.

Nagaraja[102] using bamboo fibres dipped in water for one hour, observed that bamboo fibres contributed significantly to the tensile strength of the composite, the increase being almost linear with the percentage of the fibre. Table 3.4 shows the comparison of the split strength and flexural capacities between bamboo fibre reinforced beams and the corresponding plain concrete beam. The results indicated that an appreciable increase in the composite modulus of elasticity occurred with increase of fibre percentage. The gain in strength indicated an optimum fibre percentage of around 3% for split tensile strength and 2% for flexural strength. The typical load deflection curves for these optimum percentages are shown in Figure 3.27.

3.3.5.2 *Bamboo pulps.* At the Asian Institute of Technology, the use of bamboo pulp as suitable substitute for asbestos fibres was investigated[46,101]. Pulp refers to those fibres which were obtained by cooking bamboo chips in 20% NaOH solution at approximately 170°C with a steam pressure of 8274 N/mm^2 for 6 h. About 2% of sulphur by weight of bamboo was added to increase the efficiency of pulping. The pulp must be washed before use. The average length and diameter of the pulp used are 2.7 mm and 0.03 mm respectively.

The mechanical and physical properties of composites with bamboo pulp reinforcement, where the pulp content was varied, were obtained in axial compression, direct tension, flexure and torsion. The ultimate strength in compression, lower than unreinforced cement paste, indicated that the pulps were ineffective in compression and that their presence can be construed as

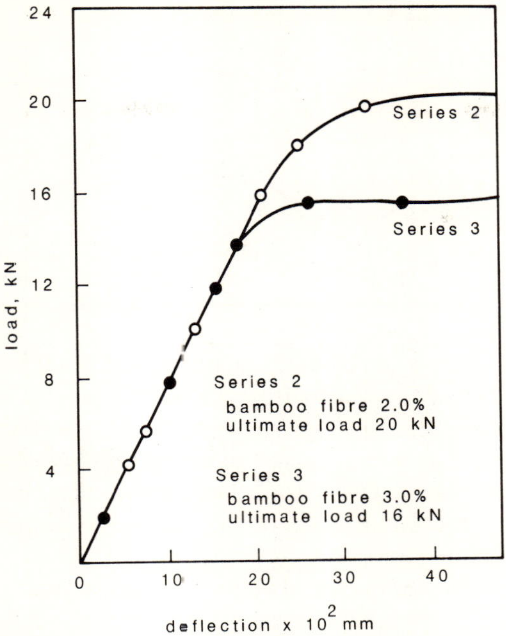

Figure 3.27 Typical load–deflection curves of bamboo fibre reinforced beams[102].

voids in the matrix which make the composite weak in compression. The presence of the pulps, however, prevents sudden failure and total disintegration of the composite.

In direct tension tests, it was observed that the presence of the pulp significantly increased the first crack strength of the composites. The flexure test showed that bamboo pulp boards satisfy the ASTM requirements for bending. The anti-clastic test indicated that the pulp reinforcement was effective in resisting shear stresses.

Pakotiprapha *et al.*[46,98,99] concluded that with cement paste binder, the resulting composite yields building boards which can be used for walls, ceilings and roofs. The boards are also impervious, non-combustible and possess good impact properties. The cost analysis also showed that bamboo pulp boards are cost-competitive compared with asbestos fibre board.

3.3.5.3 *Bamboo strips.* Nagaraja[102] investigated slabs, simply supported beams and continuous beams reinforced with various percentages of bamboo strips. To improve the bond with concrete, the bamboo strips were provided with rectangular groves at 100-mm centres along one of the edges of the strip.

The bamboo strip reinforced slab specimens were loaded at four points, one at each middle point of the half diagonals of the slab. Dry as well as wet bamboo was used as reinforcement; the ultimate loads of the dry bamboo

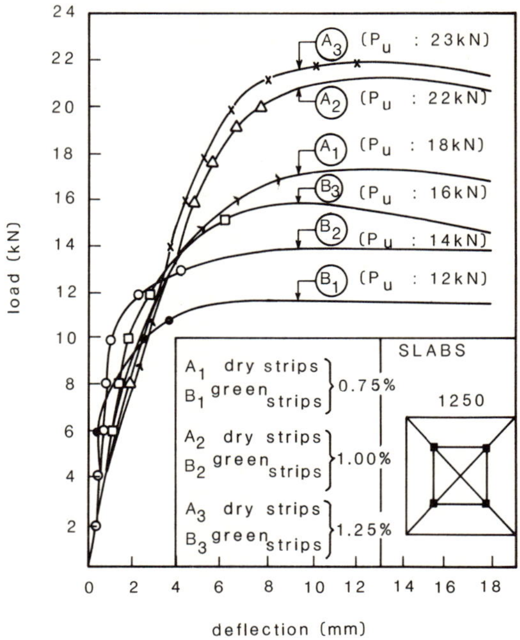

Figure 3.28 Load deflection curves of bamboo strip reinforced slabs[102].

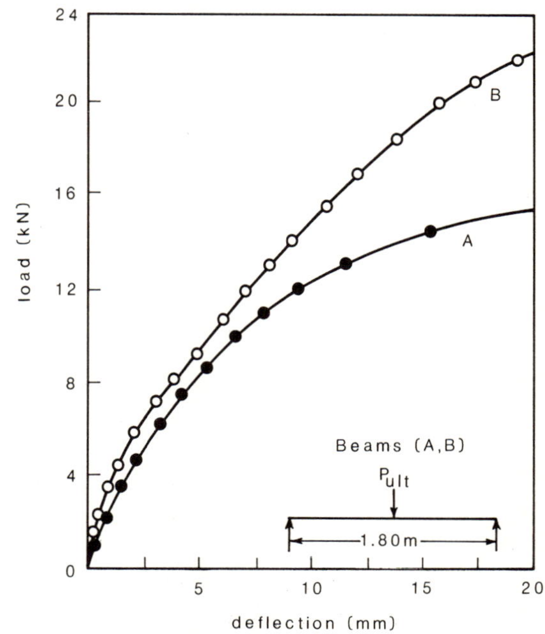

Figure 3.29 Load deflection curves of bamboo reinforced beams[102]. *A*: bamboo reinforcement 2.18%, ultimate load 15.85 kN. *B*: bamboo reinforcement 4.55%, ultimate load 24.14 kN.

specimens were higher than those with green bamboo. The ultimate loads calculated from the yield-line mechanism compared very satisfactorily with observed failure loads. Figure 3.28 shows the load deflection curves of the bamboo reinforced slabs.

For the simple beam, no bond slip and rupture of the reinforcement was observed. The beam deflection increased slowly in the initial stages until the appearance of the first crack, then deflection increased rapidly until failure (Figure 3.29). For the two-span continuous beams, significant redistribution of moments from the support section to the midspan was determined by careful analysis of the sequence of crack development at the critical sections and the rate of increase in the width of cracks at these sections. Slab and beam prototypes were also observed for long-term loading (one year), and their behaviour was found to be satisfactory.

Two dwelling units each with a plinth area of about 30 m^2 were built, using dry bamboo strips as reinforcement for the roof slab and lintels. The maximum span of the slab was 4 m, with 2.5% bamboo reinforcement. The buildings were periodically inspected for cracks and any other form of distress. After 10 years, the slabs and lintels are without any structural problems[102].

3.4 Recommended practices

Bamboo has been used as reinforcement for concrete and mortar in many countries. Admittedly, bamboo reinforcement has certain obvious limitations; however, this should not limit its increased adoption provided precautions are taken. The recommendations of different researchers based on their experience are compiled here as a guide for the use of bamboo as reinforcement.

3.4.1 *Selection of bamboo*

(i) Use only bamboo showing a pronounced brown colour. This will ensure that the plant is at least three years old[59,65]. A relatively heavy bamboo with uniform colour indicates superior strength and soundness[50].

(ii) Select the longest large diameter culms available[65]. As far as possible, straight long bamboo without any deformation and cracks should be selected[50].

(iii) It should be free of any decay, fungus growth or holes due to white ants[50].

(iv) Do not use whole culms of green, unseasoned bamboo[65]. Whole culms should be limited to distribution reinforcement only[60].

(v) Avoid bamboo cut in spring or early summer. These culms are generally weaker due to increased fibre moisture content[50,65].

(vi) All main reinforcing should consist of bamboo splints with the nodes exposed upwards to increase bond; thus select bamboo species with the greatest number of nodes[50,63].

Table 3.5 Cross-sectional Areas (ref. 65)

Whole culms		$\frac{3}{4}$in (19.1 mm) splints	
Diameter in (mm)	Area in^2 (mm^2)	Thickness in (mm)	Area in^2 (mm^2)
$\frac{3}{8}$ (9.50)	0.008 (5.16)	$\frac{1}{8}$ (3.18)	0.094 (60.65)
$\frac{1}{2}$(12.70)	0.136 (87.74)	$\frac{1}{4}$ (6.35)	0.188 (121.29)
$\frac{5}{8}$(15.88)	0.239 (154.19)	$\frac{3}{8}$ (9.50)	0.282 (181.94)
$\frac{3}{4}$(19.10)	0.322 (207.74)	$\frac{1}{2}$(12.70)	0.375 (241.94)
1 (25.40)	0.548 (353.55)	$\frac{5}{8}$(15.88)	0.469 (302.58)
2 (50.80)	1.920 (1238.71)	$\frac{3}{4}$(19.10)	0.563 (363.22)

3.4.2 *Preparation of bamboo*

(i) Splints are generally more desirable than whole culms as reinforcement; larger culms should be split into splints of approximately 19 mm wide[18,65]. Table 3.5 shows approximate net area provided by whole culms and by 19-mm wide splints[65].

(ii) Limit the cross-sectional area of bamboo to 5% of the total reinforced cross-section area[21].

(iii) All culms should be cut approximately 50 mm above ground level, stacked vertically for drying and dried to approximately 20% moisture content before surface treatment[63]. When possible, the bamboo should be allowed to dry and season for 3–4 weeks before using[59,65]. The culms must be supported at regular spacings to reduce warping[65].

(iv) After seasoning and cutting into required sizes, bamboo should be lightly filed to induce roughness on the surface[59,62]. Bond can also be increased by depositing fine sand over double coats of resin alcohol mixture and white paint, by short lengths of wire wound round the bamboo at intervals or by notches cut in the middle of each bamboo link[9,18,60].

(v) Bamboo, either dry or wet, can be permanently bent if heat is applied while applying pressure. This procedure can be used for forming splints into U-shaped stirrups and for putting hooks on reinforcement for additional anchorage[65].

(vi) When seasoned bamboo, either split or whole, is used as reinforcement, it should receive a waterproof coating to reduce swelling when in contact with concrete. The type of coating examples are asphalt emulsion, native latex, coal tar, paint, dilute varnish and sodium silicate – will depend on the materials available[68].

3.4.3 *Construction procedure*

(i) In general, techniques used in conventional reinforced concrete construction need not be changed when bamboo is used as reinforcement. Same mix designs can be used. Concrete slump should be as low as workability will allow. High early strength cement is preferred to minimize cracks caused by swelling of bamboo when seasoned bamboo cannot be waterproofed[65].

(ii) Bamboo reinforcement should not be less than 38.1 mm (1.5 inches) from the face of the concrete surface[50,65]. In China, the following are used as concrete covers: 20–30 mm for slabs, 25–35 mm for beams, and 35–40 mm for columns[9].

(iii) The clear spacing between bamboo rods or splints should not be less than the maximum size aggregate plus 6.35 mm (0.25 inch)[65]. Spacing limitations rather than the risk of over-reinforcement govern the bamboo ratio; this is normally[62] not greater than 4%.

(iv) Reinforcement should be evenly spaced and lashed together on short sticks placed at right angles to the main reinforcement. When more than one layer is required, the layers should be tied together. Ties should preferably be made with wire for primary members and with vegetation strips for secondary members[65].

(v) When using whole culms, the top and the bottom of the stems should be alternated in every row and the nodes should be staggered. This will ensure a fairly uniform cross-section throughout the length of the member[65].

(vi) Bamboo should be securely tied down before placing concrete. It should be fixed at regular intervals of 0.91–1.22 m to prevent it from floating up in the concrete during placement and vibration[65].

(vii) When splicing is essential, splints should be overlapped for at least 457 mm. Splicing at area of high stress should be avoided. In no case should more than 30% of the reinforcement be spliced at one section. For an effective splicing, the spliced ends should be sand-coated and well tied together[50, 65].

Minimum lengths l of overlap for a given thickness t for tension and compression members are[10]:

For tension reinforcement:

seasoned bamboo $l = 10t$

green bamboo $l = 16t$

For compression reinforcement:

seasoned bamboo $l = 8t$

green bamboo $l = 12t$

Minimum overlap should in no case be less than 300 mm[10].

3.5 Conclusions

Bamboo as reinforcement is gaining acceptance in different applications in developing countries. Its wider use in lieu of other materials will be greatly influenced by the understanding of its mechanical properties and physical behaviour, and the ability of engineers to quantify these rationally, using simple accepted engineering principles.

Numerous studies have been published in the last few years and confidence in the material is building up. However, there is a need for testing standards. These will make comparison of results possible. This is specially significant considering the random nature of bamboo. Further refinement of the methods of analysis and design is also necessary. Nevertheless, sufficient design information is now available and adequate field experience has now been established to enable the safe design and construction of many bamboo reinforced mortar and concrete structures.

References

1. Sharma, Y.M.L. (1980) Bamboos in Asia-Pacific Region. *Bamboo Research in Asia, Proceedings of a Workshop*, Singapore, May 1980, 99–120.
2. Lindayen, T.M., Valbuena, R.R., and Tamolong, F.N. (1969) Erect bamboo species in the Philippines. *Philippine Lumberman* **15**, 44–48.
3. Tamolong, F.N., Lopez, F.R., Semana, J.A., Casin, R.F., and Espiloy, Z.B. (1980) Properties and utilization of Philippine erect bamboos. *Bamboo Research in Asia, Proceedings of a Workshop*, Singapore, May 1980, 189–200.
4. Laxamana, M.G. (1985) Drying of some commercial Philippine bamboos. *FPRDI Journal* **XIV**, 8–19.
5. *Bamboo.* Division of Silviculture, Royal Forest Department, Ministry of Agriculture and Cooperatives, Bangkok, 1985, 35 pp. (in Thai).
6. Ramyarangsi, S. (compiler) (1984) *Bamboo Research Project, Annual Report.* Division of Silviculture, Royal Forest Department, Ministry of Agriculture and Cooperatives, 22 pp.
7. Ramyarangsi, S. (compiler) (1985) *Bamboo Research Project, Annual Report Year II*, Division of Silviculture, Royal Forest Department, Ministry of Agriculture and Cooperatives, 44 pp
8. Chu, H.K. (1914) Bamboo for reinforced concrete. Thesis, Massachusetts Institute of Technology, U.S.A.
9. Engineering development in U.S.S.R. *Civil Engineering and Public Works Review* **53** (1985) 1005.
10. Masani, N.J., Dhamani, B.C., and Singh, B. (1977) *Studies on Bamboo Concrete Composite Construction.* Controller of Publications, Delhi, 39 pp.
11. Datta, K. (1936) Versuche über die Verwendung von Bambus in Betonbau (Investigation on the use of bamboo in concrete). *Der Bauingenieur* **17**, 17–27.
12. De Simone, D. (1939) Norme, materiali e strutture pe la limitazione dell'imjuego del ferro nelle construzionei. *Annali dei Lavori* **77**, 1189–1223 (English translation by Holme, A. (1940) Substitutes for steel in reinforced concrete, *Royal Engineers Journal* **54**, 237–244).
13. Glenn, H.E. (1950) *Bamboo Reinforcement in Portland Cement Concrete.* Clemson Agricultural College, South Carolina, 171 pp.
14. Saucier, K.L. (1964) Precast concrete elements with bamboo reinforcement. *Tech. Rep.* **6–646**, US Army Engineer Waterways Experiment Station, CE, Vicksburg, Mississippi, 33 pp.
15. Smith E.F., and Saucier, K.L. (1965) Design, analysis and construction of precast concrete elements with bamboo reinforcement. *Misc. Paper* **6–705**, US Army Engineer Waterways Experiment Station, CE, Vicksburg, Mississippi, 33 pp.
16. Saucier, K.L., and Smith, E.F. (1967) Design, analysis and construction of precast concrete elements with bamboo reinforcement. *Misc. Paper* **6–926**, US Army Engineer Waterways Experiment Station, CE, Vicksburg, Mississippi, 17 pp.
17. Cox, F.B., and Geymayer, H.G. (1969) Expedient reinforcement for concrete for use in Southeast Asia: Report 1—Preliminary tests of bamboo. *Tech. Rep.* **C-69-3**, US Army Engineer Waterways Experiment Station, CE, Vicksburg, Mississippi, 123 pp.

18. Cox, F.B., and Geymayer, H.G. (1970) Bamboo reinforced concrete. *Misc. Paper* **C-70-2**, US Army Engineer Waterways Experiment Station, CE, Vicksburg, Mississippi, 15 pp.
19. Charters, D. (1976) Model study of bamboo hydraulic structures for erosion abatement. *New Horizons in Construction Materials*, Vol. 1, ed. H.Y. Fang, Envo Publishing, Lehigh Valley, 511–523.
20. Youssef, M.A.R. (1976) Bamboo as a substitute for steel reinforcement in structural concrete, Part I. *New Horizons in Construction Materials*, Vol. 1, ed. H.Y. Fang, Envo Publishing, Lehigh Valley, 525–534.
21. Mehra, S.R., Uppal, H.L., and Chadda, L.R. (1951) Some preliminary investigations in the use of bamboo for reinforcing concrete. *Ind. Concr. J.* **25**, 20–21.
22. Chadda, L.R. (1956) Bamboo reinforced soil cement lintels. *Ind. Concr. J.* **30**, 303–304.
23. Mehra, S.R., Ghosh, R.K., and Chadda, L.R. (1957) *Bamboo-Reinforced Soil-Cement as a Construction Material.* Central Road Research Institute, New Delhi, India, 18 pp.
24. Narayana, S.K., and Rehman, P.M.A. (1962) Bamboo-concrete composite construction. *J. Inst. Engrs. India* **42**, 426–440.
25. Purushotham, A. (1963) A preliminary note on some experiments using bamboo reinforcement in cement concrete. *J. Timber Dryers and Preservers Assoc. India* **9**, 3–14.
26. Mehra, S.R., Ghosh, R.K., and Chadda, L.R. (1965) Consideration as material for construction of bamboo-reinforced soil-cement with special reference to its use in pavement. *Civ. Eng. and Publ. Wks. Rev.* **60**, 1457–1461, 1643–1645, 1766–1768.
27. Datye, K.R. (1976) Structural uses of bamboo, *New Horizons in Construction Materials*, Vol. 1, ed. H.Y. Fang, Envo Publishing, Lehigh Valley, 499–510.
28. Kalita, U.C., Khazanchi, A.C., and Thyagarajan, G. (1977) Bamboocrete low-cost houses for the masses. *Ind. Concr. J.* **51**, 309–312.
29. Datye, K.R., Nagaraju, S.S., and Pandit, C.M. (1978) Engineering applications of bamboo. *Proc. Int. Conf. on Materials of Construction for Developing Countries*, Bangkok, August 1978, 3–20.
30. Dietz, A.G.H. and Moavenzadeh, F. (1977) Innovative uses of materials for housing in developing areas. *Proc. Int. Sem. on Low-Cost Housing*, Madras, 42.
31. Kalita, U.C., Khazanchi, A.C., and Thyagarajan, G. (1978) Bamboo-crete wall panels and roofing elements for low-cost housing. *Proc. Int. Conf. on Materials of Construction for Developing Countries*, Bangkok, August 1978, 21–36.
32. Widjaja, E.A., and Risyad, Z. (1983) Anatomical and mechanical properties of utilized bamboos in Indonesia. In *Proc. Int Bamboo Workshop* (in press).
33. Shimada, H. (1939–40) A study of the physical and chemical properties of bamboo poles. *Trans. Arch. Inst. Japan*, April (translated by Dr. Kuchi Sagawa, Mississippi Medical Center, Jackson, Mississippi).
34. Espinosa, J.C. (1930) Bending and compressive strength of the common Philippine bamboo. *Philippine J. Sci.* **41**, 121–134.
35. Purugganan, V.A. *et al.* (1957) *Research Study on the Use of Bamboo as Reinforcement in Portland Cement Concrete.* Division of Materials Testing and Physical Research, Bureau of Public Highways, Manila, Philippines.
36. Bamboo reinforcement for concrete, *Civ. Eng. and Publ. Wks Rev.* **53** (1958) 627.
37. Abang Abdullah, A.A. (1983) Utilization of bamboo as a low cost structural material. *Appropriate Building Materials for Low Cost Housing, African Region*, E. & F.N. Spon, London, 177–182.
38. Abang Abdullah, A.A., and Abang Abdul Rahim, A.A. (1984) Basic strength properties of a few selected Malaysian bamboos. *J. Inst. Engs. Malaysia*, **34**, 68–71.
39. Abang Abdullah, A.A. (1984) Development of basic mechanical tests for Malaysian bamboo. *Pertanika* **7**, 2–7.
40. Abang Abdullah, A.A. (1984) Research on Malaysian bamboos. *Bull. IEM*, September, 13–14.
41. Hamid, A. (1973) Structural aspects of bamboo reinforced soil cement. Thesis No. 523, Asian Institute of Technology, Bangkok, 105 pp.
42. Ali, Z. (1974) Mechanical properties of bamboo reinforced slabs. Thesis No. 676, Asian Institute of Technology, Bangkok, 41 pp.
43. Durrani, A.J. (1975) A study of bamboo as reinforcement for slabs on grade. Thesis No. 801, Asian Institute of Technology, Bangkok, 56 pp.

44. Abu Sadeque, A.H.M. (1975) Behavior of bamboo reinforced tied columns. Thesis No. 816, Asian Institute of Technology, Bangkok, 58 pp.
45. Apirathvorakij, V. (1976) Design and performance evaluation of a rice storage bin for rural areas. Thesis No. 950, Asian Institute of Technology, Bangkok, 52 pp.
46. Pakotiprapha, B. (1976) A study of bamboo pulp and fiber cement paste composites. Dissertation No. D20, Asian Institute of Technology, Bangkok, 59 pp.
47. Singh, R.V. (1977) A study of bamboo as reinforcement for concrete columns subjected to eccentric loads. Thesis No. 1046, Asian Institute of Technology, Bangkok, 81 pp.
48. Cook, D.J., Pama, R.P. and Singh, R.V. (1978) The behaviour of bamboo-reinforced concrete columns subjected to eccentric loads. *Mag. Concr. Res.* **30**, 145–151.
49. Ali, Z. and Pama, R.P. (1978) Mechanical properties of bamboo reinforced slabs. *Proc. Int. Conf. Materials of Construction for Developing Countries*, Bangkok, August 1978, 49–66.
50. Pama, R.P., Durrani, A.J. and Lee, S.L. (1976) A study of bamboo as reinforcement for concrete pavements. *1st Conf. of the Road Engineering Association of Asia and Australia*, Bangkok, 1976, 45–96.
51. Bukkanasuta, A. (1986) Laboratory pull-out tests using bamboo and tensar geogrids including a case study. Thesis No. GT-85-25, Asian Institute of Technology, 113 pp.
52. Narayanamurty, D. and Mohan, D. (1972) *The Use of Bamboo and Reeds in Building Construction.* United Nations, New York, 95 pp.
53. Janssen, J. (1980) The mechanical properties of bamboo used in construction. *Bamboo Research in Asia, Proc. Workshop*, Singapore, May 1980, 173–188.
54. Fang, H.Y., and Mehta, H.C. (1978) Sulfur-sand treated bamboo rod for reinforcing structural concrete. *New Uses of Sulfur–II, Adv. Chem. Ser.* **165**, American Chemical Society, 241–254.
55. Atrops, J.C. (1969) Elastizität und Festigkeit von Bambusrohren. *Der Bauingenieur*, **44**, 220–225.
56. Grosser, D., and Liese, W. (1974) Verteilung der Leitbundel und Zellarten en Sprossachsen verschiedenen Bambusarten. *Holz als Roh- und Werkstoff* **32**, 473–482.
57. Mentzinger, R.J., and Plourde, R.P. (1966) Investigation of treated and untreated bamboo as reinforcing in concrete. Thesis, Villanova University.
58. Mansur, M.A., and Aziz, M.A. (1983) Study of bamboo-mesh reinforced cement composites. *Int. J. Cement Comp.* and *Lightweight Concr.* **5**, 165–171.
59. Perry, S.H., Kankam, J.A. and Ben-George, M. (1986) The scope for bamboo-reinforced concrete. *10th Int. Congr. FIP*, New Delhi, February 1986, 205–214.
60. Kowalski, T.G. (1974) Bamboo-reinforced concrete. *Ind. Concr. J.*, April, 119–121.
61. Kankam, J.A., Perry, S.H., and Ben-George, M. (1986) Bamboo-reinforced concrete one-way slabs subjected to line loading *Int. J. Cement Comp. and Lightweight Concr.* **XX**, 1–9.
62. Kankam, J.A., Ben-George, M., and Perry, S.A. (1986) Bamboo-reinforced concrete two-way slabs subjected to concentrated loading. *Structural Engineer* **64B**, 85–92.
63. Johansen, K.W. (1962) Yieldline theory. Cement and Concrete Association, London.
64. ACI Committee 544 (1978) Measurement of properties of fibre reinforced concrete. *Proc. Amer. Concr. Inst.* **75**, 283–289.
65. Brink, F.E., and Rush, P.J. (1966) *Bamboo Reinforced Concrete Construction.* US Naval Civil Engineering Laboratory, California, 17 pp.
66. Manga, J.B. (1983) The feasibility of bamboo as reinforcement for ferrocement housing walls. *J. Ferrocement* **13**, 345–349.
67. Wirjosumarto, F. *et al.* (1982) *Application of Ferrocement and Related Composite Materials in Indonesia.* Pusat Teknologi Pemtangunan, Institut Teknologi Bandung, Bandung, 220 pp.
68. Bresler, B., and Gilbert, P.H. (1961) Tie requirements for reinforced concrete columns. *Proc. Amer. Concr. Inst.* **58**, 555–570.
69. Watt, S.B. (1978) *Ferrocement Water Tanks and their Construction.* Intermediate Technology Publications, London, 118 pp.
70. Sashi Kumar, K., Sharma, P.C., and Robles-Austriaco, L. (1984) Review of design considerations and construction techniques for ferrocement water resources structures. *J. Ferrocement* **14**, 49–64.
71. Sashi Kumar, K., Sharma, P.C., and Robles-Austriaco, L. (1979) Rainwater collection tank (Indonesia). *J. Ferrocement* **9**, 47–48.
72. Winarto (1981) Rainwater collection tanks constructed on self-help basis. *J. Ferrocement* **11**, 247–254.

73. Winarto (1982) Rural rainwater cistern design in Indonesia. *Proc. Int. Conf. on Rainwater Cistern Systems*, Hawaii, June 1982, 294–298.
74. Rolloos, H. (1979) *Tangki Air Hujan Bambusemer*. Pusat Teknologi Pembangunan Institute Teknologi Bandung, Bandung, 37 pp.
75. Kaufman, M. (1983) *From Ferro to Bamboo*. Yayasan Dian Desa, 48 pp.
76. Aristanti, C. (1986) Dian Desa's rainwater catchment program—its implementation and development. *Rainwater catchment status and research priorities in the Southeast Asian Region, Proc. Regional Seminar and Workshop*, 29 November–3 December 1983, Khon Kaen, 132–144.
77. Kerkvoorden, R. van (1982) Rainwater collector for villages in West Java, Indonesia. *Proc. Int. Conf. on Rainwater Cistern Systems*, Hawaii, June 1982, 299–307.
78. Robles-Austriaco, L., Pama, R.P., and Valls, J. (1981) Ferrocement for the water decade. *J. Ferrocement* **11**, 229–245.
79. Tankrush, S. (1981) Bamboo-cement water tank—a solution to water shortage problem in Thailand. *J. Ferrocement* **11**, 255–258.
80. Vadhanamkkit, C., and Pannachet V. (1987) Investigations of bamboo reinforced concrete water tanks. *Proc. 3rd Int. Conf. on Rainwater Cistern Systems*, Khon Kaen, January 1987 C13-1/C13-6.
81. Fricke, T. (1982) *Bamboo-Reinforced Concrete Rainwater Tanks*. AT International, Washington, 50 pp.
82. Chindaprasert, P., Hovichitr, I., and Wirojanagud, P. (1986) A study and development of rainwater tank, Thai-Australia Village Water Supply Project. Khon Kaen University, Khon Kaen, 31 pp.
83. Thiensiripipat, N. (1986) Bamboo reinforced concrete water tanks, *Rainwater catchment: status and research priorities in the Southeast Asian Region, Proc. Regional Seminar and Workshop*, Khon Kaen, 29 November–3 December 1983, 50–67.
84. Sriruenthong, P. (1986) Rainwater catchment tank—bamboo reinforced concrete. *Rainwater catchment: status and research priorities in the Southeast Asian Region, Proc. Regional Seminar and Workshop*, Khon Kaen, 29 November–3 December 1983, 75–87.
85. Prempridi, T. (1986) Current research and practices in rainwater catchment and storage in Thailand. *Rainwater catchment: status and research priorities in Southeast Asian Region, Proc. Regional Seminar and Workshop*, Khon Kaen, 29 November–3 December 1983, 114–131.
86. Paro, F.J. (1986) Rainwater catchment project. *Rainwater catchment: status and research priorities in the Southeast Asian Region, Proc. Regional Seminar and Workshop*, Khon Kaen, 29 November–3 December 1983, 174–179.
87. Hazbun, J.A. (1983) Use of alternative methods and materials in ferrocement rainwater storage tank. *J. Ferrocement* **13**, 169–176.
88. Innovative water cistern. *J. Ferrocement* **13** (1983) 370–371.
89. Innovative rural applications (Indonesia). *J. Ferrocement* **10** (1980) 164–166.
90. Pompe, C., Kerkvoorden, R. van, and Sirivoyo, H. (1982) Ferrocement applications in the West Java Rural Water Supply Project. *J. Ferrocement* **12**, 51–61.
91. Pompe, C. (1983) A total-water-system for 500 people–the 'MCKS-500'. *J. Ferrocement* **13**, 161–168.
92. Smith, R.B.L., Boon-Long, S., Loo, Y.C., Nutalaya, S., and Pataragetvit, S. (1971) Hermetic storage of rice for Thai Farmers. *Thai J. Agric. Sci.* **4**, 143–155.
93. Bigg, G.W. (1975) Bamboo reinforced ferrocement grain storage silo. *J. Struct. Eng.* **2**, 173–182.
94. Chadda, L.R. (1956) The use of bamboo for reinforcing soil-cement eliminating shrinkage cracking in walls. *Ind. Concr. J.* **30**, 200–201.
95. Nainan, P.R., and Kalam, A.K.A. (1977) Bamboo-reinforced soil-cement for rural use. *Ind. Concr. J.* **51**, 382–389.
96. Pon, K.S., Arunagiri, S., and Ravichandran, S. (1980) Bamboo-reinforced soil-cement composite beams. *Proc. Silver Jubilee Sem., Modern Trends in Rural Engineering*, Muzaffarpur, India.
97. Ghosh, R.K., Phull, Y.R., and Chadda, L.R. (1968) Construction of experimental road length near Rohtak using bamboo-reinforced soil-cement as underlay and base course. *20 Years Design and Construction of Roads and Bridges*, Vol. 1, Ministry of Transport and Shipping (Road Wing), Govt. of India, New Delhi.

98. Pakotiprapha, B., Pama, R.P., and Lee, S.L. (1983) Analysis of a bamboo fibre-cement paste composite. *J. Ferrocement* **13**, 141–159.
99. Pakotiprapha, B., Pama, R.P., and Lee, S.L. (1983) Behavior of a bamboo fibre-cement paste composite. *J. Ferrocement* **13**, 235–248.
100. Smith, P.O., Amedoh, A. Nawa-Acheampong, H., and Bailey, W.A. (1979) Bamboo fibre reinforcing material in concrete. *Appropriate Technology* **6**, 8–10.
101. Pakotiprapha, B., Pama, R.P., and Lee, S.L. (1976) Development of bamboo pulp boards for low-cost housing. *Proc. IAHS Int. Symp. on Housing Problems*, Vol. 2, South Carolina, 1096–1115.
102. Nagaraja, R. (1986) Strength and behaviour of concrete elements reinforced with bamboo fibres and strips. *RILEM Symp. FRC86, 3rd Int. Symp. on Developments in Fibre Reinforced Cement and Concrete*, Vol. 1, Sheffield, July 1986, 13–17.

4 Durability of natural fibres in concrete

HANS-ERIK GRAM

Abstract

Natural fibre concrete stored outdoors becomes embrittled with time. This embrittlement is caused by a decomposition of the fibre in the alkaline pore water of the cement matrix. The rapidity with which, for instance, the sisal fibre in concrete is decomposed depends on ambient temperature and the intensity of the transport of alkaline pore water to the embedded fibre and of decomposed products from the fibre. This transport of the pore water in concrete is initiated by, for example, drying out and moistening the concrete respectively, but also by temperature variation and variations in the relative humidity of the ambient air.

A simple method for accelerated ageing of sisal fibre concrete is presented.

Countermeasures against decomposition of sisal fibres include different fibre impregnations and changes of the cement matrix.

The embrittlement of sisal fibre concrete can be delayed by protectively impregnating the fibres, but the impregnation does not result in complete protection.

Sealing the pore system of the cement matrix with wax, zinc stearate or sulphur may also improve the durability of the fibre.

The embrittlement of sisal fibre concrete can be almost completely avoided by reducing the alkalinity of the concrete pore water. This reduction of the alkalinity is achieved by replacing part of the ordinary portland cement with fine-grained pozzolanas, for instance silica fume, rice husk ash or diatomite. Tests on specimens stored outdoors in different climates confirm the results obtained in the laboratory.

4.1 Introduction

Interest in using natural fibres of vegetable origin as reinforcement of thin concrete elements such as roofing sheets and wall panels, has grown during the last ten to fifteen years. One reason for this is that roofing in developing countries is a vital necessity for the great majority of the population. This problem is often solved by importing iron sheets or asbestos fibres for asbestos cement sheets. The increasing economic burden many less developed countries have to carry makes it harder to meet vital needs, such as roofing, with imports. Efforts to find appropriate solutions based on locally available raw materials have thus become more and more important. Many different solutions for roofing, based on natural fibres such as sisal and coir in a cement mortar, have been found. Some are successful and some not. One of the reasons for failure has been a change of properties of the fibre with time – the fibre is decomposed in the cement matrix. The mechanisms behind the deterioration of the fibres and different countermeasures will be discussed in this chapter.

4.2 Historical background

Natural fibres have been used to reinforce inorganic materials for thousands of years. Examples include straw for bricks, mud and poles, plaster and reeds. During this century other fibres such as bamboo, wood as cellulose fibres, wool or chips, bast fibres, leaf fibres, and seed and fruit fibres have been used in cement-based products.

An extensive survey of the literature regarding natural fibres in cement-based products has been made by Gram[1,2] and the results can be summarized as follows.

Bamboo has been used as reinforcement of concrete slabs. The durability problems observed when using bamboo derive from its moisture movements and the concomitant cracking in the composite, and from the impaired adhesion between the bamboo and the concrete. The moisture movements in bamboo are said to decrease to a harmless level if the bamboo is treated with a water-repellent agent. The adhesion between the bamboo and the concrete is also improved when this treatment is combined with sprinkling sand on to the newly treated surface.

Considerable attention has been devoted to cellulose fibre. During World War II, efforts were made in countries which did not have domestic supplies of asbestos to replace asbestos fibre in whole or in part with cellulose fibres in the production of asbestos cement. The experience gained indicates that this was sometimes successful and sometimes not. Once again, moisture movement in the cellulose fibres gave rise to problems. Another explanation may be impaired frost resistance which can be expected to play an important part in areas with a cold climate. During the 1970s the injurious effects to health of asbestos fibres were mapped out with increasing clarity, and asbestos cement producers began to look for a replacement for asbestos fibre. In Finland, asbestos cement producers have succeeded in replacing asbestos fibre with cellulose fibre combined with a small quantity of polypropylene fibre. The moisture movement problem has not, however, yet been solved in a satisfactory manner. In Denmark, Great Britain and Australia, producers of asbestos cement products have developed autoclaved cellulose fibre cement products and their durability is reported to be excellent.

Woodwool slabs, built up of long wood shavings covered with a thin layer of cement paste, have been manufactured for more than 50 years. The loadbearing capacity of these slabs is so low that they are used only as insulation material, but the slabs are highly durable. The only problem encountered in connection with this product occurs in its production. Woodwool can contain substances which affect the cement reactions. Consequently, only certain wood species constitute suitable raw materials, and sometimes only when they have been subjected to pretreatment.

Cement particle boards have been manufactured for almost 30 years and they have been successfully used in outdoor structures, for example as roof coverings. Once again, the choice of wood species is affected by its content of components which can disturb the cement reactions. Research in this sector has been oriented towards finding methods for treating the wood so as to permit the use of a larger number of wood species as raw material for cement particle boards. In the literature indications are provided that the durability of the cement particle boards is improved if the surface is treated with a water-repellent agent.

At the end of the 1960s, investigations began on the possibility of using organic natural fibres as reinforcement in thin concrete sheets. These investigations soon indicated that it was possible to manufacture products of natural fibre concrete which had properties comparable with those of the asbestos cement products. An appropriate method for manufacturing roof sheets of natural fibre concrete was rapidly developed, and spread to countries in Central America, Africa and Asia through IT Building Materials Workshop in Great Britain and others. The method, which involves reinforcing cement or concrete products with natural fibres such as sisal, coir and jute, for example, has been applied in at least 28 countries. When combining these fibres with the cement matrix the fibres are utilized in two ways. On one hand, the fibre in the fresh concrete makes it possible to mould a product in a simple manner. On the other, the fibre increases the toughness of the material so that the product can withstand handling and a structural load. The durability aspect is important in connection with the loadbearing property of the composite. Several researchers have reported an embrittlement in the composite with time. The reasons are said to depend on the poor resistance of the natural fibre in alkaline environments. Gram[1] found that sisal fibre concrete became embrittled within 10 months when stored outdoors in Tanzania. Lola[3] found that fibre concrete roofing sheets produced in Guatemala, Sri Lanka and the Dominican Republic were partly badly cracked. Gram *et al.* evaluated the experience from 20 fibre concrete roofing projects in 19 developing countries[4]. Producers of roofing sheets report higher damage rates due to cracking than do tile producers. Smaller sheets seem to behave better than bigger sheets. Willi[5] reports reroofing 3000 m^2 roofing sheets with tiles in Nicaragua because of problems with the sheets. Obviously the fibres in the sheet are not capable of reinforcing the mortar and controlling crack development due to poor durability. This low durability of natural fibres in an alkaline environment and ways to improve it are discussed in the following sections.

4.3 Sisal fibre concrete

4.3.1 *The constituents of natural fibre concrete*

The durability of natural fibre is affected by the composition of the fibre as well as by the environment surrounding the fibre. The sisal fibre is chosen as an example representing the natural fibres of vegetable origin.

4.3.2 *Structure of sisal fibres*

Sisal is a bast fibre which is extracted from the leaves of *Agave sisalana*. The fibre content of the leaves, which can have a length of up to 1.5 m, amounts to between 3.5 and 5%. The fibre diameter is less than 0.2 mm and the appearance of the fibre is illustrated schematically in Figure 4.1.

A sisal fibre in cross-section is built up of about 100 fibre cells (each with a length of 2–5 mm). The fibre cells are linked together by means of middle lamellae, which consist of hemicellulose, lignin and pectin. Figure 4.2 presents a schematic sketch of a fibre cell.

The fibre cell consists of a number of walls built up of fibrillae. In the outer wall (primary wall) the fibrillae have a reticulated structure. In the outer secondary wall (S1), which is located inside the primary wall, the fibrillae are arranged in spirals with a spiral angle of 40° in relation to the longitudinal axis of the cell. The fibrillae in the inner secondary wall (S2) have a sharper slope, 18°. The innermost wall (the tertiary wall) is thin and has a reticulated arrangement of fibrillae. The fibrillae are, in turn, built up of micro-fibrillae

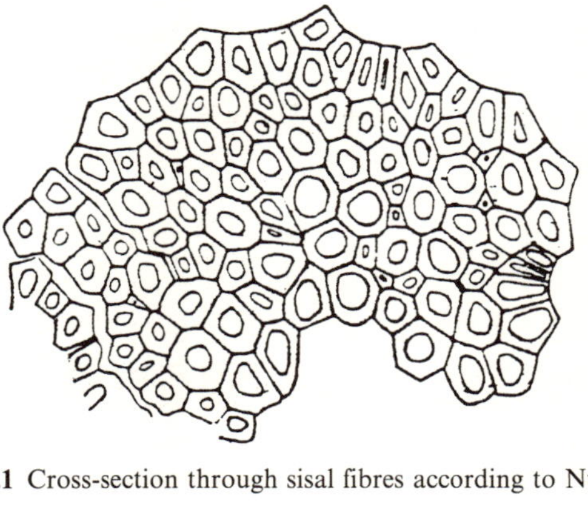

Figure 4.1 Cross-section through sisal fibres according to Nutman[6].

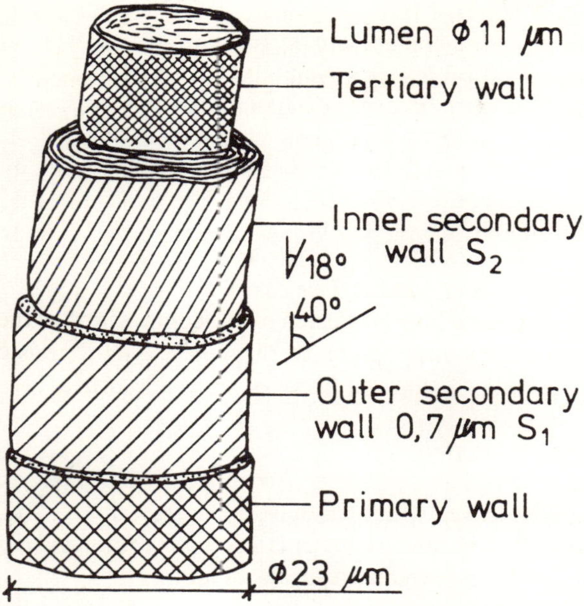

Figure 4.2 Schematic sketch of a fibre cell with approximate dimensions.

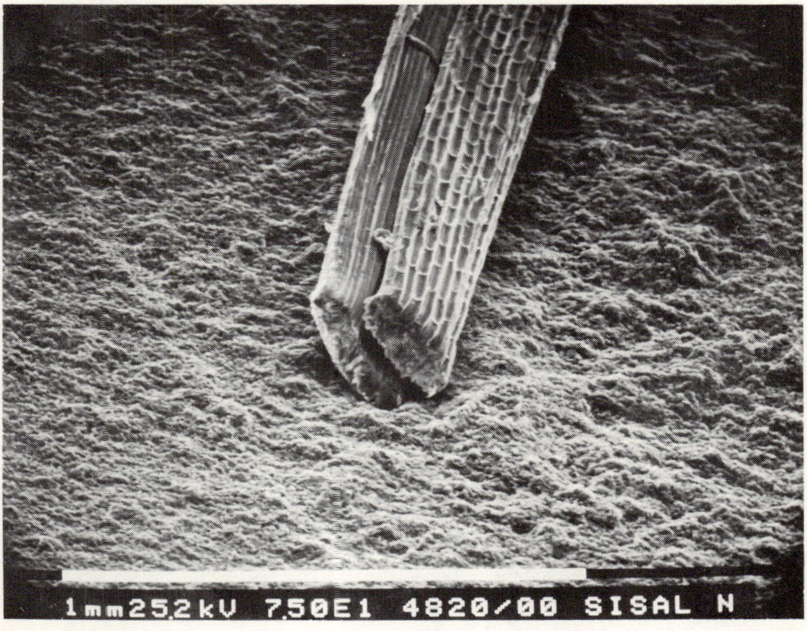

Figure 4.3 Scanning electron micrograph of a sisal fibre. The fibre is built up of fibre cells linked to each other by means of the primary wall and the middle lamella.

with a thickness of about 20 nm. The microfibrillae are composed of cellulose molecular chains with a thickness of 0.7 nm and a length of a few μm (the degree of polymerization for sisal is about 25 000).

About 12% of the fibre consists of hemicellulose and this mainly occurs in the outer layer of the fibre cell wall, in other words the primary wall, and in the middle lamella. The hemicellulose consists of polysaccharides, mainly xylose, and has an acid character, which may have a negative effect on the hydration of cement. The degree of polymerization of hemicellulose lies between 50 and 200.

The pectin (about 1%) occurs in the middle lamella and probably has a cohesive function.

The lignin (about 10%) mainly occurs in the middle lamella. The middle lamella contains 70% lignin. Figures 4.3 to 4.5 show scanning electron micrographs of the microstructure of the sisal fibre.

4.3.3 *Structure of concrete*

Concrete is built up of aggregate particles with varying particle sizes which are bonded together by means of a cement gel. The cement gel is formed from the cement paste, a mixture of cement and water. The most commonly used binder is ordinary portland cement (opc) and the mean diameter of the cement particles is approximately 10 μm. The chemical properties of concrete are mainly determined by the composition of the binder and the structure of the

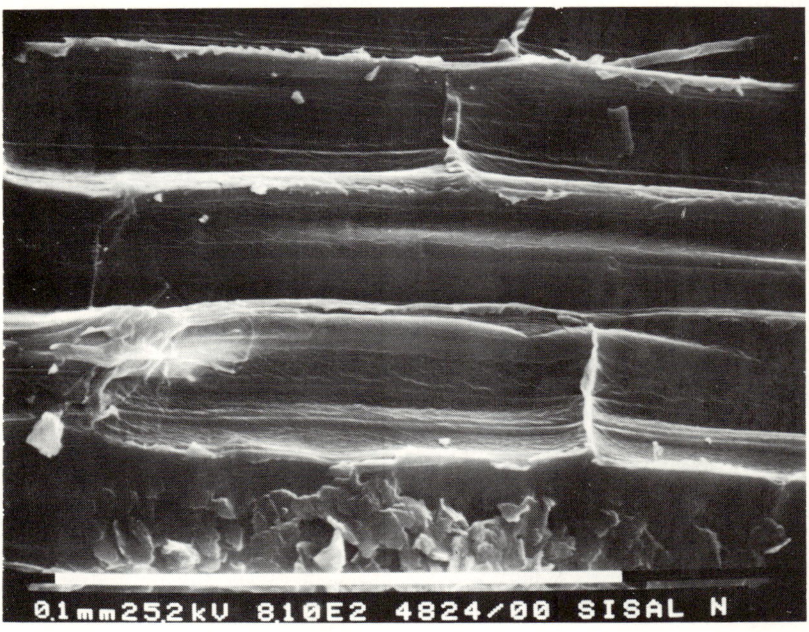

Figure 4.4 Close-up of sisal fibre. The lumen and the tertiary wall are exposed.

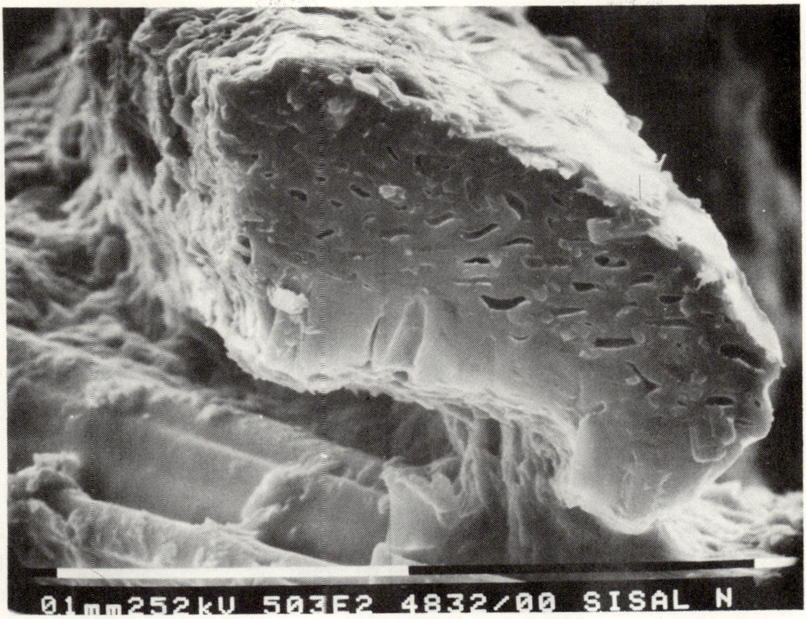

Figure 4.5 Cross-section through a cut sisal fibre with the lumen of fibre cells visible (compare Figure 4.1).

cement gel. A reaction is started between cement and water when the two constituents are mixed. This reaction, in which the cement gel is formed, can be subdivided into five stages.

During the *initial stage*, which lasts about 10 minutes (in laboratory conditions), the most easily dissoluble substances in the cement are liberated. The following ions can be detected in the water phase: Ca^{2+}, OH^-, SO_4^{2-}, K^+ and Na^+. A thin membrane, which is rich in SiO_2, is formed around the calcium silicates in the cement, and a membrane of ettringite crystals is formed around the calcium aluminates.

During the following stage, the *dormant stage* which can last up to two hours, the thin membranes around the cement minerals are slowly thickened. The concentration of Ca^{2+}, OH^-, K^+ and Na^+ in the liquid phase rises while SO_4^{2-} decreases in pace with the thickening of the membrane, as shown in Figure 4.6.

When the concentration of OH^- and Ca^{2+} has increased to a certain level, the chemical compounds which built up the membrane around the cement minerals are converted to large crystals and the cement minerals begin to hydrate again. Now the converted and permeable membrane no longer prevents hydration of the cement minerals and we have reached the *acceleration stage*. The crystals grow further and further out from the surface of the cement mineral and bind together the individual cement particles. The

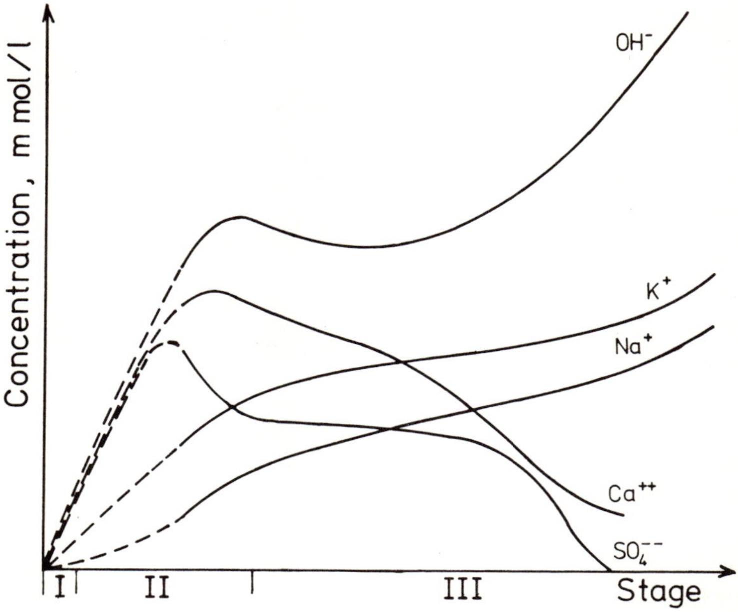

Figure 4.6 Ion concentration in a water solution in contact with hydrating cement during the first 24 h period, according to Vivian[7].

cement binds, and crystals of calcium hydroxide are precipitated in the immediate vicinity of the particles. The formation of colloidal mineral binders, cement gel and calcium hydroxide, begins. Hydration takes place with considerable heat development during the acceleration stage, which lasts for about 24 h.

During the *deceleration stage*, which lasts about 28 days, the hydration sequence is retarded mainly due to the fact that the concentration of Ca^{2+} in the liquid phase decreases. This is, in turn, due to the fact that the diffusion resistance in the crystal structure around the cement minerals increases in pace with the growth of the crystals. In previous stages the hydration mainly led to the formation of crystals in a direction out from the surface of the clinker particles. The crystals now develop inwards into the cement particle. In the deceleration stage the Ca^{2+} ions in the liquid phase have been forced to precipitate in pace with the increasing OH-ion concentration. As a result, the calcium hydroxide crystals have developed still further. The ion concentration in the liquid phase of K^+ and Na^+ increases constantly during the first four stages, with the result that all potassium and sodium ions are regarded as liberated after 28 days.

During the last stage, the *stabilization stage*, the intensity of the hydration sequence decreases. The crystal lattice which was built up in the earlier stages is refined and densified. At the same time, the liquid diffusivity in the lattice

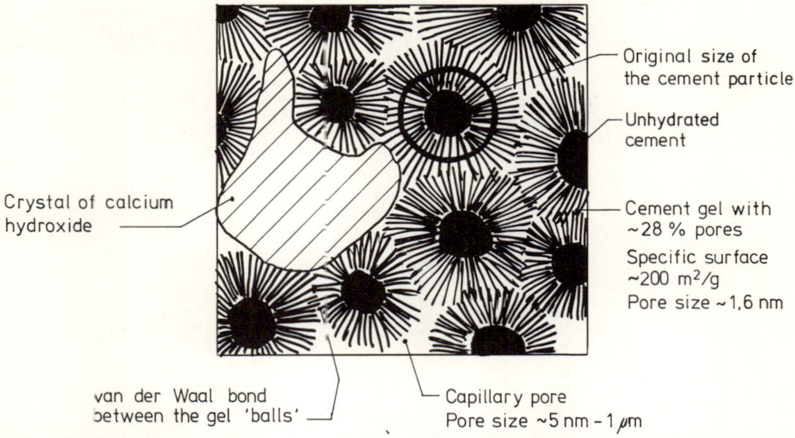

Figure 4.7 Schematic illustration of the structure of partly hydrated cement paste, according to Fagerlund[9], modified after discussions with Grudemo[10].

decreases and continued hydration is rendered more difficult. According to Czernin[8], cement mineral particles with diameters larger than 60 μm can never be fully hydrated.

The cement gel is thus built up of an interlaced compound of small, flat, curved plate-shaped and fibre-shaped crystals with thicknesses varying between 10 nm and 50 nm, usually surrounding an unhydrated clinker nucleus. Figure 4.7 presents a schematic illustration of the structure of a partly hydrated cement paste. The porosity of the cement gel is approximately 28%.

Larger voids are formed between the individual balls of gel. These voids are called capillary pores and their diameter varies between 5 nm and 1 μm. The proportion of capillary pores and their size depends on the weight ratio between the water and cement and the degree of hydration of the cement paste. If the water–cement ratio is reduced or if the degree of hydration is increased, the proportion of capillary pores is reduced.

When mixing concrete, air is always mixed into the cement paste. This results in pores of diameter in excess of 5 μm; a maximum size of one millimetre is common. Pores of this size interval can also be formed in the phase boundaries between cement paste and aggregate particle or reinforcement material, inter alia, due to water segregation in the fresh concrete state.

The porosity of concrete is normally less than 20% by volume. Gel pores and the smaller capillary pores are filled with water, while larger capillary pores and the surfaces of the voids are covered with a layer of water molecules whose thickness depends on the prevailing relative air humidity.

As can be seen in Figure 4.6, the ion concentration in the cement paste's liquid phase is changed with increasing hydration. Longuet *et al.*[11,12] have studied the chemical composition of pore water in concrete up to an age of 2 years (see Table 4.1).

Table 4.1 Changes with increasing age in the chemical composition of pore water solution in concrete manufactured of high-alkali portland cement according to Longuet[12]. Water–cement ratio 0.5; curing temperature 20°C; 95.5% clinker, 4.5% gypsum.

Compound	Content in cement (%)	Age and concentration in porewater solution (mg/kg)							
		5 h	2 days	7 days	28 days	3 months	6 months	13 months	2 years
SiO_2	20.51	16				0.3	200	22	126
Al_2O_3	5.14							0	65
SO_3	2.76	5300	70	31	—	0	222	344	1470
CaO	61.67	772	127	131	118	174	0	0	0
Na_2O	0.24	910	2240	3560	4960	4120	4840	4430	5110
K_2O	1.21	12010	16610	23650	26200	25000	25700	26100	27670
OH^-		176	435	623	717	625	670	635	671
pH		13.10	13.50	13.68	13.45	13.21	13.91	13.66	

Cements which have a high content of Na_2O and K_2O ($Na_2O + 0.6\ K_2O > 1.5\%$) are called high-alkaline cements. Low-alkaline cement ($Na_2O + 0.6\ K_2O < 0.5\%$) results in a pH value of about 13.1 in the pore water of a one-year-old concrete, according to Longuet[12].

Carbon dioxide, available in air, reacts with the hydration products of a cement. This process is called carbonation and begins in the surface layer of the concrete. A carbonation front gradually moves inwards via cracks and the pore system. Carbonation causes shrinkage in the concrete with concomitant crack formation, greater density and strength and a reduction of the pH value in the pore water to levels below 9. The mechanisms behind the carbonation sequence are partly simple, but are also partly unknown.

The carbonation process is slow, and the rate of carbonation is determined by the carbon dioxide concentration in the ambient air, the permeability of the concrete matrix, the moisture of the concrete, the temperature and the number of changes in moisture and temperature. Depending on these factors, the carbonation front in concrete moves inwards at a rate of 0.01–3 mm per year. The carbonation process is promoted by high relative air humidity in the pore system of the matrix, high carbon dioxide concentration in the ambient air, and high temperatures.

4.4 Durability of natural fibres in concrete

4.4.1 *Decomposition of sisal fibre in alkaline environments*

4.4.1.1 *Assumed model for decomposition.* It is generally held that decomposition of cellulose in an alkaline environment can take place in accordance with two different mechanisms. One is the peeling-off mechanism which occurs at the end of the molecular chain. The end group, which is reductive, then reacts with OH^- and forms isosaccharinic acid (CH_2OH) which is unhooked from the molecular chain. End groups are liberated in this way all the time. The probability of the end group forming metasaccharinic acid instead, which is not unhooked and which is stable in an alkaline solution, is 1:50, according to Mjöberg[13]. Since the degree of polymerization in sisal cellulose is high, about 25 000, the peeling-off mechanism is fairly harmless in itself. Peeling-off occurs within a wide temperature interval but the rate does not become markedly high until a temperature of 75°C or above.

The other form of cellulose decomposition is alkaline hydrolysis. This causes the molecular chains to divide, and the degree of polymerization decreases. Since the division of the molecular chain entails the exposure of new reductive end groups, the peeling-off mechanism can be started. Alkaline hydrolysis does not take place at a high rate until the temperature is in excess of 100°C.

Decomposition of hemicellulose in an alkaline environment is the same as for cellulose. The degree of polymerization for hemicellulose lies between 50

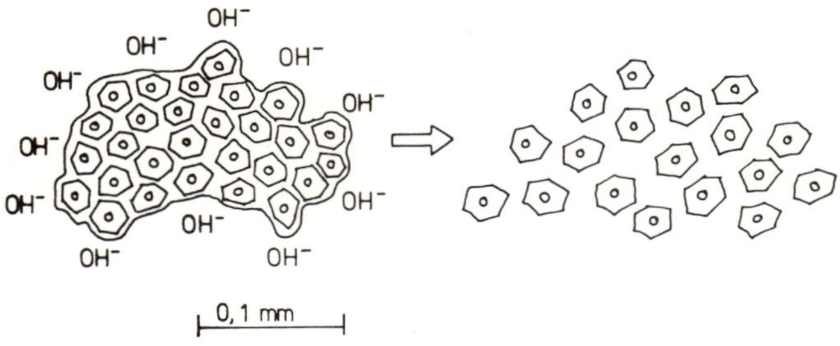

Figure 4.8 Schematic sketch of the decomposition of sisal fibres in concrete. The middle lamella is dissolved by the alkaline pore water in the concrete and the lumen is filled by calcium hydroxide.

and 200. Consequently, the peeling-off mechanism becomes the dominating decomposition mechanism.

Lignin is composed of large three-dimensional molecules, the detailed structure of which is not known. Lignin consists of aromatic substances, is easily broken down in an alkaline environment and is coloured yellow and brown when oxidized. Lignin begins to soften at 70°C–80°C. At 120°C it is partly liquid.

The primary cause of the change in the characteristics of sisal fibre in concrete is assumed by the author to be a chemical decomposition of the lignin and the hemicellulose in the middle lamella. The alkaline pore water in the concrete dissolves the lignin and the hemicellulose and thus breaks the link between the individual fibre cells (Figure 4.8). The long sisal fibre breaks down into numerous small units and loses its reinforcing capacity in concrete. The lumen of the fibre cells is filled with calcium hydroxide and the fibre loses its flexibility.

4.4.1.2 *Sisal fibres stored at room temperature in chemical solutions with various pH values.* Conditioning sisal fibres at room temperature in buffer solutions of different pH values has shown that fibres conditioned in solutions with a pH value in excess of 12 are coloured yellow[1]. This yellow discoloration of the fibres indicates a reaction between the buffer-solution OH^- ions and the lignin in the sisal fibre. Fibres stored in water become coloured red and black with time, thus indicating that bacteria break down the components in sisal fibre when the fibre is stored in neutral solution.

The pH alone does not determine whether the sisal fibre is decomposed or not. A significant reduction in strength of a fibre with time can be noted for fibres conditioned in a saturated lime solution with a pH value of 12.6. The yellow discolouring of fibres noted in solutions with a pH value in excess of 12 does not generally entail a reduction in strength. It is possible that the calcium ions in the saturated lime solution contribute to different decomposition effects than do the sodium and potassium ions in the buffer solutions with pH values of 12.5 and 13.

Fibres conditioned in tap water also lose strength with time, probably due to microbiological action. Wet fibres in solutions with a pH value in excess of 12 could be pulled apart fairly easily by finger force. This was not possible with fibres in other buffer solutions. It thus seems likely that the hemicellulose and the lignin in the middle lamella are reorganized during the drying.

Sisal fibres conditioned in a saturated lime solution at a temperature of 50°C proved to have lost all their strength after 12 days[1]. When dried, the fibres disintegrated into a brown powder. The individual fibre cells had been released from each other, probably through a considerable decomposition of the lignin and the hemicellulose in the middle lamella. Sisal fibres conditioned in a concrete pore solution (pH 13.7) at a temperature of 70°C for 7 days, retained about 30% (tested dry) of their original strength[1]. Wet fibres could easily be pulled apart by finger force. The occurrence of calcium ions in the saturated lime solution appears to have had a more powerful decomposing effect on the natural fibre than the occurrence of sodium and potassium ions in the prepared concrete pore water.

4.4.1.3 *Observations on sisal fibres embedded in concrete.* When a roof unit of sisal fibre concrete manufactured in Tanzania was subjected to load testing at an age of 10 months, the load caused a brittle failure[1]. When the failure surface was investigated, the fibres located at the surface appeared to be intact. Their pullout length was about 50 mm and, in a subjective assessment, they were regarded as flexible and strong just as the fibres usually are in bending beams which have been stored and tested in a laboratory environment. Figure 4.9 presents a schematic sketch of the fibres at the failure surface.

The fibres in the centre of the cross-section, on the other hand, have become yellow. The pullout length of the fibres here was 2 mm or less, and they easily broke off at their anchorage in the concrete.

When the carbonation depth was determined, the undamaged fibres were found to lie in carbonated concrete, in other words in concrete in which the pH value of the pore water is under 9. The short yellow fibres lay in that part of the concrete in which the pH value of the pore water exceeded 13. The intensity of the yellow colour was stronger than that observed in fibres conditioned in chemical solutions at room temperature, possibly with the exception of the fibres which were conditioned in saturated lime solution.

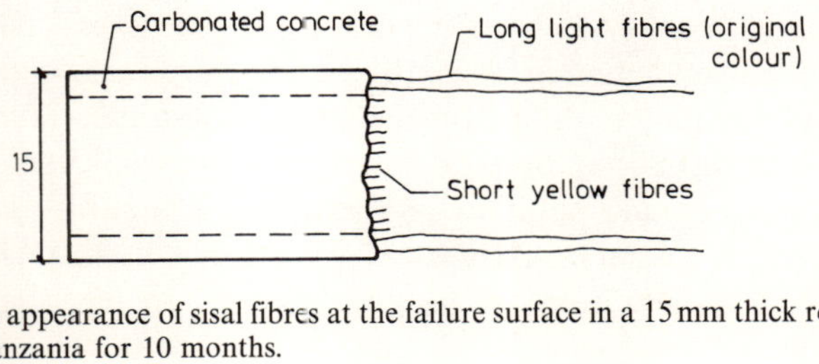

Figure 4.9 The appearance of sisal fibres at the failure surface in a 15 mm thick roof sheet stored outdoors in Tanzania for 10 months.

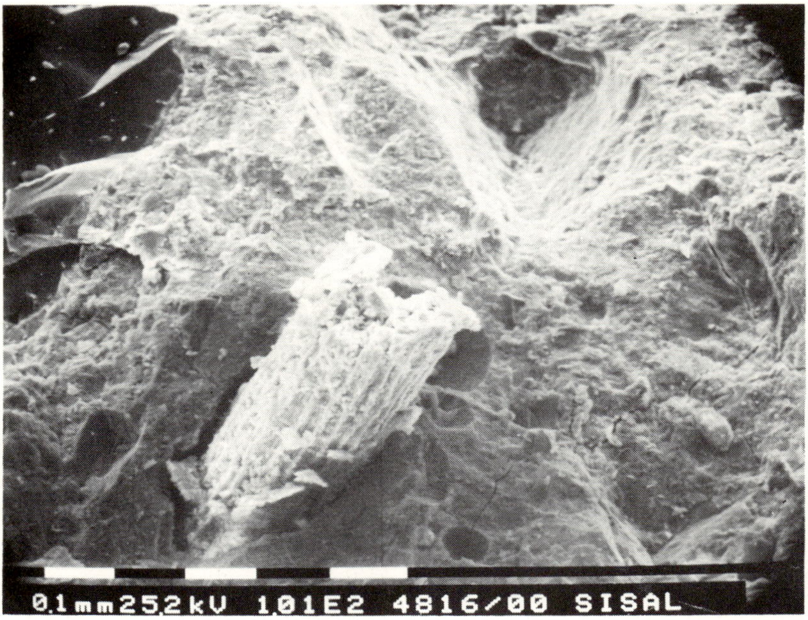

Figure 4.10 Sisal fibre at the failure surface in cement mortar stored outdoors in Stockholm for 9 years.

Figure 4.11 Cross-section through a cut sisal fibre from a 9-year-old specimen. The fibre seems to have been 'mineralized'. No lumen is visible.

A corrugated roofing sheet manufactured in 1978 and stored outdoors in Stockholm until 1987 has been investigated. The material is brittle and the pullout length of the sisal fibres is less than 0.5 mm. As can be seen in Figure 4.10, cement hydration products are visible on the fibre surface. In Figures 4.10 and 4.11 it can be seen that the lumen of the fibre cells seems to be filled with hydration products. The fibre is no longer flexible, and its reinforcing capacity is lost.

4.4.2 *Embrittlement of the composite due to ageing*

4.4.2.1 *Embrittlement – test method and criterion.* Specimens measuring 55 × 200 × 8 mm, cut out of larger sheets manufactured with a mortar composed of binder, aggregate (0–2 mm) and water in the weight ratios 1:2:0.5 respectively, have been used in the tests reported in ref. 1. The part by volume of the embedded fibre was 2%. Two layers of 220 mm long fibres uniformly distributed and aligned parallel to the edge of the mould (280 mm) were worked into thin layers of mortar so that each individual fibre was surrounded by mortar. The specimens were moist-cured for one week (20°C) and then kept in laboratory conditions (20°C, 50% RH) for another 3 weeks.

The behaviour of the small beams during bending according to the load set-up in Figure 4.12 was used in the evaluation of the toughness of the composite. The resulting stress–strain curve falls into two main different categories, as illustrated in Figure 4.13. The shape of the stress–strain curve depends on factors such as the strength of the matrix, the modulus of elasticity of the matrix and the fibre, the tensile strength of the fibre, the bond between the matrix and the fibre, the fibre length and diameter, the fibre orientation as well as the volume of incorporated fibres. The position of the first crack, point *A* (limit of proportionality), is mainly dependent on the properties of the matrix.

After the development of the first crack, the fibre properties govern the appearance of the stress–strain curve. One simple way to evaluate the reinforcing effect of the fibre and consequently its durability or embrittlement, and its capacity for enhancing the fracture resistance of the composite, is to determine the post-cracking stress at point *B* in curve 1 (modulus of rupture) or in curve 2 (the stress at 1% strain) as shown in Figure 4.13. The parameter *B* thus represents a stress value at at least 1% strain and may be considered as a toughness parameter.

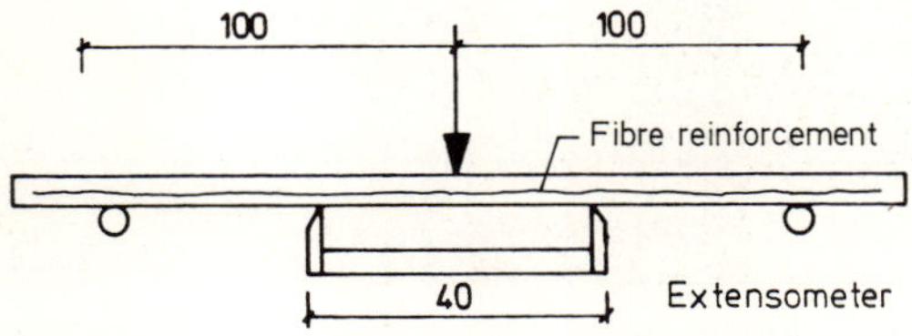

Figure 4.12 Load set-up when bending fibre concrete beams.

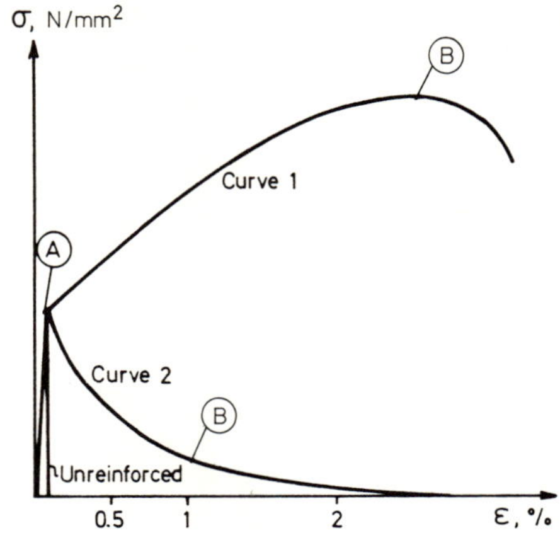

Figure 4.13 Principal layouts for stress–strain curve during tests of flexural strength of natural fibre concrete beams.

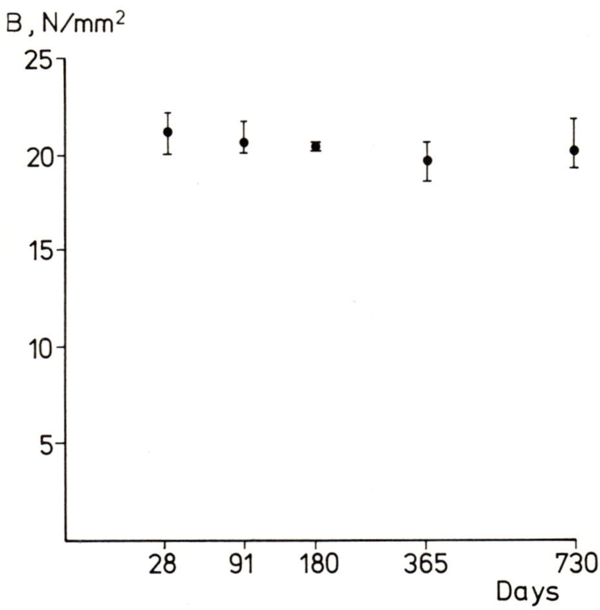

Figure 4.14 Toughness parameter *B* (mean value of three beams and highest and lowest value) as a function of age in specimens stored at 20°C and 50% RH[1].

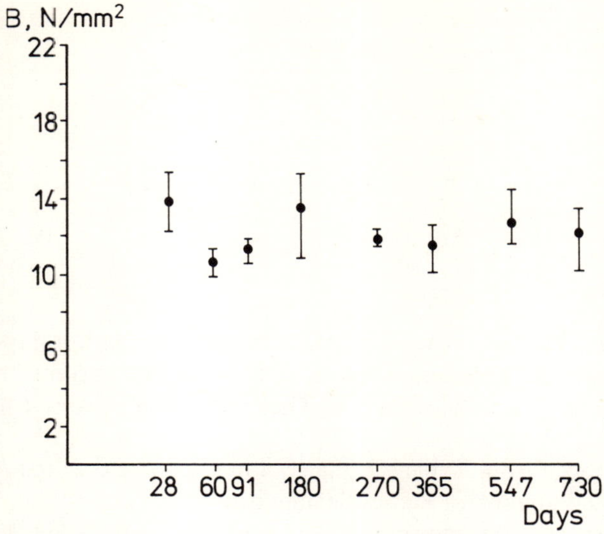

Figure 4.15 Toughness parameter *B* (mean value of three beams, and highest and lowest value) as a function of age in specimens conditioned in a water bath at a temperature of 50°C. Specimens tested wet[1].

A tough and ductile composite has a stress–strain relation according to curve 1 with a high value of the parameter *B* and the ultimate tensile strain. When the fibres are damaged the stress–strain curve is gradually changed from curve 1 to curve 2, or even to the same curve, as in the unreinforced case.

4.4.2.2 *Specimens stored dry.* Specimens stored in a laboratory room at 20°C and 50% relative humidity for up to two years do not lose their toughness[1] (see Figure 4.14).

4.4.2.3 *Specimens stored wet.* Specimens stored in a water bath at 50°C for up to two years have been tested (Figure 4.15). As can be seen, sisal fibre concrete conditioned in a water bath at 50°C retains its toughness for at least two years[1].

4.4.2.4 *Specimens stored alternately wet and dry.* Specimens with a thickness of 15 mm and reinforced with cut sisal fibres (30 mm) have been conditioned for up to two years[1], alternately for one week in water at 20°C and for three weeks in a conditioning room at 20°C and a relative humidity of 50%. T toughness parameter *B* developed as follows: 2.4 N/mm² at the age of 28
days, 1.5 after one year and 0.4 after two years. The fibres lose t
capacity within 1–2 years.

4.4.2.5 *Specimens stored outdoors.* Natural fibre concrete util
in, for example, a roof sheet is subjected to alternating wet a

Table 4.2 Toughness parameter *B* for specimens after various lengths of time conditioned outdoors in Ethiopia, Thailand, Tanzania and Sweden.

	Storing time and parameter B(N/mm^2)				
Locality	28 days	200 days	1 year	2 years	4 years
Addis Ababa	15.0	1.4	1.9	—	—
Bangkok	10.5	—	0.5	0.7	—
Dar es Salaam	12.5	0.3	—	—	—
Stockholm	30.7	6.9	—	—	4.8

temperature in the roof sheet can reach 70–80°C during hot days. Table 4.2 presents the results from tests on sheets which were conditioned outdoors in Dar es Salaam, Tanzania; Bangkok, Thailand; Addis Ababa, Ethiopia; and Stockholm[1,14,15].

As can be seen from Table 4.2, the sisal fibre concrete stored outdoors in Addis Ababa, Bangkok or Dar es Salaam is seriously embrittled. Within one year the toughness parameter *B* has dropped to a value below 2 N/mm^2. In practice, the ability of the sisal fibre to impart tensile strength to the concrete had completely ceased after a comparatively brief conditioning period outdoors in a tropical climate. Sisal fibre concrete conditioned outdoors in Stockholm for 4 years was not embrittled to the same extent.

4.4.2.6 *Specimens subjected to accelerated ageing.* It was previously shown that sisal fibre concrete conditioned in a dry or wet constant environment at 20°C

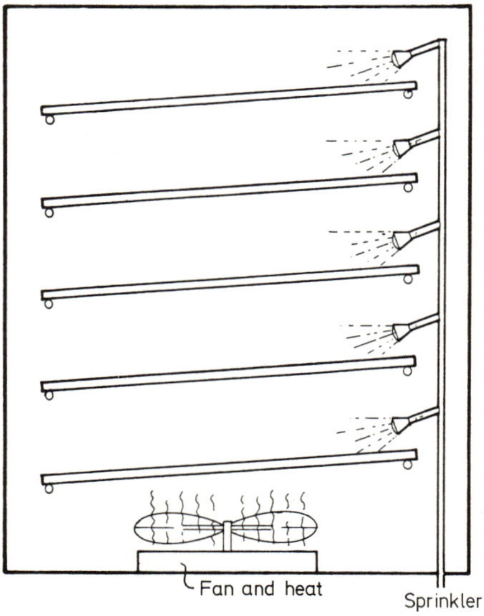

Figure 4.16 Schematic sketch of a climate cubicle used for accelerated ageing.

or 50°C does not become brittle with time. It is likely that the transport of OH^- ions or Ca^{2+} ions to the fibre proceeds very slowly when the environment is kept constant. When sisal fibre concrete is subjected to alternating wet and dry conditioning in the laboratory or outdoors, it is embrittled. An accelerated ageing method should therefore be based on changes in the moisture state of the specimen, allowing the capillary pore system of the cement matrix to be alternately filled with and emptied of alkaline water. This can be achieved in a simple climate cubicle[1] shown in Figure 4.16.

The specimens in the climate cubicle are subjected to moistening and cooling by spraying them with water from the sprinkler nozzles. The water, which is transported into the cubicle during 30 minutes (1.5 litre/minute and specimen), maintains a temperature of 10°C. The period of 30 minutes has been selected on the basis of the capillary suction capacity of the mortar. After water spraying, the heater and fan start and the temperature is increased to 105°C. The heat is switched on for 5.5 h before the water is sprayed into the climate box again.

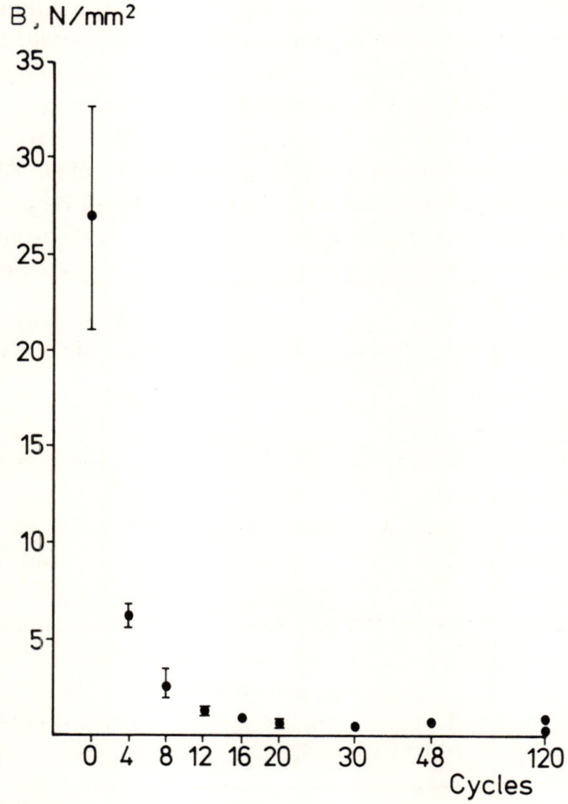

Figure 4.17 Toughness parameter *B* for sisal fibre concrete beams (mean value for three beams) as a function of the number of cycles in the climate cubicle.

Figure 4.17 presents the results from experiments with sisal fibre concrete conditioned between 0 and 120 cycles (15 days) in the climate cubicle[1]. As can be seen, the value for the toughness parameter *B* drops from 28 N/mm^2 to less than 2 N/mm^2 in 12 cycles. The considerable alternation between wet and dry in the climate cubicle thus gives a marked embrittlement in the sisal fibre concrete within a few days.

4.4.2.7 *Conclusions.* The decomposition of cellulose in an alkaline environment follows two mechanisms: peeling-off and hydrolysis. The peeling-off mechanism starts to show severe effects at temperatures above 75°C and the hydrolysis at temperatures above 100°C. The peeling-off mechanism is relatively harmless to cellulose because of its high degree of polymerization but has a considerable effect on hemicellulose with its low degree of polymerization and lignin. One possible explanation to the embrittlement of sisal fibre concrete is a chemical decomposition of lignin and hemicellulose in the middle lamella which bonds the fibre cells together, combined with a 'mineralization' of the fibre by calcium hydroxide deposited in the lumen of the fibre cells. Sisal fibres in carbonated concrete, where the alkalinity is low and where there is no calcium hydroxide, are not decomposed chemically or biologically.

Degradation of sisal fibre in an alkaline cement matrix is very slow in constant dry or wet environments. If the composite is subjected to changes in its moisture state, the degradation of the fibre can take place. An accelerated ageing method based on changes in the moisture state gives a marked embrittlement in the sisal fibre concrete within a few days. This method can be used for seeking countermeasures against the fibre deterioration.

4.5 Methods of improving the durability of sisal fibres in concrete

4.5.1 *Actions dealing with the fibre*

4.5.1.1 *Bundled fibres.* It might be thought that embedding a bundle of fibres, instead of individual fibres, in concrete would reduce the tendency towards embrittlement in the composite. Outer fibres, which are in direct contact with the matrix, can be broken down in time while the inner fibres retain their strength. To evaluate this theory, a mesh of fabric consisting of sisal fibres with about 30 fibres in the weave threads was embedded in a concrete matrix.

As can be seen from Figure 4.18, the toughness parameter *B* achieved when the reinforcement was in the form of mesh was not at all as high. The parameter *B* occurs at a far greater deformation, in this case at a strain of about 6‰ compared with a normal value of 2–3‰. The displacement of *B* is due to the fact that the individual fibres in the mesh have a far longer anchorage length, since they are not in continuous contact with the matrix; at the same time, they are not tensioned to the same extent. As can also be seen from Figure 4.18, the

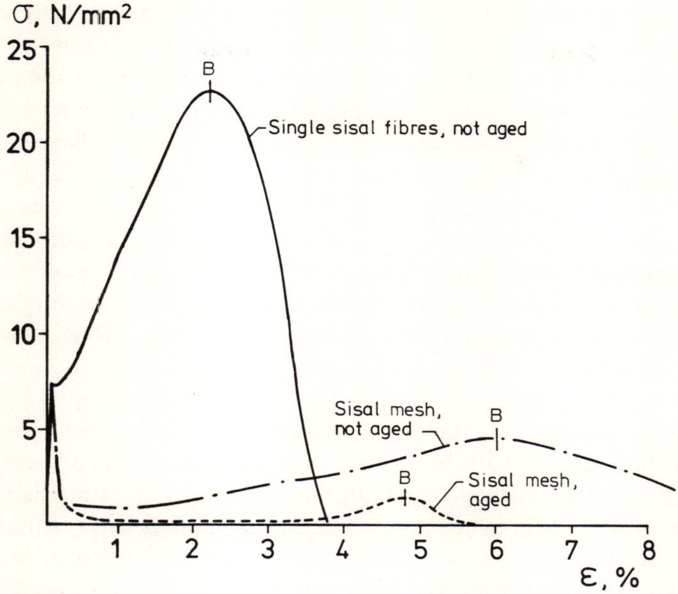

Figure 4.18 Stress–strain curves for mortar reinforced with sisal mesh which had been subjected to 0 and 12 cycles respectively in a climate cubicle, and non-aged mortar with individually embedded fibres.

toughness is reduced after ageing. Alkaline pore water will also penetrate the fibres in the mesh which are not in direct contact with the matrix.

4.5.1.2 *Fibre impregnation with blocking agents.* Decomposition of sisal fibres in concrete results because the fibre is attacked chemically by alkaline pore water in the concrete. One way to avoid or delay this decomposition could be to impregnate the fibre with agents which react with certain fibre components and build up compounds which are difficult to dissolve in an alkaline environment. Different impregnation agents, such as sodium silicate ($Na_2SiO_3.9H_2O$), sodium sulphite ($Na_2SO_3.H_2O$), magnesium sulphate ($MgSO_4$) and other magnesium, iron or copper compounds, barium salts, anti-oxidants, hydroxylamine (NH_2OH), sulphite salts, sodium sulphate (Na_2SO_4), nitrites, sodium boron hydride ($NaBH_4$), hydrosulphuric acid (H_2S), anthraquinone ($C_6H_4COC_6H_4CO$), various chrome or arsenic compounds, anti-oxidation agents used in the plastics and rubber industry, sulphamine acid (H_2NSO_3H), pentachlorphenol (Cl_5C_6OH), boron triethyl ($(C_2H_5)_3B$) and dieldrin ($C_{12}H_8Cl_6O$), have been discussed as possible solutions. So far none of the investigated impregnating agents tested according to ref. 1 improved the durability of the fibre in a cement matrix.

4.5.1.3 *Fibre impregnation with water-repellent agents.* Another way to avoid or delay decomposition of natural fibre in the concrete's alkaline environment might be to impregnate the fibre with a water-repellent agent. If the alkaline pore water can be prevented from penetrating to the fibre, it is not likely that the fibre will be decomposed chemically. Treatments with epoxy and polyester resins, sulphur, bituminous products, linseed oil with turpentine, white lead with varnish, cashew nut oil, calcium stearate, PVA, amide wax, silicon oil, carboxymethyl cellulose, tar, rubber latex, polyvinyl dichloride, chromium stearate, asphalt, fluorine-carbon-hydrogen stearate, sodium stearate and stearic acid have been tested[1,16–20]. The findings may be summarized as follows. Impregnation may reduce the bond between the fibre and the matrix. Impregnation with water-repellent agents may slow down the embrittlement process to some extent, but it will not prevent the fibre from being decomposed.

4.5.1.4 *Fibre impregnation with blocking and water-repellent agents.* Sisal fibres were first impregnated with blocking agents, for instance boric acid, borax, formine, potassium nitrate, potassium nitrite, magnesium sulphate, sodium chromate, sodium metasilicate and sodium nitrite, and then impregnated with water-repellent agents such as chromium stearate, sodium stearate, stearic acid, PVC_2, and fluorine-carbon-hydrogen stearate. The results from these tests[1] are summarized in Table 4.3.

As can be seen from Table 4.3, impregnating sisal fibres with blocking and water-repellent agents causes the value for the toughness parameter B for non-aged specimens to be lower than for specimens with unimpregnated fibre. Impregnating the sisal fibres reduces the embrittlement, however, after ageing in a climate cubicle. Fibres impregnated with formine and stearic acid, potassium nitrate and stearic acid, sodium chromate and fluorine-carbon-hydrogen stearate or borax and chromium stearate give values for B after 120 cycles which exceed 3 N/mm^2.

Tests of specimens stored outdoors in Stockholm and in Dar es Salaam have

Table 4.3 Toughness parameter B after an increasing number of cycles in a climate cubicle for specimens reinforced with sisal fibres impregnated with blocking and water-repellent agents.

	Number of cycles and B(N/mm^2)			
Impregnating agent	0	12	60	120
Unimpregnated sisal	32.8	2.8	0.9	0.4
Borax and chromium stearate	16.1	—	—	3.3
Formine and stearic acid	13.6	9.7	9.5	5.6
Potassium nitrate and sodium stearate	15.5	1.6	2.8	1.4
Potassium nitrate and stearic acid	14.0	—	6.4	3.5
Potassium nitrite and stearic acid	10.2	5.9	2.5	2.3
Sodium chromate and fluorine-carbon-hydrogen stearate	18.2	—	3.9	3.7
Sodium metasilicate and stearic acid	12.3	—	2.8	1.9
Sodium nitrite and chromium stearate	19.4	—	4.1	1.5

Table 4.4 Toughness parameter *B* for specimens reinforced with fibres impregnated with blocking and water-repellent agents. The specimens were stored outdoors in Dar es Salaam (*Dar*) or Stockholm (*Sth*).

Impregnating agent	Locality	Number of days and B(N/mm^2) 0	200	1460
Unimpregnated	Sth	10.7	6.9	4.8
Unimpregnated	Dar	12.4	0.3	—
Formine and stearic acid	Dar	11.4	6.9	6.8
Barium nitrate and stearic acid	Sth	18.2	12.2	5.9
Barium nitrate and stearic acid	Dar	14.1	9.7	7.1

also been made (see Table 4.4). The results in Table 4.4 confirm the results achieved with specimens aged in a climate cubicle. Impregnating sisal fibres with formine and stearic acid or barium nitrate and stearic acid has a favourable effect, and leads to a retardation of the embrittlement tendencies in the composite. After four years the fibres still reinforce the composite.

4.5.2 *Actions dealing with the cement matrix*

4.5.2.1 *Effects of a sealed matrix.* The transport of alkaline pore water to the fibre, and the removal of neutralized water with decomposed products from the fibre which occur when the temperature and moisture conditions of the composite are changed, might be reduced or eliminated completely by sealing the pore system in the matrix. This would delay or perhaps completely eliminate an embrittlement of natural fibres in concrete.

The permeability of concrete can be influenced by choosing aggregate with a higher proportion of fines, or by increasing the cement content. A denser concrete is also obtained when the water–cement ratio is lowered, in which case the proportion of capillary pores and the size of capillary pores are decreased, thus decreasing the pore volume available for moisture transport.

The concrete can also be made less permeable to water by admixing polymers, silicons or stearates. Another method is to admix small balls of wax

Table 4.5 Toughness parameter *B* after different number of cycles in a climate cubicle for specimens with and without a wax-sealed pore system and reinforced with sisal fibres.

Wax (% of cement weight)	Number of cycles and B(N/mm^2) 0	12	60	120
0	23.1	3.6	0.7	0.2
6.5	17.4	14.5	13.8	18.9
15	19.8	13.2	—	5.1

in the fresh concrete and then, when the concrete has hardened, to heat the concrete so that the wax melts and flows out into the pore system, thus sealing it. Impregnating concrete with polymers gives an efficient sealing of the pore system. Impregnating with sulphur, tar, asphalt or oil also constitutes a possible means of preventing moisture movements in the pore system in the concrete.

Methods tested[1] with the aim of making the cement matrix less permeable included: the admixture of fine-grained particles of magnesium trisilicate; reduction of the water–cement ratio from 0.5 to 0.37, 0.31 or 0.25; changing the binder from ordinary portland cement to finely-ground blastfurnace slag which is activated with powerfully alkaline admixtures; and the admixture of polymer microparticles. None of these methods proved capable of completely eliminating the embrittlement of sisal fibre concrete.

One method tested for changing the pore system of the matrix which had a positive effect on the reduction of the embrittlement of sisal fibre concrete in laboratory tests was the admixture of wax beads (Table 4.5).

As can be seen from Table 4.5, the value of *B* decreases after 120 cycles to about 5 N/mm^2 for specimens with wax amounting to 15% of the cement weight. Specimens with wax amounting to 6.5% of the cement weight do not lose as much of their toughness, and the value for *B* after 120 cycles even exceeds the value for the non-aged specimen.

Another method tested[21,22] for sealing the pore system was the admixture of zinc stearate (Table 4.6). As can be seen from the table, sealing of the pore system with zinc stearate seems to have a positive effect on the durability of the composite. The impregnation of the cement mortar with plasticized sulphur had a positive effect[20] on the durability of specimens aged in the laboratory.

Results from tests on specimens stored outdoors in Addis Ababa, Bangkok, Dar es Salaam and Stockholm are presented in Table 4.7. As can be seen, sealing the pore system improves the durability of the composite when conditioned outdoors in Stockholm or Addis Ababa. In Bangkok and Dar es Salaam, where the climate is hotter and more humid, the sealed material is still embrittled within one year.

Table 4.6 Toughness parameter *B* after different number of cycles in a climate cubicle for specimens with and without zinc stearate and reinforced with sisal fibres.

Zinc stearate (% of cement weight)	Number of cycles and *B*(N/mm^2)			
	0	12	24	120
0	22.0	2.9	2.7	0.1
0.7	29.5	7.7	3.1	7.6
2.0	26.0	17.1	11.2	10.4
5.0	20.7	17.1	9.2	11.8

Table 4.7 Toughness parameter *B* for specimens with or without a sealed pore system reinforced with sisal fibres and stored outdoors in Addis Ababa, Bangkok, Dar es Salaam and Stockholm.

Sealing (% of cement weight)		Locality	Number of days and B(N/mm^2)					
			0	200	365	540	730	1460
6.5	Wax beads	Stockholm	17.4	12.3	—	—	—	4.3
4	Wax beads	Dar es Salaam	16.4	1.8	—	—	—	0.8
8	Wax beads	Dar es Salaam	11.5	1.5	—	—	—	—
4	Wax beads	Bangkok	10.9	—	1.2	—	1.0	—
2	Wax beads	Addis Ababa	7.6	5.1	7.4	1.7	—	—
6	Wax beads	Addis Ababa	9.1	8.2	10.3	7.5	—	—
10	Wax beads	Addis Ababa	6.7	7.4	16.4	10.1	—	—
0.7	Zinc stearate	Addis Ababa	11.3	5.2	6.7	2.1	—	—
2	Zinc stearate	Addis Ababa	10.9	9.8	10.3	10.5	—	—
5	Zinc stearate	Addis Ababa	10.3	13.1	14.4	9.9	—	—

4.5.2.2 *Effects of reduced alkalinity*. The pH value for the pore water in ordinary concrete exceeds 13. One way to avoid or delay decomposition of natural fibres in the concrete's alkaline environment might be to reduce the alkalinity of the pore water to a pH value below 12–12.5.

It is well known that the alkalinity of the pore water drops to pH values of less than 9 in carbonated concrete. One method of creating a more favourable environment for the fibres should therefore be to carbonate the composite rapidly when it has reached a certain age and strength, for example at 14–28 days. Attempts to carbonate the composite rapidly have been carried out[1]. Specimens were placed in a sealed container and the air was evacuated from the container. The container was then filled with carbon dioxide and the specimens were stored for a number of days in the container. When later determining the penetration of the carbonation front in the specimens, it could be seen that the concrete had not carbonated at all. There were several interacting reasons for this. The water–cement ratio of the investigated matrix was 0.5. This resulted in a comparatively impermeable concrete in which the carbonation process normally proceeds very slowly. Water is liberated during the carbonation process. As a result, the relative air humidity in the pore system in the matrix increases. The higher the relative air humidity rises above 50%, the more slowly the carbon dioxide is transported in the pore system. New hydration products, which re-alkalize the pore water, are formed during the continued hydration of the matrix, which is also favoured by the increasing relative air humidity in the pore system in the matrix.

Further experiments in this field could give interesting results. One way to avoid embrittlement of sisal fibre concrete could be to place the fibre reinforcement in the surface layer, where the concrete is first carbonated and where it also provides the most benefit from the structural viewpoint.

The alkalinity of the pore water in the matrix can be reduced somewhat by replacing part of the cement with granulated blastfurnace slag or fly ash. The pH value of the pore water drops[1] from 13.2 to 13.0 for slag cement (70% slag) and to 12.8 for fly ash cement (67% fly ash). The replacement of part of the ordinary portland cement with slag or fly ash, however, has no or only very limited effect on reducing the embrittlement of the composite. The pH value is not sufficiently reduced.

The alkalinity of the pore water in the matrix can be reduced considerably by using high alumina cement as a binder instead of ordinary portland cement. The pH value of the pore water in a high alumina cement matrix is normally between 11.5 and 12.0. The tendency of the composite to become brittle is reduced, but not completely eliminated, by using high alumina cement as a binder instead. The toughness parameter *B* drops[1] from 30 MPa to 13.5 after 120 cycles of ageing in a climate box.

The pH value of the pore water in the matrix is also reduced by replacing some of the ordinary portland cement with a highly reactive, ultrafine silica fume. The pH value of the pore water decreases from 13.2 for a matrix with only ordinary portland cement as binder to 12.9 and 12.0 for matrices in which 17% and 33% respectively of the binder is replaced with silica fume.

Efforts to reduce the alkalinity of the cement matrix by adding different amounts of silica fume, so that decomposition of the sisal fibre is prevented, have been made[1] (Figure 4.19). As can be seen, replacing the cement with silica fume does not entail any reduction in the value of *B* in non-aged specimens. The interaction between the fibre and matrix is improved in a number of cases

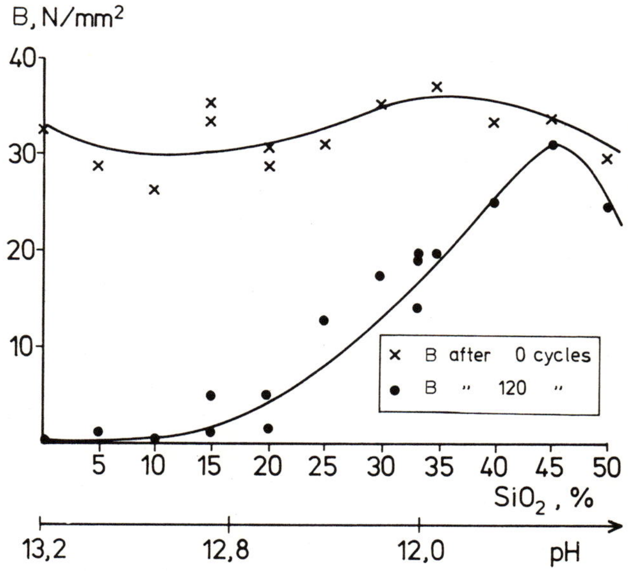

Figure 4.19 Toughness parameter *B* (mean value for three specimens reinforced with sisal fibres) after 0 and 120 cycles in a climate cubicle as a function of the percentage silica fume of the binder weight.

Table 4.8 Toughness parameter *B* for specimens with various percentages of silica fume and reinforced with sisal fibres, and conditioned outdoors in Addis Ababa, Bangkok, Dar es Salaam or Stockholm.

		Number of days and B(N/mm^2)				
Silica fume (%)	Locality	0	200	540	730	1460
15	Stockholm	35.6	12.1	—	—	—
20	Stockholm	30.8	10.6	—	—	9.4
17	Dar es Salaam	24.0	12.7	—	—	8.7
33	Dar es Salaam	32.0	24.6	—	—	—
25	Bangkok	23.0	—	—	8.9	—
40	Bangkok	24.4	—	17.9	17.0	—
25	Addis Ababa	12.3	11.6	15.9	—	—
40	Addis Ababa	12.1	10.9	10.6	—	—

when silica fume is admixed. Specimens in which up to 20% of the binder weight consisted of silica fume showed comparatively insignificant or moderate increases in the value of *B* after 120 cycles compared with specimens without an admixture of silica fume. When 30–50% of the cement is replaced with silica fume, a dramatic improvement is obtained in the toughness of the composite after ageing.

Results from tests on specimens with silica fume stored outdoors in Addis Ababa, Bangkok, Dar es Salaam and Stockholm are presented in Table 4.8. These confirm the results achieved with specimens aged in a climate cubicle.

However, reduction of the alkalinity in a cement matrix by using high alumina cement or by replacing part of the ordinary portland cement with silica fume is not appropriate for developing countries. The silica fume can be substituted there with locally available pozzolanas such as rice husk ash or pumice, scoria and diatomite[14,15,20–23] (Table 4.9).

4.5.3 *Combinations of fibre impregnation and changes of the cement matrix*

One way to delay or completely avoid embrittlement in the composite would be to use fibres which have been impregnated with blocking and water-repellent agents in a matrix in which the alkalinity of the pore water has been

Table 4.9 Toughness parameter *B* for specimens with various percentages of cement replacement with natural pozzolanas and reinforced with sisal fibres and conditioned outdoors in Addis Ababa or Bangkok.

		Number of days and B(N/mm^2)				
Pozzolana replacement (%)	Locality	0	180	365	540	730
Pumice, 40	Addis Ababa	7.5	6.0	10.2	8.2	—
Scoria, 40	Addis Ababa	12.2	14.0	11.8	14.4	—
Diatomite, 20	Addis Ababa	9.1	6.3	7.5	8.2	—
Rice husk ash, 40	Bangkok	22.7	—	12.4	8.3	9.6

Table 4.10 Toughness parameter *B* for specimens with 17% of cement replaced with silica fume and reinforced with sisal fibres impregnated with blocking and water-repellent agents. The specimens were conditioned outdoors in Dar es Salaam for up to four years.

	Number of days and $B(\mathrm{N/mm^2})$		
Impregnating agent	0	200	1460
Formine and stearic acid	13.1	9.1	15.0
Barium nitrate and stearic acid	15.1	9.6	14.8

reduced. Barium nitrate + stearic acid and formine + stearic acid were selected[1] from amongst the most promising impregnation combinations. Fibres impregnated with these agents were embedded in the matrix in which part of the ordinary portland cement had been replaced with silica fume (Table 4.10). As can be seen from the results in the table the toughness of the composite is still excellent after four years of exposure in a moist and tropical climate.

4.6 Conclusions and future prospects

From the materials technology viewpoint, two important advantages are obtained by combining natural fibres and cement mortar. One is that admixing fibres in the mortar makes it possible to give the fresh concrete the desired shape in a simple manner. The fibres hold the fresh concrete together. The second advantage derives from the reinforcing capacity of the fibres. Fibres increase the toughness and impact resistance of hardened concrete and the material can be utilized in structural applications. Durability must be taken into consideration if the purpose is to utilize natural fibre as a loadbearing reinforcement in concrete regardless of whether the application refers to the deadweight of a roof sheet or the loadbearing capacity of a beam. Sisal fibre concrete stored outdoors in a tropical climate has, for example, proved to become markedly embrittled within a period as brief as 200 days. This embrittlement of the material is, of course, harmless as long as it is used in applications where the stress is lower than the tensile strength of the matrix.

The embrittlement is due to the fact that the alkaline pore water in the concrete reacts with components in the sisal fibre so that the loadbearing capacity and toughness of the fibre are decreased. The rapidity with which sisal fibres in concrete are decomposed is mainly dependent on the intensity of the transport of alkaline pore water to the embedded fibres and the removal of the decomposed products from the fibre. The transport of the pore water in concrete is initiated by, for example, drying out and moistening the concrete respectively, but also by temperature variations and variations in the relative humidity of the ambient air. The decomposition of the fibre is also dependent on the temperature. The higher the temperature of the pore water, the more rapidly the decomposition proceeds.

The embrittlement of sisal fibre concrete can be avoided almost completely. The long-term properties of the composites are improved by markedly reducing the alkalinity of the concrete pore water.

Replacing about 45% of the cement with fine-grained, reactive silica fume reduces the pH value of the pore water to such low levels that the sisal fibre concrete retains its toughness after it has been subjected to accelerated ageing. Experiments with specimens stored outdoors confirm laboratory experiments in this respect.

Another way of reducing the alkalinity of the pore water is to replace part of the ordinary portland cement with natural pozzolanas such as rice husk ash, finely ground pumice, scoria or diatomite. The natural pozzolanas are not as effective as silica fume in reducing the embrittlement of the composite, but they do have a positive effect on the durability.

Carbonation of concrete results in a low pH value for the pore water. The durability of the fibre is thus improved by placing the fibre reinforcement in the outer layer of the composite, in the part which is first carbonated.

It is also possible to reduce the embrittlement of sisal fibre concrete by sealing the pore system in the concrete matrix. Admixing small beads of wax or powder of zinc stearate in the fresh mortar, or impregnating the hardened product with sulphur, show promising results in the laboratory but varying results in outdoor exposure tests.

The embrittlement of sisal fibre concrete can also be delayed by protectively impregnating the fibres, but attempts to prevent a chemical decomposition of fibre components completely through impregnation were not successful. In most cases, the impregnation of the fibre resulted in poorer composite properties due to reduced fibre strength and changed interaction between the fibre and the matrix.

A combination of changed matrix through the use of pozzolanas and fibre impregnations also has favourable effects on the durability of the sisal fibre in concrete.

What are the prospects for natural fibres in concrete? To answer this question one might compare natural fibres with another fibre sensitive to alkalis – the glass fibre. During the last 20 years glass fibres with improved alkali resistance have been developed, and products made of glass fibre cement have been and are used extensively in the industrialized world. The durability of alkali-resistant glass fibres and sisal fibres has been studied[24], and indications are given that natural fibres in a cement-based matrix – with countermeasures taken against fibre deterioration – have a better durability than do alkali-resistant glass fibres. With further research and development this might result in interesting applications of natural fibre concrete.

References

1. Gram, H.-E. (1983) Durability of natural fibres in concrete. *CBI Research fo 1.83*, Swedish Cement and Concrete Research Institute, Stockholm, 255 pp.

2. Gram, H.-E. Persson, H. and Skarendahl, Å. (1984) Natural fibre concrete. *Sarec Report R2:1984* (Swedish agency for research cooperation with developing countries), Stockholm, 139 pp.
3. Lola,C.R. (1985) Fibre reinforced concrete roofing sheets. *Technology Appraisal Report*, AT International, Washington DC, USA, 28 pp.
4. Gram, H.-E. *et al.* (1987) *FCR–Fibre Concrete Roofing.* SKAT (Swiss Center for Appropriate Technology), St Gallen, Switzerland, Intermediate Technology Publications, London, 185 pp.
5. Willi, K. (1987) Nicaragua: a country predestinated for Fibre Concrete Roofing. *FCR–News No 1*, FCR Advisory Services, SKAT, 1–4.
6. Nutman, F.J. (1937) *Empire J. Exp. Agric.* **5**, 79.
7. Vivan, H.E. (1976) Alkalis in cement and concrete. *Proc. Symp., London, Sept. 1976*, Cement and Concrete Association, Wexham Springs, Slough, 15.
8. Czernin, W. (1962) *Cement Chemistry for Civil Engineers.* Crosby Lockwood, Dublin, 139 pp.
9. Fagerlund, G. (1980) The structure of cement paste. *Betonghandboken*, Material, Stockholm, Svensk Byggtjänst, 177 (in Swedish).
10. Grudemo, Å. (1983) Private communication, Swedish Cement and Concrete Research Institute, Stockholm.
11. Longuet, P., Burglen, L. and Zelwer, A. (1973) La phase liquide du ciment hydraté. *Revue de Matériaux* **676**, 35–41.
12. Longuet, P. (1976) La protection des armatures dans le béton armé élaboré avec des ciments de laitier. *Silicates Industriels* **7/8**, 321–328.
13 Mjöberg, J. (1983) Private communication. Division of Cellulose Technology, Royal Institute of Technology, Stockholm.
14. Gram, H.-E. and Nimityongskul, P. (1987) Durability of natural fibres in cement–based roofing sheets. *Proc. Symp. Building Materials for Low-Income Housing*, Bangkok, January 1987, 328–334.
15. Andersson,R. (1987) Roofing tiles made of mortar reinforced with natural fibres. Report from a visit to Ethiopia during August 1987. *Report* **87074**, Swedish Cement and Concrete Research Institute, Consultant Section, Stockholm, 60 pp.
16. Geymayer, H.G. and Cox, F.B. (1970) Bamboo reinforced concrete. *ACI Journal*, October, 841–846.
17. Fang, H.Y. and Fay, S.M. (1978) Mechanism of bamboo-water-concrete interaction. *Proc. Int. Conf. on Materials of Construction for Developing Countries*, Bangkok, August 1978, 37–48.
18. den Uijl, J.A. (1976) Betong met bamboe gewapend. *Cement* **28** (6), 278–281 (In Dutch).
19. Mohan, D. and Rai, M. (1977) New building materials and techniques for rural housing, *Proc. Int. Conf. on Low Income Housing–Technology and Policy*, Asian Institute of Technology, Bangkok, 1977, 1077–1093.
20. Ziraba, Y.N., Baluch, M.H. and Azad, A.K. (1985) Use of plasticized sulphur in sisal-fibre concrete. *Durability of Building Materials* **3**, 65–76.
21. Andersson, R. and Gram, H.-E. (1986) Roofing sheets made of mortar reinforced with natural fibres. [Joint project, Swedish Cement and Concrete Research Institute, Department of Civil Engineering, University of Ottawa, and the Faculty of Technology, Addis Ababa University. Report of activities undertaken by the Swedish part during the second half of 1985.] Rept. **8607**, Swedish Cement and Concrete Research Institute, Consultant Section, Stockholm, 32 pp.
22. Andersson, R., Gram, H.-E. (1986) Roofing sheets made of mortar reinforced with natural fibres. Report of activities by CBI during the first half of 1986. Swedish Cement and Concrete Research Institute, Consultant Section, Stockholm, 24 pp.
23. Berhande, Z. and Gram, H.-E. (1988) Effect of cement composition on the durability of natural fibres in mortar. *4th Int. Conf. on Durability of Building Materials and Components*, Singapore, November 1987. (in press).
24. Bergström, S.G. and Gram, H.-E. (1984) Durability of alkali-sensitive fibres in concrete. *Int. J. Cement Comp. and Lightweight Concr.* **6** (2), 75–80.

5 Natural or modified cellulose fibres as reinforcement in cement composites

Z. FÖRDÖS

Abstract

The relatively high cost of man-made fibres (glass, steel, plastics, etc.) used in fibre reinforced cement-based composites, combined with a reduction in the usage of asbestos fibres due to shortage and health hazard, make it desirable to evaluate natural fibres, such as cellulose fibres, as possible substitutes. In recent years there has been considerable worldwide research into developing cellulose fibre reinforced cements and mortars. A short review is given of the development and use of natural fibre reinforced cement-based composites, particularly as a material for roof sheets in developing countries. This chapter describes, in addition, some results of a research project in Denmark to study the use of cellulose fibres as reinforcement for cement-based matrices. Some relevant physical, chemical and mechanical properties of cellulose fibres are discussed, and guidelines have been prepared to assist users in the design of cellulose fibre reinforced cement-based composites.

The effect of cellulose fibres on the dry and wet strength of fibre reinforced products and the production process are discussed. Air-cured and autoclaved cement pastes and mortars, reinforced with refined wood-pulp fibres, have been tested in bending (in both wet and dry conditions) and the fracture surfaces have been examined by scanning electron microscopy.

5.1 Introduction

Extensive research and development in the understanding and applications of fibre concrete materials are still taking place all over the world. These activities include, amongst other things, the development of new, stronger fibres, better fibre reinforced composites and new substitutes for products which have some disadvantages.

Since the end of the 19th century, asbestos-cement has had a wide range of applications, such as corrugated roofing elements, pipes and tiles. Because of the well-known health risks involved in the use of asbestos fibres, together with a possible future shortage of asbestos, there has, in recent years, been considerable research into the development of new high-performance reinforcing fibres. Krenchel[1] and Fördös[2] mention various types of fibres that might prove suitable as a substitute for asbestos and examine the advantages and disadvantages of the different products. However, the cost of these composites restricts their use to special applications.

Although natural fibres generally have poor mechanical properties compared with high performance man-made fibres, they have the advantage of low density, low cost and low energy demand during manufacture. The relatively high cost of man-made fibres also makes it desirable to evaluate natural fibres as possible substitutes for asbestos fibres.

In Denmark, the asbestos-cement industry has used cellulose fibres, both as filter agent and as a partial replacement of asbestos fibres, to improve the filtration time of the sheets and, simultaneously, their impact resistance; also, during World War II, waste cellulose fibres were used to produce cellulose fibre cement sheets, 'Cembrit'.

Another aspect of interest in the use of natural fibres is for making lowcost construction elements for, houses. Vegetable fibres are available in relatively large quantities in many developing countries. Some natural fibres such as flax, hemp etc. sometimes show surprisingly good mechanical properties. The last few years have seen intensive worldwide effort in the development and usage of cellulose fibres in cement-based composites.

The purpose of this work is to summarize recent research, particularly in Denmark, aimed at developing non-asbestos fibre reinforced cement materials based on cellulose fibres. The aim of, and the emphasis in, this report is therefore those properties of cellulose fibres and cellulose fibre reinforced cement-based materials which have special relevance to their design and application and in particular for their durability. Since Denmark has limited vegetable fibre resources, this report will be (mainly) restricted to naturally occurring commercial cellulose fibres, mainly derived from wood sources.

5.2 Natural fibres

The use of natural fibres as reinforcing material has been known since mankind started making straw reinforced sunbaked clay or mud bricks. Since the end of the 19th century, asbestos-cement composites have had a wide range of applications. The annual consumption of asbestos is still enormous – 20 million tonnes – but it will be necessary to find a substitute fibre, because of the well-known health risk involved in the use of asbestos fibres, combined with a possible future shortage of asbestos.

There is currently a great deal of interest in developing the technology for using natural organic fibre materials in cement composites. Among the most promising of such materials are cellulose fibres. These are produced mechanically or chemically, either from wood or other vegetable fibre sources.

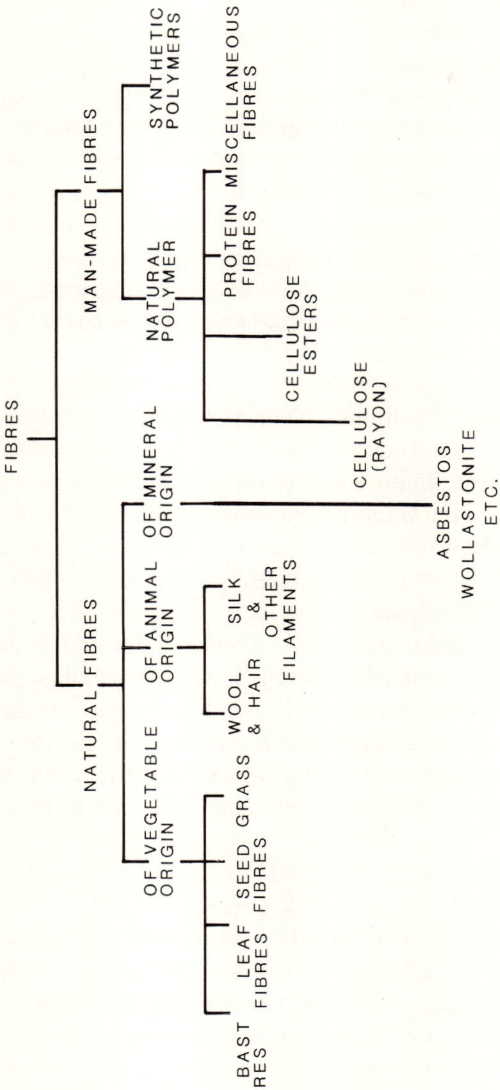

Figure 5.1 Classification of fibres.

Research and development in this field have been reviewed by Cook[3] and Gram[4,5].

Natural organic fibres exist in reasonably large quantities all over the world, and natural vegetable fibres are produced in most developing countries. Fibres may be classified as either natural or man-made, and the natural fibres further divided into different groups (Figure 5.1).

5.2.1 *Wood fibres*

For practical purposes, woods are grouped into two classes, softwoods and hardwoods. These two terms, however, are not very definitive, since some softwoods are harder than certain hardwoods. The term 'softwoods' refers to the conifers, which have needle-like or scale-like leaves. They are nearly all evergreens. The term 'hardwoods' refers to the broad-leaved trees, which are nearly all deciduous. The fibres of softwoods and hardwoods also possess great differences in length, and are fundamentally different in their anatomy and morphology.

5.2.1.1 *Chemical composition.* Wood consists primarily of cellulose, hemicellulose and lignin. Softwoods and hardwoods are slightly different in chemical composition and react differently with certain chemicals. Softwood, the term arbitrarily applied to the wood of needle-bearing trees, consists of about 43% cellulose, 28% hemicellulose and 29% lignin; hardwood, the term arbitrarily applied to the wood of broad-leaved trees, consists of about 43% cellulose, 35% hemicellulose and 22% lignin.

Cellulose is a crystalline linear polymer of glucose. The glucose units string together in long chains that tend to lie parallel to one another in a regular array (Figure 5.2). Hemicellulose is the various polysaccharide polymers of wood, other than cellulose. Lignin is an amorphous polymer, consisting of aromatic units (benzene rings), that serves to cement the fibres of wood together.

5.2.1.2 *Fibre structure.* A typical wood fibre wall contains four main layers as shown in Figure 5.3: the primary wall, and three layers of the secondary wall, the outer (S_1), middle (S_2) and inner (S_3) layers. The thickness of layers, other than the S_2 layer, remains relatively constant from one fibre to another. Therefore, the difference in wall thickness between in a thin-walled spring-

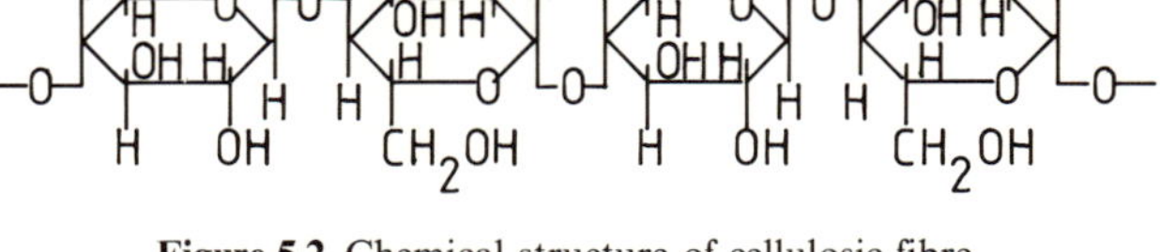

Figure 5.2 Chemical structure of cellulosic fibre.

wood and a thick-walled summerwood fibre can be considered to be the difference in thickness of the S_2 layer.

Fibres in wood are cemented together by the lignin. When the wood is chemically pulped, the middle lamella is softened and the lignin removed in order to separate the fibres.

The fibrils in the primary wall are irregularly oriented.

5.2.2 *Natural fibres of vegetable origin other than wood*

Natural fibres of vegetable origin other than wood include bast, seed, leaf and grass fibres.

(i) *Bast fibres* (*flax, hemp etc.*). Flax and hemp are the most common bast fibres. They are quite similar in their general appearance and dimensions[6]. The bast fibres of *flax* are obtained in bundles from the bark of the stem. Separation of these fibres from other cellular elements of the stem is achieved by retting, a process of rotting and then running the stems through breaker rolls. The broken matter is separated from the fibres by a beating process.

A single flax fibre has a cylindrical shape (Figure 5.4). A very fine lumen, thick walls, and clear transverse dislocations are characteristic of flax fibres.

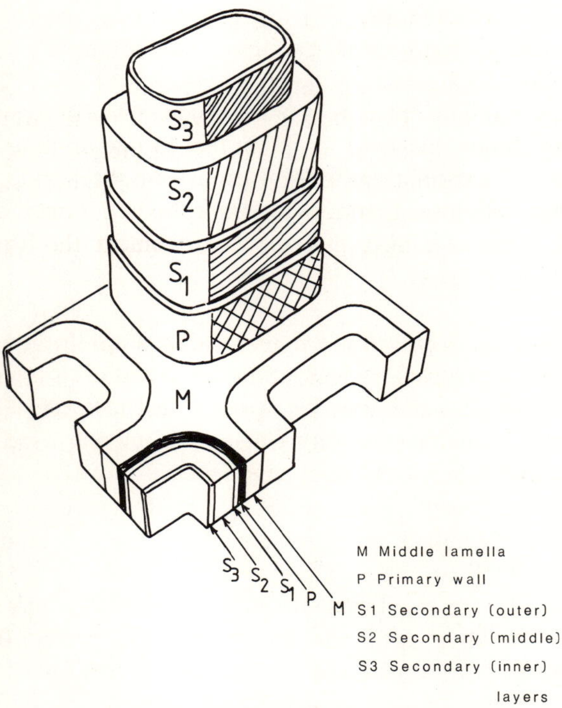

Figure 5.3 Model of wood fibre.

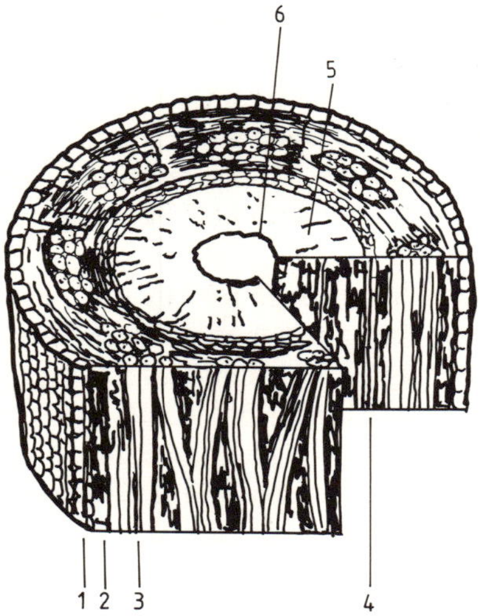

Figure 5.4 Schematic sectional pattern of flax straw. 1, Epidermis; 2, parenchyma; 3, bundles of bast fibres; 4, cambium; 5, woody element; 6, pith.

The cross-section of a fibre is round to polygonal. Fibre length varies from 9 to 70 mm, and fibre width from 5 to 38 μm.

Hemp, like flax, is an annual plant. It grows from 2 to 3 m in height and about 12 mm in diameter. The fibres are obtained from the inner part of the stem. The fibre shape is cylindrical, the fibre walls are thick, while the cross-section of a fibre is polygonal with rounded edges. Fibre length varies from 5 to 55 mm, and fibre width from 10 to 51 μm.

Jute fibres are obtained in strands from the bark of the jute plant, grown in Pakistan, India and Brazil.

(ii) *Seed fibres* (*cotton, coir, Java kapok*). *Cotton* is a long, single cell fibre, ribbon-like in shape, with a large continuous lumen throughout its length. The cotton fibre has a primary wall which consists of tiny fibrils. The secondary wall is composed of layers of cellulose. Fibre length varies from 10 to 40 mm and fibre width from 12 to 38 μm.

Coir fibres are obtained in bundles from the husk that surrounds the nut of the coconut plant. The fibres are extremely resistant to wear and decay. Fibre length varies from 0.3 to 1 mm, and fibre width from 12 to 24 μm.

The *kapok* fibre is obtained from the inside of the seed husk of the kapok tree. It consists of a single cylindrical cell, having a wide lumen, and thin walls. The cross-section is generally circular or oval, and the walls of kapok fibres are highly lignified. Fibre length varies from 8 to 30 mm and fibre width from 10 to 30 μm.

(iii) *Leaf fibres.* Manila hemp or abaca is a perennial plant from the Philippines and Indonesia. In height, it ranges from 5 to 10 m and it has a diameter of about 50 mm (maximum). Its stem is surrounded by leaf sheaths. The fibres are obtained in the form of stands from these sheaths. Individual fibres are of fairly uniform diameter, and are rather thin-walled. The longitudinal shape is approximately cylindrical. Fibre length varies from 2 to 12 mm and fibre width from 16 to 32 μm.

Sisal fibres are obtained from the agave plant, which is cultivated in Africa, America and in the West Indies. The plant has a short stem with leaves 1.2–2 m long. Individual fibres of sisal are nearly cylindrical in shape. Cell walls are thick and have some fine cross-markings. Fibre length varies from 1 to 8 mm and fibre width from 8 to 40 μm.

Pineapple fibres are obtained from the leaves of the pineapple plant, which is cultivated in Hawaii and the West Indies. Pineapple fibres are readily distinguishable from other fibres because they are extremely fine. Chemically, pineapple fibres have a higher percentage of cellulose (~82 weight%) than other leaf fibres[6]. Fibre length varies from 2 to 10 mm and fibre width from 3 to 14 μm.

Caroa fibres are obtained from the caroa plant, which grows wild in Brazil. Caroa fibres look like pineapple fibres. Chemically, caroa fibres have a lower percentage of cellulose (~60 weight%) and higher lignin content (~13%) than pineapple fibres. Fibre length varies from 2 to 6 mm and fibre width from 8 to 12 μm.

Mauritius hemp is cultivated in Brazil, Ceylon, Africa and on the islands of Mauritius. Fibre length varies from 2 to 6 mm and fibre width from 18 to 32 μm.

New Zealand flax is indigenous to New Zealand. The fibres are obtained from the leaves and are mainly used for bagging fabrics and paper. The cell walls are thick. Fibre length varies from 2 to 15 mm and fibre width from 5 to 27 μm.

(iv) *Grass fibres.* Cereal straws are used for paper production in areas where wood is not abundant. For example, in Europe straw pulps are used in the production of various grades of paper. Fibre length varies from 0.5 to 3 mm and fibre width from 7 to 40 μm.

Esparto grass grows wild in Spain and in Africa, and the pulp of the fibres is used, for example, in Europe for the manufacture of high-quality papers. Fibre length varies from 0.5 to 2 mm and fibre width from 7 to 14 μm.

Rice straw elements look like esparto grass, but the fibres of rice are longer and more slender than those of esparto. Fibre length varies from 0.7 to 4 mm and fibre width from 5 to 14 μm.

Bamboo is a tropical plant, cultivated in Asia. The cell elements of bamboo pulps are similar to those of other grasses. Fibre length varies from 1.5 to 5 mm and fibre width from 7 to 30 μm.

The chemical composition of the various plant fibres is quite similar.

5.3 Cellulose fibres

Woods or plants are composites built up of a variety of organic matter and materials. Cellulose constitutes chemically the bulk of many naturally occurring fibrous substances, including wood, bast, leaf, seed, and grass fibres. The discrete cellulose fibres are embedded in, bonded and held together in different ways by a continuous organic matrix, the lignin (Figures 5.3, 5.4 and 5.5).

Cellulose is a material of particular importance because of its widespread natural occurrence and its past and present practical use. The cellulose, normally used for reinforcement, consists of wood fibres either as milled, single fibres or in the form of pulp or paper.

The dimensions of the individual fibres vary considerably from type to type and depend on the origin of the fibre source, on the method which is used to remove the fibre and on the following pulping process. Thus, different paper-pulp qualities are available worldwide. The price of the paper-pulp cellulose is also very low compared with most other man-made fibre materials, but the low modulus of elasticity and the durability of the natural cellulose are two main

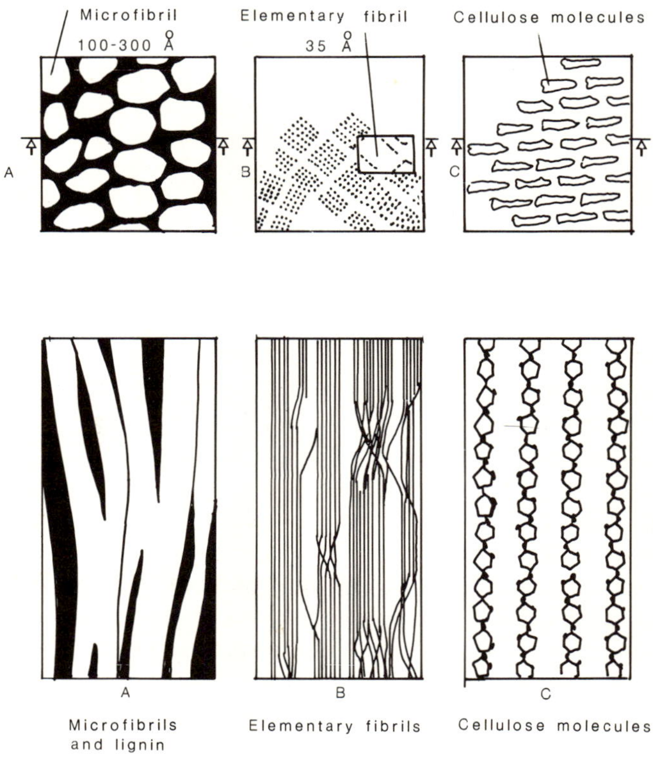

Figure 5.5 Model of cellulose microfibrils.

factors which cast doubt on their direct application as a reinforcement of cement matrices. Therefore, for decades, this type of fibre has mainly been used as a filter agent in the production of other fibre reinforced materials, such as asbestos-cement, polypropylene fibre reinforced cement and glass fibre reinforced cement.

Scientific and technological necessity have, however, provided a great deal of knowledge about the structure of cellulose. Cellulose exists in several crystalline forms, characterized by their distinctive X-ray diagrams. The generally accepted unit cell dimensions for these modifications are given in Table 5.1. Cellulose I is the form which occurs abundantly in natural fibres such as wood, cotton, ramie, and jute[7].

Table 5.1 Crystallographic data for the different forms of cellulose[7]

Type	System	*a*(Å)	*b*(Å)	**c*(Å)	γ^0
Cellulose I	Monoclinic	8.20	7.90	10.3	96–98
Cellulose II	Monoclinic	8.02	9.03	10.3	117
Cellulose III	Monoclinic	7.74	9.96	10.3	122
Cellulose IV, X	Orthorhombic	8.12	7.99	10.3	90

5.4 Fibre processing

The method of removal of the fibres from wood and the grade of the refining has a profound effect on the properties of the cellulose fibres. Pulping is the process by which wood is reduced to a fibrous mass. Although pulp can be obtained in many ways, the commercial processes are generally classified as mechanical, thermo-mechanical, semi-chemical and chemical.

Most pulp is made from wood fibres, but a large number of other fibres do find their way into pulp production, including both vegetable and man-made fibres. In areas where wood supply is short, use is made of locally available raw materials, such as straw, grass, bagasse and bamboo. More pulp is produced from softwood than from hardwood; the ratio is roughly 4:1.

5.4.1 *Mechanical pulping*

Mechanical pulping is frequently called the groundwood process, because in its original form it entails literally grinding the fibres out of the wood. Refiner groundwood is generally preferred over stone groundwood, because it contains more long fibres and yields a stronger end-product. Sometimes the wood is thermally or chemically pre-softened to reduce the power needed for grinding. Mechanical separation of the fibres, although giving a high yield of pulp, produces an inferior product because of the high lignin content and partially collapsed fibres.

5.4.2 *Thermo-mechanical process* (*TMP*)

The thermo-mechanical process, conducted at temperatures greater than the glass transition temperature of the lignin binder, gives a high yield of lignin-coated, uncollapsed fibre. The fibres may be further refined to produce paper pulps with partially collapsed lignin-coated fibres.

5.4.3 *Semi-chemical pulping*

The most important high-yield semi-chemical pulping processes are the cold-caustic process and the neutral-sulphite process. All cold-caustic pulps are inferior to kraft pulps in physical properties, but are stronger than softwood groundwood pulps. In the neutral-sulphite process, the most successful of the semi-chemical pulping operations, hardwood chips are impregnated under pressure with the cooking liquor. The cooked pulp is defibreized mechanically with refiners.

Young eucalyptus, while unsuitable in its untreated state, will produce a satisfactory groundwood pulp if it is treated with sodium hydroxide before grinding.

5.4.4 *Chemical pulping*

The principle underlying chemical pulping is that the lignin in wood can be degraded and dissolved by various chemical reagents, leaving most of the cellulose and hemicellulose in the form of fibres. The two main methods of chemical pulping are the kraft or sulphate process and the sulphite process. Because lignin has an adverse effect on cellulose strength and colour, this rather stiff macromolecular material is removed to a greater or lesser extent in all chemical pulping processes.

The *kraft process* involves cooking the wood chips in a solution of sodium hydroxide, sodium carbonate and sodium sulphide. Several different grades of kraft pulp are made for different end-products. Unbleached grades contain more lignin and have a higher yield from wood than pulps destined to be bleached into grades that can be made into white paper. Kraft pulps yield strong fibres.

In the *sulphite process*, an acidic mixture of sulphur dioxide, water and a chemical base material is the means of attacking and dissolving the lignin. Here the mechanism of chemical attack removes the lignin as salts of lignosulphonic acid, and the aromatic ring structure is left largely intact. The chemical base can be ionic calcium, magnesium, sodium or ammonium. Sulphite pulps are fairly light in colour and can be bleached easily. The sulphite pulps are weaker than the equivalent sheets of kraft pulp.

5.4.5 *Dissolving pulp*

In the dissolving pulp industry, pulps made by either the kraft process or the sulphite process are further purified chemically to remove all hemicellulose

Table 5.2 Properties of fibres[2,8,9]

Fibre	Density (g/cm^3)	Diameter (μm)	Length (mm)	Tensile strength (MPa)	Modulus of elasticity (GPa)	Elongation at break (%)
Asbestos	2.6–3.4	0.02–20	1–5	3000	170–200	2–3
Carbon	1.4–1.9	8–10	< 1000, continuous	2–3500	200–500	0.7–1.0
Cellulose						
Wood, straw (dry)	1.5	15–40	0.5–4	300–500	8–10	2.10
Sisal	1.5	10–50	1000	800	30	3
Cotton	1.5	10–30	60	300–900	10	3–9
Kraft cellulose, unbleached (*Pinus*)	1.5	15–60	1.2–2.7	200–1500	40	
Kraft cellulose, bleached (*Pinus*)	1.5	15–60	1–2.7	200–1300	40	
Thermo-mechanical pulp	1–1.5	20–70	1.5–2.5	200–800		
Kraft cellulose, bleached (birch)	1.5	18–60	1.0–1.2	200–1000	5–40	
Kraft cellulose, bleached (eucalyptus)	1.5	12–30	0.9–1.2	200–1300	45	
Glass						
E-glass	2.5	5–20	Continuous	2–2500	74	2–5
Alkali-resistant	2.7	10–15	Continuous	2–2500	70	2–4
Plastics						
Polyamide (nylon)	1.15	5–25	Continuous	800–900	8	10–12
Polypropylene	0.91	20–600	Continuous	350–500	5–8	8–20
PP high tensile		20–600	Continuous	2000	200	4
Kevlar PRD	1.45	10–12	Continuous	3000	70–140	3–4
Rockwool	2.9	2–10	2–20	1–150	70–120	1–2
Steel	7.8	10–600	Continuous	400–3000	200	3–15

and leave only pure cellulose, which can be transformed into such products as viscose rayon, cellophane and acetate fibres and film.

5.5 Properties of fibres

Fibres commonly used to provide reinforcement for cement-based materials include asbestos, steel, glass, carbon and plastic. The mechanical properties of such fibres are compared with those of cellulose from various sources in Table 5.2.

The method of removal of the cellulose fibres from the wood has an important effect on both the mechanical and the chemical properties of the fibres. A typical softwood pulp, illustrated in Figures 5.6 and 5.7, is composed of the two basic fibre types, i.e. thin-walled, wide, springwood, and thick-walled, narrow summerwood. Hardwood pulps are characterized by shorter, thinner fibres, the presence of vessel elements, and a larger proportion of short ray cells, as shown in Figures 5.8 and 5.9.

The wood or cellulose fibres have relatively poor mechanical properties compared with man-made fibres. They have the advantages of low density, low cost and low energy-demand during manufacture.

The chemical composition of some cellulose fibres is shown in Table 5.3[8,9].

5.6 Special problems with cellulose

The industries which produce cellulose fibre cement-based composites are growing in importance. There are different considerations to be taken into account before selection of the 'right fibre' for production of cellulose cement based composites. These include the many different fibre types to choose from,

Table 5.3 Chemical composition (%) of various cellulose fibres[8,9]

	Cellulose	Hemicellulose	Lignin	Other
Wood				
Pinus	45	20	27	8
Birch	54	15	29	2
Eucalyptus	49	25	24	2
Kraft-cellulose				
Pinus	84	15.5	0.5	—
Birch	81	19	—	—
Eucalyptus	89	10.5	0.5	—
Cotton	94	—	—	6
Flax				
Flax, green	62.8	17.1	2.8	17.3
Flax, alkali treated	94	4	0.3	1.7

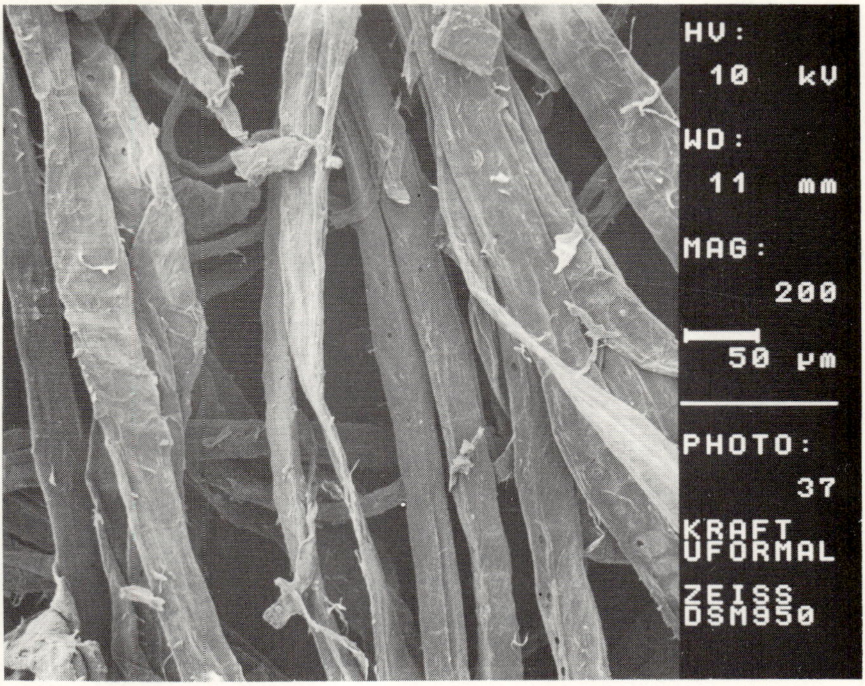

Figure 5.6 Pine sulphate, unground.

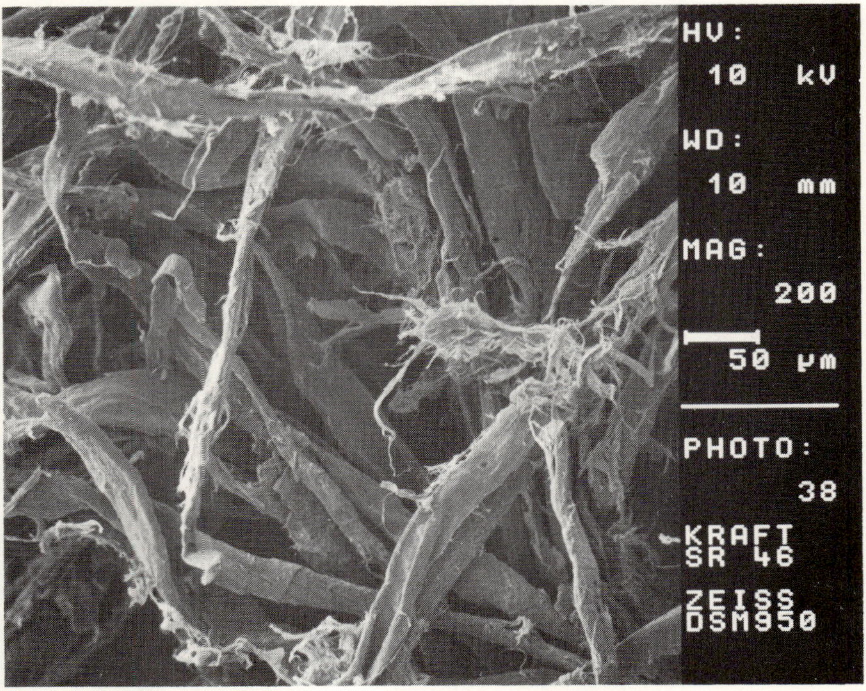

Figure 5.7 Pine sulphate, ground, 60° SR.

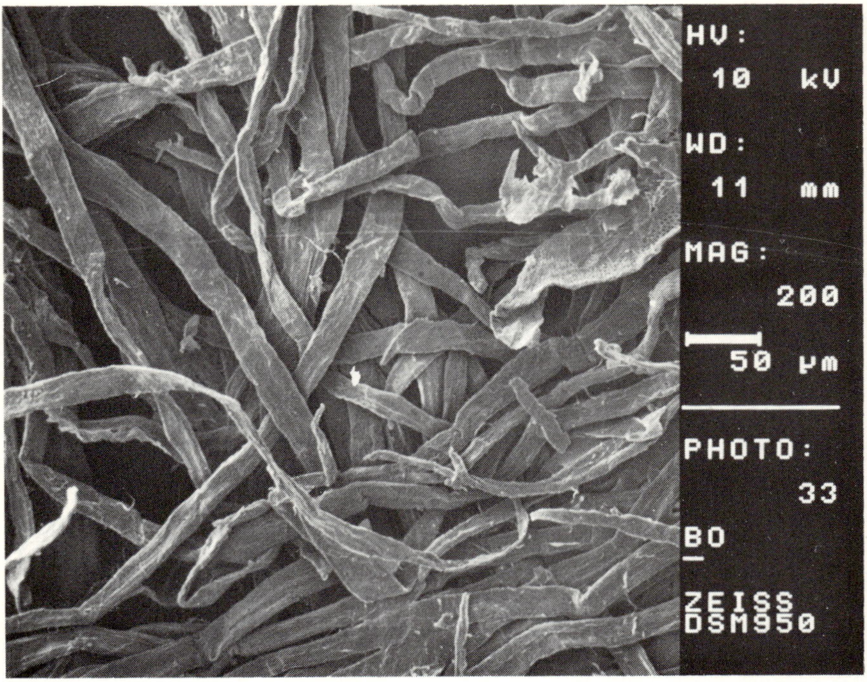

Figure 5.8 Birch sulphate, unground.

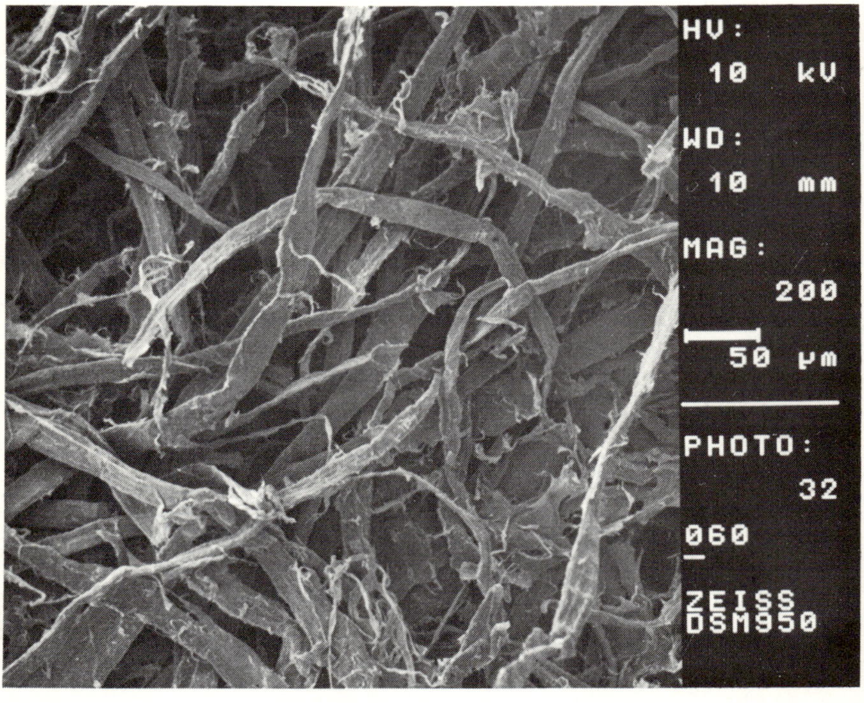

Figure 5.9 Birch sulphate, ground, 60° SR.

dependence of wood types, mechanical or chemical treatment, pulping process and possible after-treatment of the pulp.

5.6.1 *Setting and curing of cement*

It is well known that when setting cement paste comes into contact with some small species of wood, curing of the cement is inhibited, and hardwoods are generally more inhibitory to cement setting than softwoods. Weatherwax and Tarkow found that hardwood hemicellulose had a pronounced inhibitory effect on the setting of cement[10]. Sandermann found, in addition, starches, sugars, tannins, and certain phenols to be inhibitory[11,12]: lignosulphonic and hydroxylated carboxylic acids are used commercially as retarders of cement setting[13], and various sugars have an inhibitory effect on cement[14].

The kraft process produces a low yield of partially collapsed, delignified fibres. These fibres will have little effect on the setting and curing of the cement. Unbleached fibres in every case contain various amounts of lignin, and can therefore inhibit the setting and curing of cement. In some cases, it may be necessary to pretreat the pulp, because it may contain leachable matter which may retard the setting and hardening of the cement.

5.6.2 *Fibre form*

It has been pointed out earlier that the method of removal of the cellulose fibres from the wood has an important effect on both the mechanical and the chemical properties of the fibres. The mechanical treatment of wood, for example with disc refining, mainly influences the physical dimensions and form of the fibres, and has only little influence on the chemical composition of the fibres. After this treatment the fibres will be partially collapsed but unlignified.

On the other hand, the chemical treatment may influence both the fibre form (collapsed fibres) and the fibre chemical composition, as the chemical treatment mostly removes the lignin and also the hemicellulose content of the fibres.

The uncollapsed, lignified cellulose fibres might be expected to give fibre reinforced cement-based composites a lower density, strength and elastic modulus than composites containing delignified fibres, but having greater water resistance and toughness, as cellulose fibres are hollow.

5.6.3 *Strength*

Published values for elastic modulus and tensile strengths of pulp fibres show considerable variation as they depend on fibre angle and the extent and type of fibre defects[15,16]. Table 5.2 shows that account has to be taken of the fact that cellulose fibres may be hollow whereas the other fibres are solid. If a cellulose fibre collapses on drying, as do most delignified pulp fibres, the area of cell wall will closely approximate the overall fibre area.

Values for the strength and elastic modulus of uncollapsed lignified wood fibres are not available. Such fibres are relatively thin-walled, and do not collapse on drying; therefore these fibres are considered from the point of view of fibre reinforcement of composites. Their relative strength may be lower than those of delignified fibres.

Bleaching the fibres increases both the elastic modulus and flexural strength but reduces the specific work of fracture. Kraft fibres are always stronger than acid sulphite and sulphite-bisulphite fibres.

5.6.4 *Effect of season on fibre properties*

It is well known that springwood fibres have a thin-walled structure compared with thick-walled summerwood fibres. The superior strength of summerwood fibres of various species has been reported by McIntosh[17]. Also, the different constituents of wood vary with the season[10].

5.6.5 *Wet condition*

Water has a dramatic adverse effect on the elastic modulus and flexural strength of cellulose fibres irrespective of whether the fibres are bleached or not[18–21]. The uncollapsed, lignified cellulose fibres retain their strength better than delignified fibres when exposed to moisture[18].

Wet summerwood fibres show a distinctly higher breaking load than the corresponding dry fibres, but this is not the case for springwood fibres.

5.6.6 *Attack by micro-organisms*

The action of common fungi increases the permeability of the wood, and also lowers the degree of polymerization of wood constituents, thereby increasing the amounts of carbohydrate and lignin soluble in alkalis.

It is common knowledge that, in spite of their unusual durability, cellulose microfibrils are degraded by a variety of biological processes and therefore their formation is reversible[22].

5.7 Fibre reinforced cement-based composites

The main concept of fibre reinforcement is shown in Figure 5.10[2].

Fibre reinforced materials may be classified into various groups, according to the type of fibre and matrix (Table 5.4).

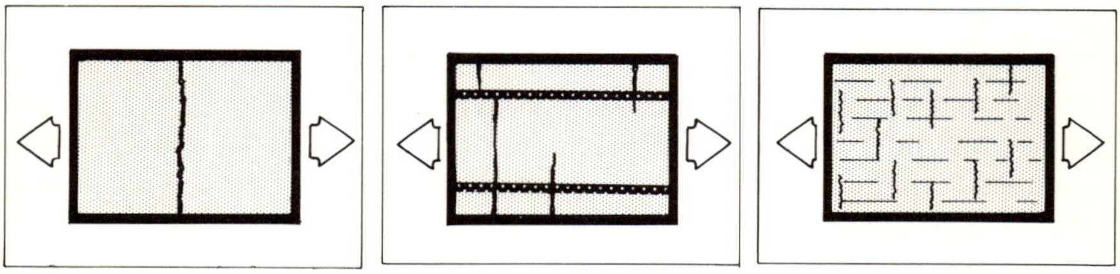

Figure 5.10 Schematic illustration of the concept of fibre reinforcement of a cement matrix.

Table 5.4 Classification of fibre composites

Type	Fibre	Matrix	Example
1	Ceramic	Plastic	Glass fibre reinforced plastics
2	Ceramic	Metal	Under development
3	Ceramic	Ceramic	Glass fibre reinforced cement paste. Advanced fibre reinforced ceramics
4	Metal	Plastic	Optical and thermal applications
5	Metal	Metal	Rocket nozzles, jet engines
6	Metal	Ceramic	Steel fibre reinforced Densit™
7	Plastic	Plastic	Kevlar reinforced plastics. Rocket nozzles
8	Plastic	Ceramic	Polymer fibre reinforced cement paste

Fibre reinforced cement-based materials, which consist of either brittle or ductile fibre introduced in a brittle matrix, belong to types 3, 6 and 8.

Fibre reinforced cement based materials differ, both in their production processes and in their behaviour from the other groups, where the matrix is ductile. The matrix material in this case is normally an organic, polymeric material, which has an elongation at rupture of the same order of magnitude as that of the most common fibres (Figure 5.11).

Short discrete fibres of glass and steel have been extensively used in cement mortar and concrete to obtain certain favourable properties in the composite. The addition of such fibres generally results in higher impact and fracture toughness, improved ductility, and increases in crack-inhibiting effect and wear resistance properties (durability) of the composite.

The properties of fibre composites are controlled by a number of factors, of which the most important are:

(i) Magnitudes of both the fibre and the matrix modulus of elasticity (E_f and E_m)
(ii) The ratio of modulus of elasticity of the fibre to that of the matrix (E_f/E_m)
(iii) Type and properties of the matrix (ductile or brittle)
(iv) Fibre content, fibre length and orientation
(v) Interfacial bond strength between the fibres and the matrix.

Fibres with a high modulus of elasticity yield a considerable increase in tensile strength over that of the hardened cement paste, whereas fibres with a low modulus of elasticity inhibit crack propagation in cement-based materials, thus improving the impact resistance of the composites.

In recent years there has been considerable research into developing high-performance reinforcing fibres. Continuous graphite, boron, aramid and steel fibres are now used as reinforcement in organic or inorganic matrix composites. However, the cost of these fibre reinforced composites will restrict their use to special applications.

The mechanical properties of materials, particular inorganic materials such as cement, concrete, glass and ceramics, are structure-sensitive. Fibre composites made of these contain cracks, voids, dislocations, and other

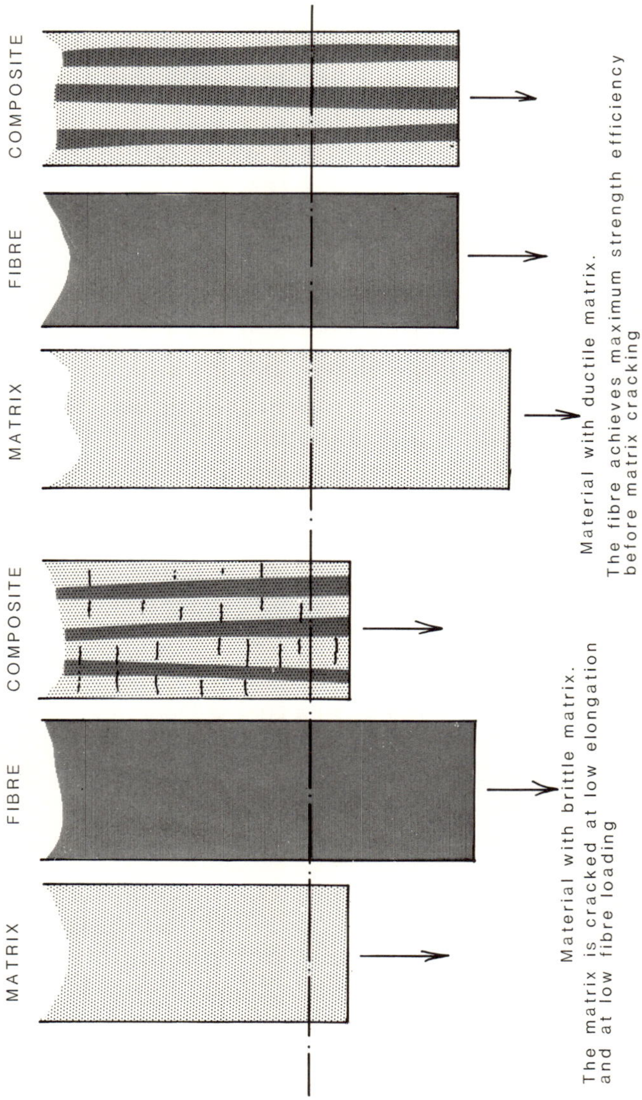

Figure 5.11 Fibre reinforced composites with brittle and ductile matrices respectively.

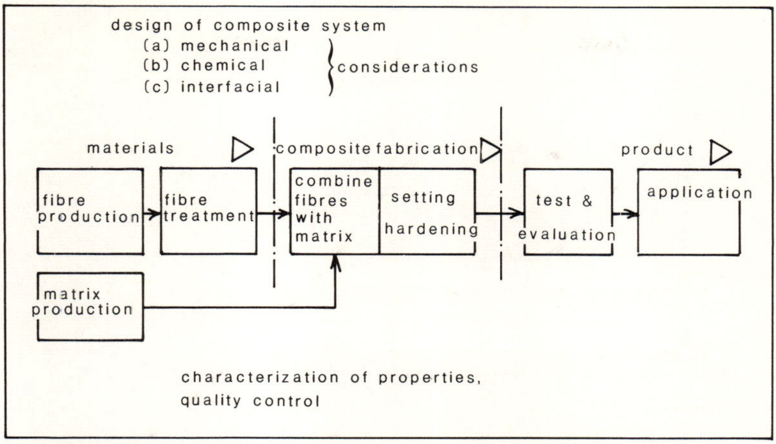

Figure 5.12 Scope of cellulose fibre reinforced cement-based composites.

imperfections both in fibre and matrix and at the interface zone. The unreinforced matrix generally fails in tension and exhibits brittle behaviour.

Many publications give comprehensive and fully up-to-date descriptions of the science and technology of fibre cements and concretes[1,2,23,24]; the only exception is, generally, that of cellulose fibre reinforced composites. The flow diagram in Figure 5.12 shows a broad coverage of this subject, including materials, fabrication, product properties such as mechanical behaviour, long-term durability, structure and applications[23].

5.8 Design of fibre cement composites

Although interest in cellulose fibre reinforced cement-based materials has grown at a greatly accelerating pace in the last few years, composites as such are not new. The main factors controlling the performance of composite materials are the physical properties of the fibres, of the matrix and of the strength of the bond between the two.

The role of the constituents in fibre composite materials is shown in Figure 5.13 and in Table 5.5[24].

Other important factors are:

(i) Fibre volume and concentration
(ii) Fibre orientation
(iii) The forms of reinforcement (fibre dimensions, aligned short fibre, continuous, combined, etc.)
(iv) Fibre distribution

5.8.1 *Volume fraction of fibre*

It is typical of composite materials, where the matrix is normally an organic, polymeric material, to have quite a high fibre content ($V_f = 0.2$–0.7). On the

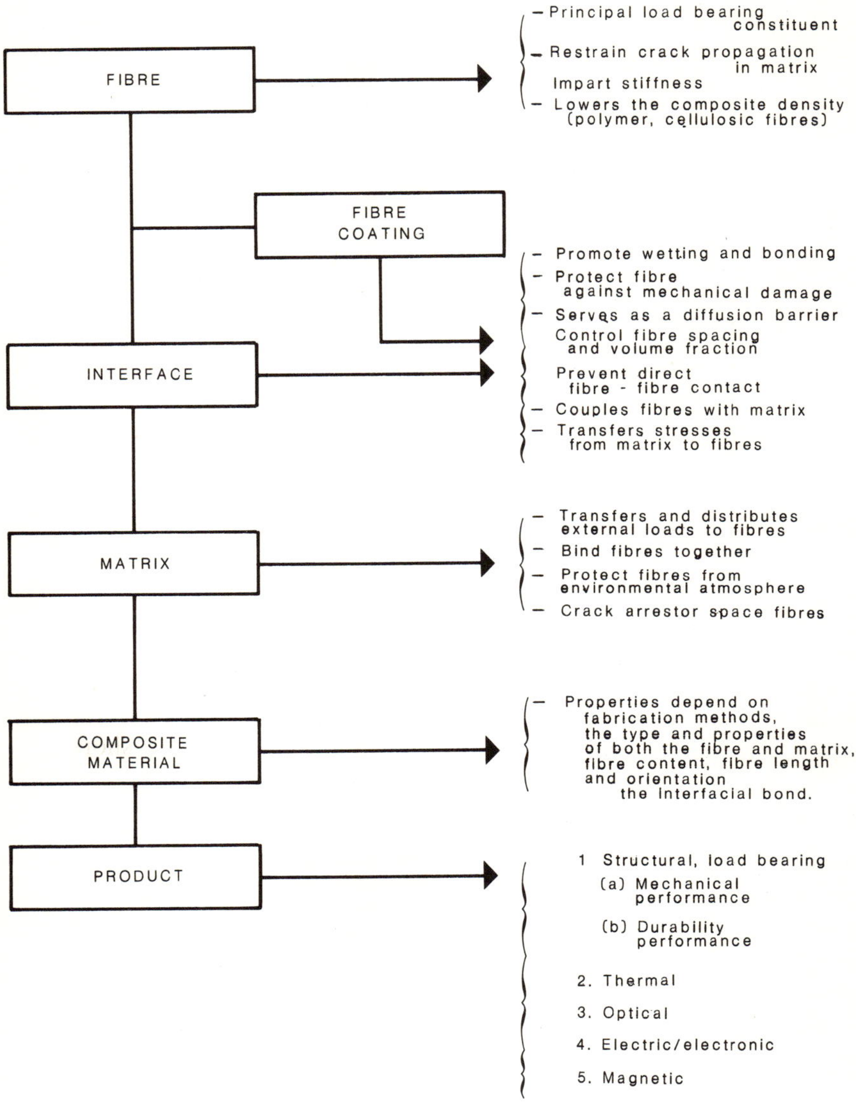

Figure 5.13 Functions of constituents in a fibre composite material.

other hand, composite materials with an inorganic cement matrix, usually have quite a low fibre content ($V_f = 0.01–0.30$).

5.8.2 *Fibre forms*

Forms in which fibres can be embedded in a matrix are varied. In many ways, the simplest form is short, randomly distributed fibres. At the opposite end of

Table 5.5 Factors contributing to tensile strength of a fibre composite material[24]

(A) *Fibre contributions*
1. Ultimate strength of the individual fibres
2. Collective (bundle) strength of the fibres
3. Strength retention during handling and fabrication
4. Volume fraction in composite
5. Strength *v.* fibre diameter or length
6. Strength *v.* temperature
7. Fracture strain (brittle *v.* ductile fibres)

(B) *Transfer of load from matrix to fibres*
(Average stress of the fibres in the matrix)
1. Stress–strain curves of the constituents
2. Fibre volume fraction
3. Matrix shear strength and shear modulus
4. Interfacial bond strength
5. Fibre critical length
6. Fibre length and aspect ratio
7. Fibre taper
8. Fibre orientation
9. Composite failure mode and statistics of the fracture
10. Constraint of the matrix near the fibre interface and ends

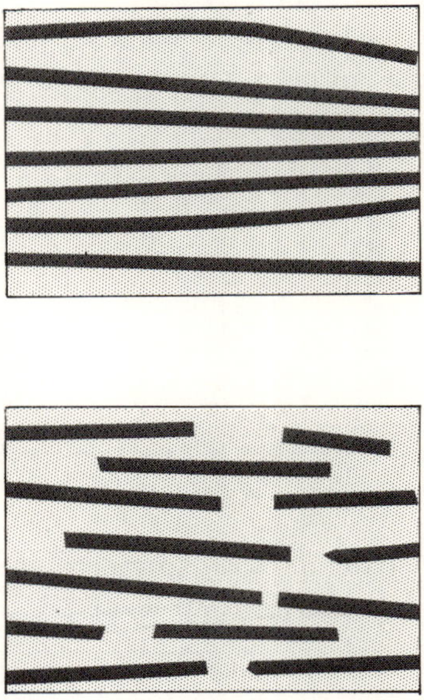

Figure 5.14 Fibre composites with short or long discrete fibres. In the case of short fibres, high bond strength between the fibres and the matrix is demanded.

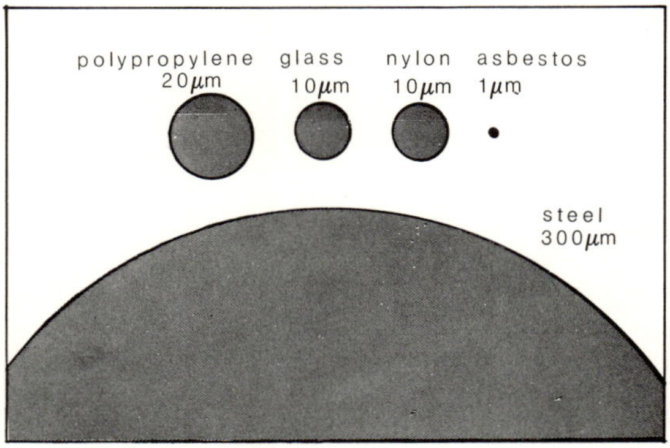

Figure 5.15 Comparison of cross-sectional dimensions of various fibres.

the scale are the long continuous fibres or filaments laid down parallel to each other to provide optimal packing and strength in that direction (Figure 5.14).

5.8.3 *Fibre diameter*

The fibre diameter is dependent on the fibre type. This is illustrated in Figure 5.15. The fibre diameter can also affect composite strength because smaller fibres are generally stronger than larger ones.

The tensile strength of glass fibres is shown in Figure 5.16 as a function of the fibre diameter.

5.8.4 *Fibre length and aspect ratio*

Fibre length can be used to classify fibre composite materials. The fibres can be continuous, discontinuous, or both, within a given composite material. How-

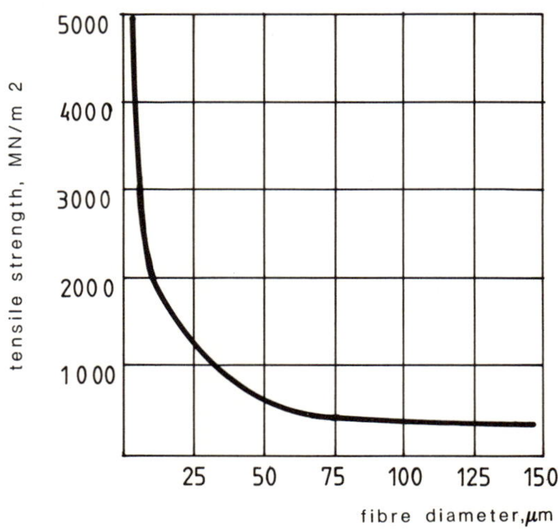

Figure 5.16 Tensile strength of glass fibres as a function of fibre diameter.

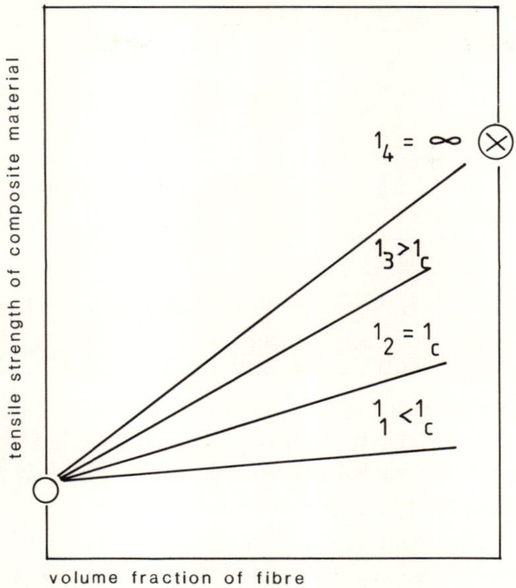

Figure 5.17 Effect of fibre aspect ratio on tensile strength of fibre reinforced composites.

ever, the fibre length or the aspect ratio can have a significant effect on the properties and on the failure modes of the composite materials. The effect of the aspect ratio and the fibre volume fraction on composite strength is illustrated in Figure 5.17. Reinforcing fibres having lengths less than the critical length l_c will tend to pull out. On the other hand, when the fibre length is greater than l_c, the fibres will fracture, rather than pull out. In the region before the matrix failure strain is reached, the length efficiency factor for an aligned short fibre composite with frictional bond at the fibre interface is nearly unity for practical composites[25].

Fibre length also affects the ease with which the fibres can be aligned and packed. In general, short fibres are more difficult to align and to pack densely than are continuous fibres.

5.8.5 *Fibre orientation*

The orientation of the fibres has a decisive influence on the mechanical properties and the behaviour of the composite material. Maximum loading of the fibres and maximum composite strength occurs when all the fibres are aligned parallel to a uniaxial tensile load. The effect of fibre orientation on the strength of a composite is shown in Table 5.6[2,26].

5.8.6 *Fibre shape*

Fibre shape can have an effect on the maximum obtainable fibre volume fraction, and can also affect composite properties.

Table 5.6 The effect of fibre orientation on strength of composite[2,26]

Fibre orientation	Efficiency
1 D aligned	100%
2 D right-angle in plane	25%–50%
2 D random in plane	25%–40%
3 D random	15%–20%

Cellulose fibres, being hollow, can in some cases confer benefits in reduced composite density and increased strength-density ratios.

5.8.7 *Constituent behaviour*

Whether the constituents are brittle or ductile can affect composite properties. If, for example, the fibres are brittle, the composite will tend to fracture in the linear portion of a stress–strain curve, even when the matrix is ductile. If ductile fibres are reinforcing a ductile matrix, the composite will show plastic deformation prior to fracture. On the other hand, if ductile fibres are reinforcing a brittle matrix, the composite will exhibit linear behaviour until the first crack, and after that, ductile behaviour (considerable elongations or deflections) may occur over a period of time.

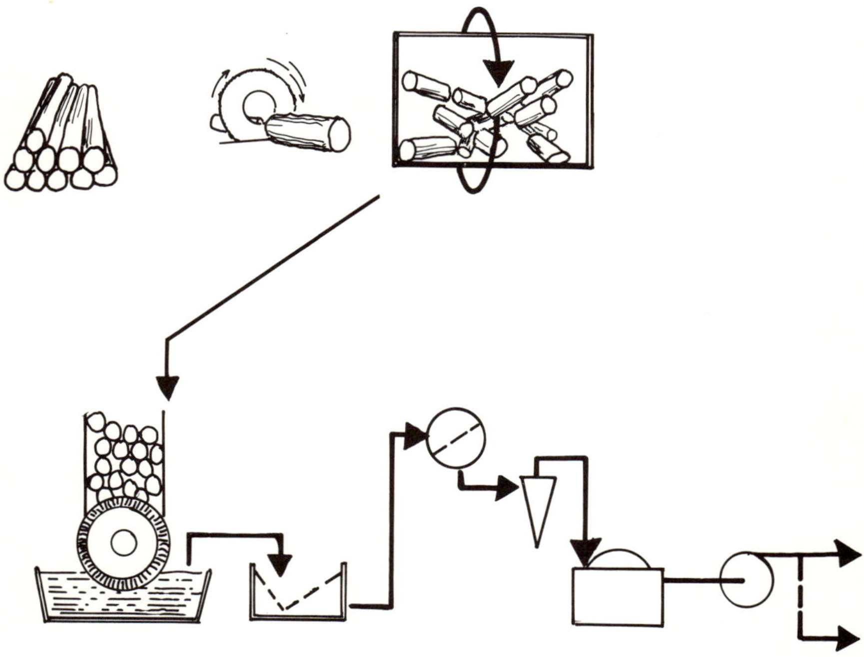

Figure 5.18 Production principle of mechanical pulping process.

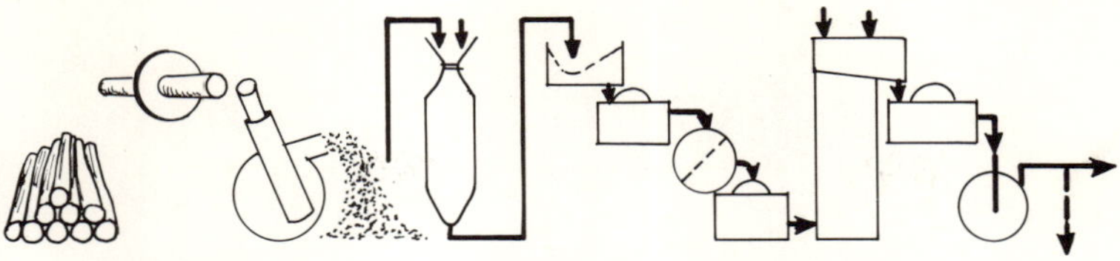

Figure 5.19 Production principle of the chemical pulping process.

The properties of a cellulose fibre cement composite will depend upon the structure and properties of the fibres from which it is made[27,28]. The principles of the mechanical and chemical production processes for cellulose pulps are shown in Figures 5.18 and 5.19[19].

Studies of natural fibres and of refining and pulping technology of wood lead our interest into different pulp qualities, and it becomes clear that the primary component, the fibre, could have functions other than load-bearing, and that the matrix not only holds the fibres together but also plays a significant role in composite behaviour. A third consideration – the interface between fibre and matrix – plays a very important role in the fracture behaviour of this composite material.

5.9 Production of cellulose fibre reinforced cement-based materials

Potential applications of cellulose fibre reinforced mortar and concretes for structural purposes are relatively limited, although products such as sheets and boards are cheap and light in weight, and can be used for both internal and external purposes. The production of roof sheets reinforced with natural fibres

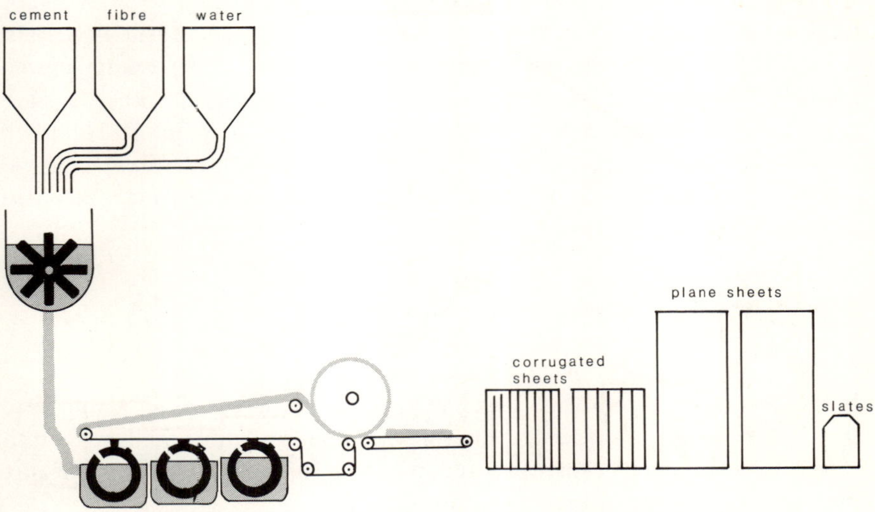

Figure 5.20 Production principle of the Hatschek process.

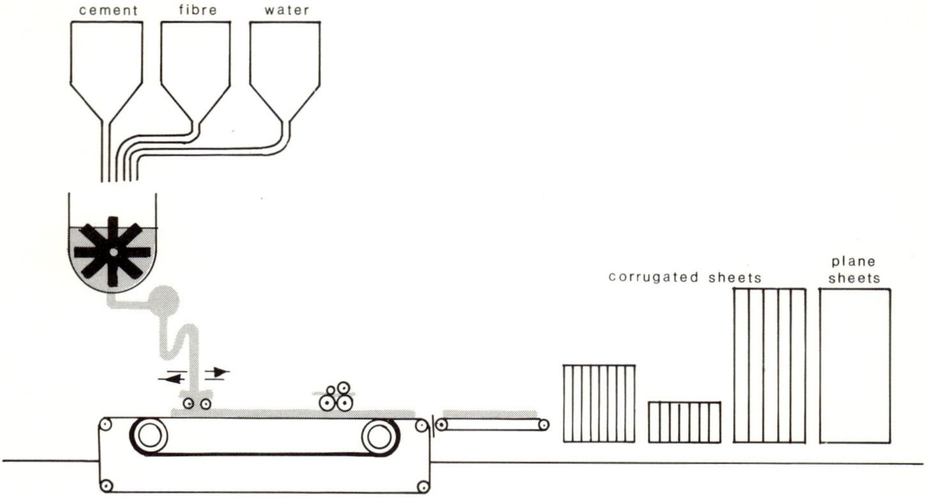

Figure 5.21 Production principle of the Magnani process.

has worldwide use thanks to the IT Building Materials Workshop, UK[29]. Specimens of long-span prefabricated roof units of sisal reinforced cement, exposed to the outdoor climate in Dar es Salaam, Tanzania, lost their favourable strength properties and became brittle after no more than a few months, thus illustrating the difficulties yet to be overcome.

Worldwide, there is a rapid development of the number and types of products manufactured from wood or natural fibres[30]. Paper, fibre boards (particle boards), laminates and sandwiches are common examples, and more recently, bast fibres from bark embedded in a polymeric matrix, and wood or natural fibres used in cement-based systems, offer new composite materials.

In this last group, the products in most cases are either manufactured manually by a moulding process followed by rolling or other surface treatment, or through industrial production by the Magnani or Hatschek processes to produce cement sheets reinforced with cellulose fibres.

The principles of the Hatschek and Magnani processes are shown in Figures 5.20 and 5.21 respectively.

5.10 Materials and experimental techniques

5.10.1 *Laboratory experiments*

In the laboratory experiments, different proportions of the cellulose pulp were thoroughly mixed with the cement and the other additions, and made into sheets by combined dewatering and pressing[31]. Table 5.7 shows the mix proportions of some of the laboratory series.

In the investigation, the cellulose content was varied from 0 to 30% by

Table 5.7 Mix compositions of some laboratory series

Material	1	2	3	4	5	6
Kraft cellulose (%)	4	8	8	8	8	8
Birch sulphate (%)	—	4	7	7	7	—
Eucalyptus (%)	—	—	—	—	—	7
Fly ash (%)	—	—	—	5	25	—
Microsilica (%)	24	24	24	24	24	24
LSC* cement (%)	72	64	61	56	(36)	61

* Low alkali sulphate resistant.

weight, the microsilica content from 0 to 40% by weight and the fly ash content from 0 to 30% by weight. In some experiments, low alkali sulphate resistant cement was replaced by other types of cement. Admixtures such as clay and hydrated lime were also used in these experiments.

The full results of these series are not presented here; however, some general remarks are made.

5.10.2 *Full-scale experiments*

Table 5.8 shows the mix proportions of some of the full-scale specimens.

Sets of specimens (laboratory: 200 × 76 × 6 mm; production: corrugated sheets 1220 × 1090 × 6 mm, cut specimens, corrugated 500 × 100 × 6 mm and flat specimens 400 × 40 × 6 mm) were then cured in air, in water for 7 days and 28 days, or in high-pressure steam autoclaves. Their bending strengths and deformations were then determined in accordance with the traditional 4-point

Table 5.8 Mix composition of some cellulose fibre cement composites produced at full process scale

Material	1	2	3	4	5	6
Kraft cellulose (%)	8	8	8	3.2	6	
Birch sulphate (%)	4	7	7	6.4	4	4
Flax (%)				6		6
Fly ash (%)	16	10				
Microsilica (%)	20	24	24	30.3	20	20
Hydrated lime (%)		5		28.1		
Other admixtures, e.g. wollastonite (%)				32	10	10
LSC* cement (%)	52	46	61		60	60
Stack compression pressure (MPa)	—	—	23	21	19	19
Process	Magnani		Hatschek			

*Low-alkali sulphate-resistant.

test method. Samples of dried composites were weighed and placed in water for a specified period of time. Upon removal, their surfaces were dried. They were then weighed to determine the amount of water absorbed. A scanning electron microscope (SEM) was used to examine the fracture surface of samples. All fracture surfaces were coated with gold under vacuum and examined in a JEOL-3 SEM at 25 kV.

5.11 Results and discussion

Results of this investigation are shown in Tables 5.9 and 5.10 and in Figures 5.22–5.27. The results are discussed below.

5.11.1 *Laboratory results*

Table 5.9 compares performance data for composites containing different types of cellulose fibres and admixtures, and prepared by different techniques. Materials in column (a) were produced at a low compression pressure, and those in column (b) at a higher compression pressure. The varying effects of the nature of the pulp and the fibre content were investigated in a separate series with different composition. The strength values of cellulose fibre composites using different types of pulp are given in Figure 5.22. Figure 5.23 shows the variation in bending strength as a function of fibre content for laboratory specimens.

5.11.2 *Full-scale test results*

The bending strength and densities of factory-produced specimens were determined on 6 specimens per test, and the data are shown in Table 5.10. Test pieces of specimens were cut from the autoclaved boards and two values of the modulus of rupture MOR of the sheets were measured: MOR-l (with rupture parallel to the direction of production) and MOR-t (with rupture perpendicular to the direction of production). Typical bending stress–strain curves for Magnani-produced asbestos-cement and cellulose fibre reinforced

Table 5.9 Properties of cellulose cement-based composites, laboratory series

Series	1	2	3	4a	4b	5a	5b	6a	6b
Filtration time (s)	1500	430	270	260	260	225	225	175	175
Solid material loss (%)	0	4.4	3.3	3.8	3.8	3.5	3.5	4.1	4.1
Dewatering pressure (MPa)	1.5	1.5	1.5	1.5	10.0	1.5	10.0	1.5	10.0
Density (g/cm^3)	1.55	1.31	1.21	1.22	1.45	1.14	1.39	1.28	1.52
Autoclaving temperature (°C)	140	140	140	—		—		—	
MOR, wet (MPa)	13.9	18.8	22.6	19.0	28.6	18.9	27.1	22.0	31.6
MOR, dry (MPa)				26.9	32.7	25.4	33.5	29.8	37.9

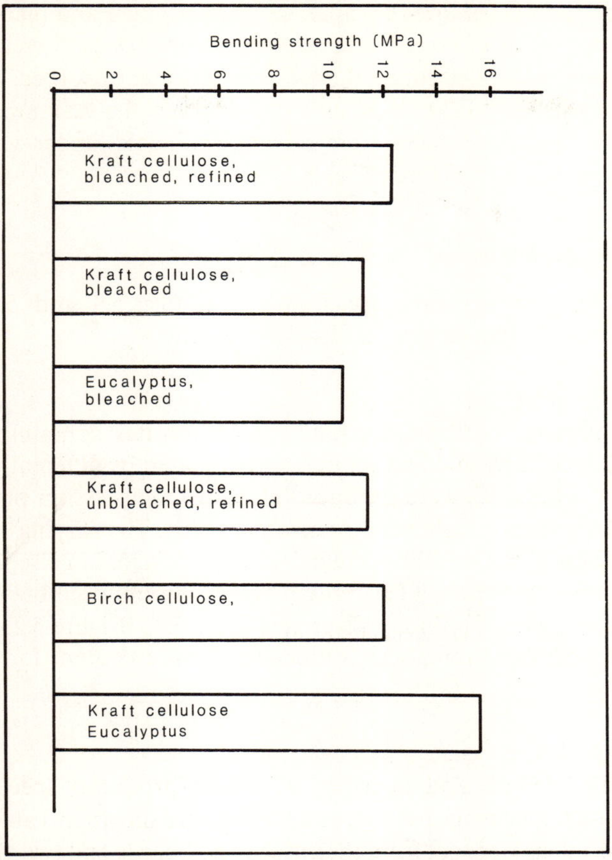

Figure 5.22 Bending strength values of laboratory-prepared cellulose fibre cement specimens with 10 wt % fibre content (wet specimens). Matrix consists of 20 wt % silica and LSC cement.

Table 5.10 Bending strength and density of full-scale trial test specimens

Series	1	2	3	4	5	6
Density (g/cm^3)	1.13	1.04	1.3	1.2	1.4	1.5
MOR, wet (MPa)	14.0	13.6				
MOR, dry (MPa)	20.0	17.7				
MOR-l, wet (MPa)			22.5	17.2	24.4	20.4
MOR-t, wet (MPa)			36.0	23.9	38.2	30.1
MOR-l, dry (MPa)			26.6	26.0	33.8	27.6
MOR-t, dry (MPa)			44.9	40.4	47.6	34.4
MOR-l, stock (MPa)			30.9	21.8	38.8	25.3
MOR-t, stock (MPa)			55.5	37.2	49.6	33.6
Process	Magnani		Hatschek			

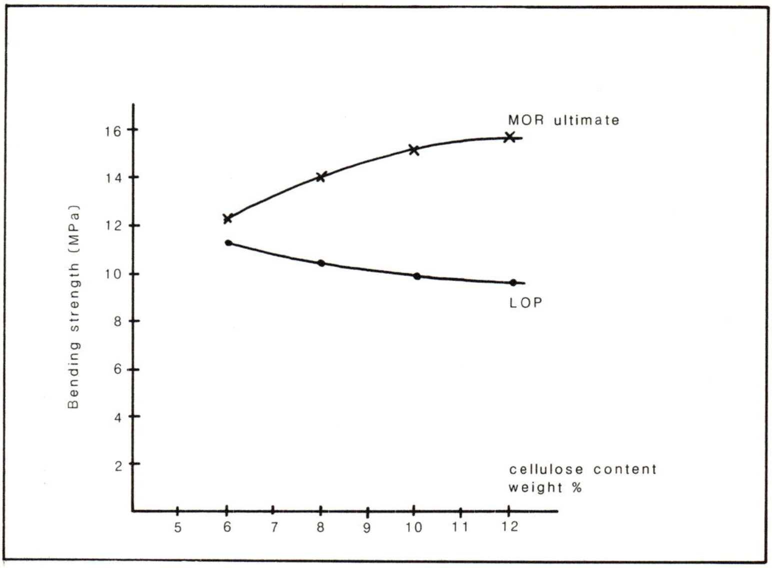

Figure 5.23 Bending strength results of laboratory specimens with varying cellulose content (kraft cellulose: eucalyptus 1:1). Matrix consists of 20 wt % silica and LSC cement (wet specimens). MOR, modulus of rupture; LOP, limit of proportionality.

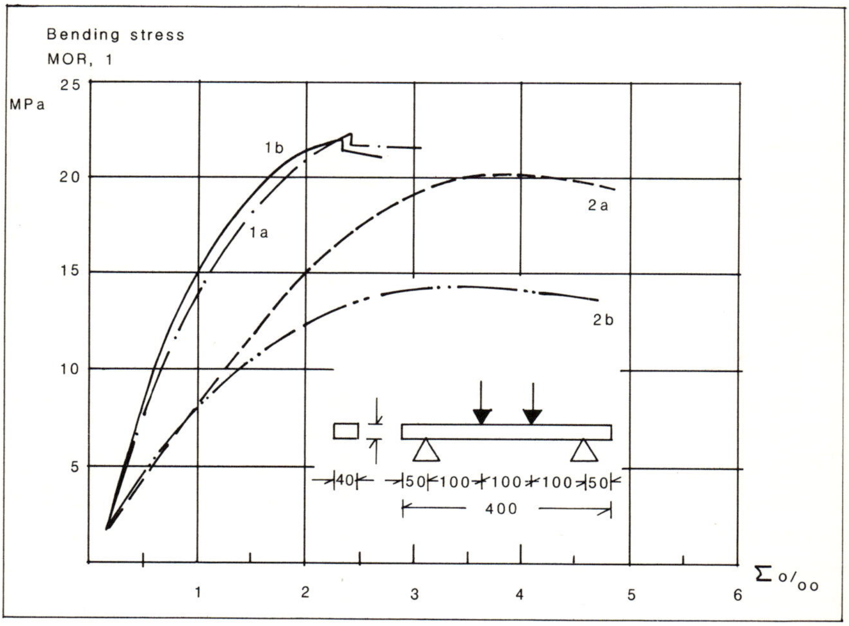

Figure 5.24 Typical stress–strain curves for Magnani-produced asbestos-cement and cellulose fibre cement composites in the wet and dry state. 1a, asbestos-cement, 65% RH; 1b, asbestos-cement, soaked; 2a, cellulose cement sheet, 65% RH; 2b, cellulose cement sheet, soaked.

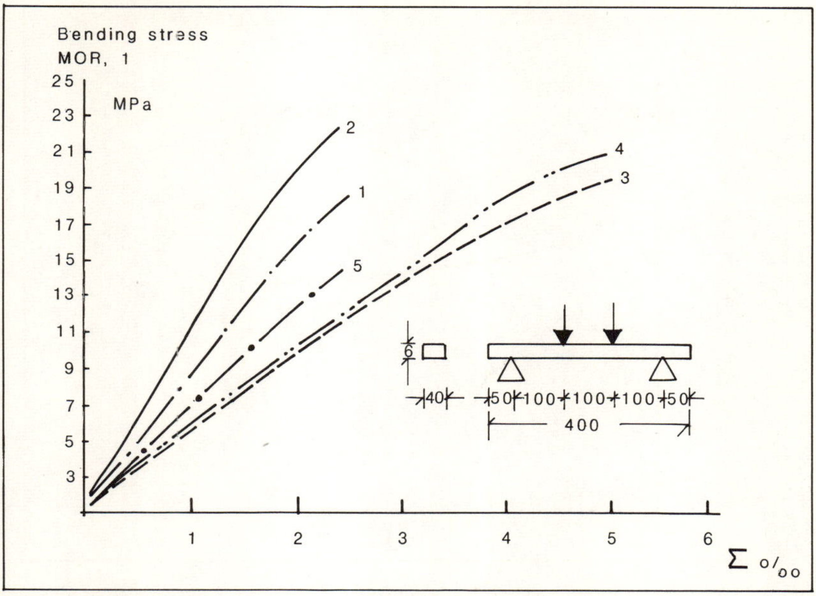

Figure 5.25 Typical stress–strain curves and reference specimens of Magnani-produced asbestos-cement and cellulose fibre cement composites (store-dry specimens). 1, asbestos-cement, 65% RH; 2, asbestos-cement, 65% RH, exposed for approx. 20 years; 3, cellulose sheet, 65% RH; 4, cellulose sheet, 65% RH, painted; 5, cellulose sheet, 65% RH, produced during World War II, exposed for approx. 40 years.

cement composites in the wet and the dry states are shown in Figure 5.24 (MOR-l).

Figure 5.25 shows typical stress–strain curves for exposed and reference specimens of asbestos-cement and cellulose fibre cement composites, produced by the Magnani process (MOR-t). For comparison, the results from a Hatschek-produced cellulose cement sheet exposed for about 40 years are included.

5.11.3 *Properties in the fresh state*

The incorporation of cellulose pulp in the cement mix decreases the workability and increases the void content of the composites.

At higher volume fractions (greater than 10%), mixing becomes difficult, and fibres tend to cluster together, resulting in inhomogeneity of the board and reduction in strength. When increasing the fibre content in the board, it is essential to increase the water ratio in the slurry to avoid fibre lumps. In most cases the optimal ratio of water to solids should be 10:1 or 20:1.

Improved forming and filtering is achieved when fibres are refined to a higher freeness. Usually such sheets also exhibit higher strength.

Thin sheets can be produced on Hatschek and Magnani machines in the

Figure 5.26 Fracture surface of a water-cured sample, showing fibre pull-out.

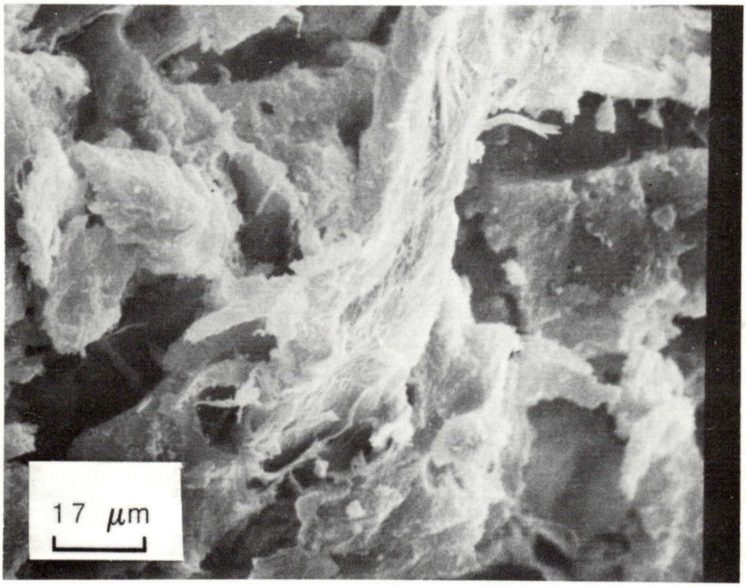

Figure 5.27 Fracture surface of an autoclave-cured sample, showing fibre damage.

same way as asbestos-cement sheets. Furthermore, the green Hatschek sheets are suitable for hand-moulding.

5.11.4 *Properties in the hardened state*

With low fibre contents ($< 3\%$ by mass) the strain capacity of the composite, as the matrix cracks, is about the same as that of unreinforced cement pastes, or mortars, particularly in the dry state. In the wet or semi-wet state the fibres can act as crack-arresters, and also absorb a significant amount of energy if a considerable proportion of the fibres are pulled out from the matrix. With higher fibre contents ($> 3\%$ by weight) both the strain capacity and bending strength of the composites increase, particularly in the dry and the semi-wet state.

The ultimate strength, however, also depends on fibre type, length and volume fraction of fibres and fibre distribution, as well as on the properties of the matrix.

The strength also depends on the type of pulp or pulping process. Kraft pulp yields a stronger pulp than thermo-mechanical pulp (TMP) or cold-caustic pulps, and higher specific fibre strength is usually related to higher composite strength. Combinations of different pulp types can in some cases be advantageous.

Water absorption, both of air-cured and autoclaved samples, increases with increasing fibre content.

Due to the hollowness of cellulose fibres, the density of the final composite is much lower than that of other fibres. A decrease in density of 25% in the composite is experienced for cellulose cement compared with asbestos-cement of equal strength level.

5.11.5 *Microstructure of cellulose fibre cement-based materials*

Cellulose fibre cement composites fail by a combination of fibre fracture and fibre pullout. SEM studies indicate that fibre fracture is the major mechanism taking place at low moisture content (dry or semi-wet). Fibre pullout increases in importance at higher moisture content (wet sample). Figure 5.26 shows an SEM photo of a fracture surface of a water-cured sample, showing fibre pullout.

Examination of the fracture surfaces of heavily autoclaved cellulose-composites shows, in some cases, that the reinforcing fibres may be damaged during the autoclave process. Figure 5.27 shows an SEM photo of a fracture surface of an autoclave-cured sample, showing fibre damage.

It is believed that the mechanism of fracture is dependent on the fibre type, fibre strength, and the moisture content of the sample. Normally, wet samples exhibit less fibre to matrix bond. Hence, more fibre pullouts are observed.

5.11.6 *Long-term durability*

Cellulose fibre cement composites have demonstrated a number of useful mechanical properties, such as high bending strength and post-cracking resistance, but the long-term performance and durability are still to be proven, although extensive accelerated ageing tests have shown promising results. However, the strain capacity of the composite decreases with time, as it does in asbestos-cement. Cellulose-cement still has the major disadvantage that both the tensile strength and the modulus of elasticity decrease considerably when the material is soaked and the fibres become wet. However, normally an increase in impact strength can be observed.

Some of the cellulose fibre reinforced cement sheets which were produced and used during World War II in Denmark are still in service and show good strength retention results after 40 years of natural weathering.

5.12 Concluding remarks

Cellulose fibre and cement-based matrix are not ideal partners to form composite materials. A great deal of consideration should be given before selection of proper materials and processes.

The investigations described in this chapter have shown that it is possible to produce cellulose-cement sheets as substitutes for a great variety of asbestos-cement sheeting, such as building boards and cladding panels. The collected data show that cellulose fibres can be incorporated in the cement matrix system and still retain considerable strength after many years (Figure 5.25).

The strain capacity of the composite decreases with time, however, as it does in asbestos-cement, and cellulose-cement still has the major disadvantage that both the tensile strength and the modulus of elasticity decrease considerably when the fibres become wet.

Recently developed cellulose-cement-based sheets have dry strength levels similar to asbestos-cement sheets. However, the cellulose-cement-based sheets are more sensitive to wetting than is asbestos-cement. A decrease in the ultimate bending strength of up to 40% can be observed, when the sheets are exposed to wet conditions. However, an increase in impact strength is normally observed. The potential for increasing the wet strength level is obvious, and further research is needed to achieve cellulose-cement-based products with improved properties.

References

1. Krenchel, H. (1974) *Fibre Reinforced Brittle Matrix Materials*. Publication SP-44, American Concrete Institute, Detroit, 45–77.
2. Fördös, Z. (1979) Fiberarmerede, cementbaserede materialer. *Beton-Teknik* 08/02/1979, Aalborg Portland, Aalborg, 16 pp. (in Danish).
3. Cook, D.J. (1980) Concrete and cement composites reinforced with natural fibres. *Concrete Soc., Proc. Symp. on Fibrous Concrete*, London, April 1980, The Construction Press, Lancaster/London/New York, 99–114.
4. Gram, H.-E. (1980) Research in the field of natural fibre reinforced concrete – an overview. *CBI Reports* **12**, Swedish Cement and Concrete Research Institute, Stockholm, 23 pp.
5. Gram, H.-E. (1983) Durability of natural fibres in concrete. *CBI Research* **1**, Swedish Cement and Concrete Research Institute, Stockholm, 255 pp.

6. Strelis, I., and Kennedy, R.W. (1967) *Identification of North American Commercial Pulpwoods and Pulp Fibres.* University of Toronto Press, 117 pp.
7. Ellefsen, Ø., and Tønnesen, B.A. (1971) *Cellulose and Cellulose Derivatives*, eds. N.M. Bikales and L. Segal, Vol. 5, Part 4, Wiley-Interscience, New York, 151–180.
8. Bach, N. (1985) Modificerede cellulosefibre som armering for fibercement, Dansk Eternit, Aalborg, internal communication, 104 pp.
9. Fördös, Z., and Tram, B. (1986) Natural fibres as reinforcement in cement-based composites. *RILEM Symp. FRC 86, 3rd Int. Symp. on Dev. in Fibre Reinforced Cement and Concrete*, July 1986, eds. R.N. Swamy *et al.*, Vol. 1, 6.
10. Weatherwax, R.C., and Tarkow, H. (1964) Effect of wood on setting of portland cement. *Forest Prod. J.* **14**, 567–570.
11. Sandermann, W., Preusser, H.T., and Schweers, W. (1960) Studien über mineralgebundene Holzwerkstoffe. *Holzforschung* **14**, 70–77.
12. Sandermann, W., and Dehn, U. von (1956) Einfluss chemischer faktoren auf die Festigkeitseigenschaften zementgebundener Holzwolleplatten. *Holz als Roh- und Werkstoff* **14**, 307–311.
13. Hobbs, C. (1964) Concrete additives from the viewpoint of the large contractor. *Chem. Ind.* **13**, 526–528.
14. Thomas, N.L., and Birchall, J.D. (1983) The retarding action of sugars on cement hydration. *Cement Concr. Res.* **13**, 830–842.
15. Page, D.H., El-Hosseiny, E., Winkler, K., and Lancaster, A.P.S. (1977) Elastic modulus of single wood pulp fibres. *Tappi* **60**, 114–117.
16. Page, D.H., El-Hosseiny, E., Winkler, K., and Bain, R. (1972) The mechanical properties of single wood-pulp fibres. Part I: A new approach. *Pulp and Paper Mag. of Can.* **73**, 72–77.
17. McIntosh, D.C. (1963) Tensile and bonding strengths of loblolly pine kraft fibres cooked to different yields. *Tappi* **46**, 273–277.
18. Laamanen, J., and Söderhjelm, L. (1977) Generella egenskaper hos de vanligaste typerna av avsalumassa. OY Keskuslaboratorio/Centrallaboratorium AB, Uppdrag F 7795, DEF 3/'Specifikationer', Internal Report, 4 pp. (in Swedish).
19. Ejderby, F., and Hult, N. (1976) *Papper och pappersmassa – en grund bok.* Sveriges Skogsindustriförbund, Markaryd, 151 pp. (in Swedish).
20. Hannant, D.J., Hughes, D.C., and Kelly, A. (1983) Toughening of cement and other brittle solids with fibres, *Phil. Trans. R. Soc. London, Ser. A* **310**, 175–190.
21. Hughes, D.C., and Hannant, D.J. (1985) Reinforcement of Griffith flaws in cellulose reinforced cement composites. *J. Mater. Sci. Lett.* **4**, 101–102.
22. Colvin, J.R. (1971) Structure and formation of the cellulose microfibril. *Cellulose and Cellulose Derivatives*, eds. N.M. Bikales and L. Segal, Part 4, Wiley-Interscience, New York, 695–718.
23. Hannant, D.J. (1978) *Fibre Cements and Fibre Concretes.* John Wiley, New York, 219 pp.
24. Rauch, H.W., Sutton, W.H., and McCreigth, L.R. (1968) *Ceramic Fibers and Fibrous Composite Materials.* Academic Press, New York, 436 pp.
25. Laws, V. (1982) Micromechanical aspects of the fibre–cement bond. *Composites* **13**, 145–151.
26. Krenchel, H. Private communication.
27. Coutts, R.S.P., and Michell, A.J. (1983) Wood pulp fibre-cement composites. *J. Appl. Polymer Sci., Appl. Polymer Symposia* **37**, 829–844.
28. Fördös, Z., Jensen, K.B., Laamanen, J., Lambrechtsen, N., and Söderhjelm, L. (1977) Användning av cellulosafibrer i cementskivor. OY Keskuslaboratorio/Centrallaboratorium AB, Internal Report, 20 pp. (in Swedish).
29. Parry, J.P.M., and Carrier, A.W., Development and testing of roof cladding materials, made from fibre reinforced cement. Stage 2, Report. *IT Building Materials Workshop*, Warley, February 1980.
30. Campbell, M.D., Coutts, R.S.P., Michell, A.J., and Willis, D. (1980) Composites of cellulosic fibres with polyolefins or cement. A short review. *Ind. Eng. Chem. Prod. Res. Dev.* **19**, 596–601.
31. Gregersen, J.C. (1983) European Patent Application No. 127, 960, 37 pp.
32. Dansk Eternit (1985) *A New Era in Building Materials and Process Technology.* Brochure from Export Dept., Dansk Eternit, 25 pp.

6 Vegetable fibre reinforced building materials—developments in Brazil and other Latin American countries

VAHAN AGOPYAN

Abstract

The main vegetable fibres suitable for building material reinforcement in Latin America are discussed, and the main characteristics of coir, henequen, malva and disintegrated newsprint fibres are described. Fibres such as those from bamboo, jute, piassava and sisal plants, which are better detailed in other chapters, are briefly mentioned and the properties of these fibres cultivated in Latin America are quoted.

The fibre reinforced materials which can be used in the production of building materials are presented mainly those based on bamboo, cane, coir, henequen and sisal fibres. The production methods and characteristics of these composites are described. Each study is appraised in terms of the properties of the final products, and the researchers' conclusions are presented. Bamboo, coir and sisal fibre reinforced materials are already in use in Latin America; data on composites reinforced with jute and piassava fibres, husk and bagasse of sugar-cane, and also mineralized wood shavings, are also considered. Flax, hemp, phormium, lechuguilla, maguey and rice husk fibres, which have also been tested for cement reinforcement, are mentioned.

The new research in progress reviewed in this chapter deals mainly with the improvement of fibre durability in cement composites based on coir, sisal, henequen and disintegrated newsprint fibres. Studies of vegetable fibres in other types of matrices have also been reviewed.

Finally, the results of this review are discussed, pointing out difficulties and achievements. Based on the available experimental results, vegetable fibre reinforced cement seems to be a very appropriate material for low-cost buildings.

6.1 Introduction

In this chapter the main available fibres in Latin America are presented. Some such fibres are considered in detail: coir, henequen, malva and disintegrated newsprint fibres. The respective composites are described, and finally, research in progress is discussed as a guideline for future developments.

Brazil and the other Latin American countries, like the other developing countries, need schools, hospitals, public service buildings and houses. This need for various buildings is greater in Latin America than in developed countries. Therefore, even in periods of economic difficulties, the building industry continues to be a major activity.

The lack of housing is the most critical problem, as millions of houses are needed to provide a decent home for individual families. The rapid increase of population has contributed to this problem. For instance, while the world population increased by 60% from 1958 to 1975, in Latin America the population doubled during this period[1]. Rapid urbanization is still in progress; 68% of Latin Americans already live in the cities in 1985, and this percentage is expected to increase to 75.7% by the end of this century, according to estimates by the United Nations. Because of this, the number and size of slums in the cities are increasing continually.

Most of the countries in this continent already have a well-established building materials industry. This industry produces traditional building materials of acceptable quality. Some sectors even export building components to Europe, North America and Japan. The production of this industry is growing, but as the building industry grows, there is a shortage of materials, and therefore prices increase. As a result, sometimes, on top of an inflated price an excess is charged, making the situation even worse. When this takes place, as occurred in 1987 in Brazil, the countries suffer badly both from the shortage of building materials and the inflation. For this reason new building materials and components suitable for the market are needed. It is also necessary to take into account that about 60% of housing is produced in the informal sector: new materials should therefore take this special use into account.

It is also necessary to point out that the cost of traditional building materials is still too high for widespread use by the people of Latin America. A considerable amount of this cost is due to the price of energy for manufacturing and to transportation costs. Such high costs are the consequence of three factors: the distance between sources of energy and producing centres; the distance between producing centres (plants) and some of the consumer areas; and finally, the tendency for economic monopoly in some types of basic materials. All the above factors are themselves consequences of the size of the subcontinent and of an industrial development not rationally organized. Energy-saving processes to produce building materials and components must be encouraged to reduce part of the costs. At the same time, new technologies must also be encouraged. These should not be sophisticated, because the materials ought to be produced locally, or even at the building sites, in order to minimize the costs of transportation.

Similar recommendations were expressed at the Second Symposium on

Building Materials for Low Cost Housing, held in Bangkok during January 1987. Sosa[2], reporting on this symposium, lists the following points which should be pursued as strategies in lower building costs:

(i) Participation by users
(ii) Use of local resources and techniques
(iii) Use of small production units which can easily be carried to other sites
(iv) Small energy demand.

In this context, fibre reinforced materials seem to be suitable materials for Latin America. There are two approaches for the development of new composites in fibre-cement. The first is based on the production of fibre-cement components free of asbestos: such components are very similar to those made with asbestos-cement, since they are mainly thin sheets. The second is related to the production of composites for new types of building components, not only for asbestos replacement; the performance of these components relies on new composite properties, and their main purpose is to replace or provide an alternative to traditional materials and components. In Latin America, researchers and producers are primarily concerned with new materials. Asbestos replacement has been a concern only for some large asbestos-cement industries, and in fact, some types of fibre-cement flat sheets are already available in the market. When considering the benefits that fibres confer on brittle materials[3] it is necessary to keep their relative costs in mind. On one hand, in Latin America the price of man-made fibres is still a considerable part of the final cost of the component. For example, in low-cost production methods, the cost of such fibres can represent as much as 25% of the final product[4]. On the other, vegetable fibres have a relatively poor performance. This cannot be interpreted as a discouraging factor for building materials in Latin America, because sometimes they are the only suitable reinforcement for low-cost buildings. A review of the role played by these vegetable fibres is presented by Aziz *et al.*[5]. Published research in Latin America on this topic started last decade. This means that few results are available. Nevertheless, some components are already being produced, mainly on building sites.

6.2 Vegetable fibres

A large part of the Latin American subcontinent is located in tropical and subtropical regions covered by a great amount of vegetation producing vegetable fibres in vast amounts. It is necessary to point out that the chemical and physical characteristics of vegetable fibres depend on the climate and soil where they grow, the processes for separation of the fibres, and even on the period of the year when the plant is cropped. Most of the fibres mentioned in this chapter are from plants which grow in uncontrolled plantations, sometimes in the wild. For this reason, the values of the properties which are reported represent average figures with high coefficients of variation, especially in the evaluation of durability. Moreover, as the fibres themselves are heterogeneous, even fibres coming from the same crop and the same place may present a high coefficient of variation for vegetable fibre properties and their products.

6.2.1 *Prospective fibres for fibre-cements*

From the large number of vegetable fibres available in Latin America it is possible to select some which have properties that make them suitable for use in building. These properties are related to mechanical behaviour (tensile strength, Young's modulus and elongation at break), physical characteristics (density, water absorption), size relations and durability. Some of the values of these properties, for the majority of the fibres available in Latin America, are not known. Moreover, existing data are mainly related to properties relevant to the textile industry. Nevertheless, Fernandes *et al.*[6] managed to select nineteen vegetable fibres as having potential use for building materials.

Table 6.1 presents 18 types of fibres. Several names are indicated for each type of fibre, because local names vary from one country to another and sometimes from one region to another in the same country. The first name presented in this table always refers to the most usual English name for the fibre; the Spanish and Portuguese names, which have no similarity to the English version, are also mentioned. To avoid misunderstanding, the full botanical name of the plant is indicated. The fibrous part of the plant is also stated, because this affects the properties of the fibre; stalk fibres are the longest, usually come in strands, and are flexible; leaf fibres are more rigid and coarse; finally, fibres from seeds or fruits are the shortest and are more resistant to alkalis. The uses presented in the table refer to those which are most often mentioned as reasons for the development of plantations.

6.2.2 *Selected fibres*

Not all the fibres presented in Table 6.1 are equally suitable as building material reinforcement. Although all have properties which make them suitable for building, they are not equal in all aspects and can be replaced by other fibres which have better performance. For example, if costs are taken into account, cotton, flax and ramie fibres are not appropriate for building, as they have a more sophisticated use in the textile industry. Due to low availability in Latin America, kenaf, Manila hemp (abaca), phormium, plantain and sunn-hemp fibres are also disregarded as general building-material reinforcement. Some of these can be used, and some studies with kenaf in Guatemala and with abaca in Ecuador are known (see section 4.1). Finally, on account of production difficulties, dishcloth gourd, elephant grass and sugar-cane bagasse fibres must be replaced by other fibres more easily produced and mixed with cement. Moreover, the studies carried out with these fibres in Latin America[7] were not very successful, probably due to difficulties of preparation of both the fibre and the composite.

Table 6.1 Prospective vegetable fibres for building

Common names of fibre or plant	Plant: Botanical name	Plant: Fibrous part	Present uses of fibres	Observations
Bamboo	*Bambusa vulgaris* Schrad.	Stalk	Paper, building	Shredding very difficult (7); used as sticks in buildings
Coir, coco	*Cocos nucifera* Linn.	Fruit	Textiles, ropes, upholstery, brushes	Low strength
Cotton	*Gossypium herbaceum* Linn.	Seed	Textiles	Expensive
Cellulose	*Pinus elliottii* and *Eucalyptus citriodora*	Stalk	Paper	Fibres from dissolved paper (8)
Dishcloth gourd, vegetable sponge, esponjilla	*Lufa aegyptica* Mill.	Fruit	None	Good durability, difficult to separate fibres (7)
Elephant grass	*Pennisetum purpureum* Schum.	Leaf	None	African species have good strength and durability (9), difficult to mix in the matrix
Flax, 'linho'	*Linum usitatissimum* Linn.	Stalk	Textiles	High durability for textile purposes (10), expensive
Henequen, maguey	*Agave fourcroydes* Lemaire	Leaf	Textiles, brushes	Similar to sisal, lower strength (10)
Jute	*Corchorus capsulanis* Linn.	Stalk	Textiles, sacking	High strength, low durability (see. ref. 11)
Kenaf, brown Indian hemp	*Hibiscus cannabinus* Linn.	Stalk	Textiles, ropes	Similar to jute, with slightly lower strengths (10, 12)
Malva, Congo jute, paka, 'malva roxa'	*Urena lobata* Linn.	Stalk	Textiles, sacking	Strengths similar to jute (10)
Manila hemp, abaca	*Musa textilis* Née.	Leaf	Ropes, paper	Salt-water resistant but slightly acid (10)
Phormium, New Zealand flax	*Phormium tenax* Forster	Leaf	Textiles, ropes	Salt-water resistant, high strength, available only in Argentina and Brazil (12)
Piassava	*Attalea funifera* Mart.	Leaf	Ropes, brushes and brooms	Not flexible, thick
Plantain	*Musa sapientum* Linn.	Leaf	None	Low durability (9)
Ramie	*Boemmiria nivea* Gaud.	Stalk	Textiles	Very high strength, not affected by humidity (12), expensive
Sisal	*Agave sisalana* Perrine	Leaf	Textiles, ropes, pulp	High strength, low durability (13)
Sugar cane	*Saccharum officinarum* Linn.	Stalk	Pulp	Use of bagasse, difficult to remove remaining sugar (7)
Sunn-hemp	*Crotalaria juncea* Linn.	Stalk	Pulp	Small production, high durability (11)

Table 6.2 Properties of bamboo sticks (14, 15)

Property	Value
Water absorption	145%
Density	1158 kg/m^3
Tensile strength	575 MPa*
	70 to 170 MPa**
Young's modulus	28.8 × 10^3 MPa
Elongation at break	3.2%

*ASTM D 3379-75 test method (16)
**Genera other than *Bambusa*.

Therefore, considering these factors, only eight types of fibres have been selected from Table 6.1. A short description of bamboo, jute, piassava and sisal fibres is presented next, with emphasis on Latin American regional data. Coir, henequen, malva and fibres from dissolved paper are also discussed. Further details for bamboo, cellulose and sisal fibres are presented in other chapters of this book.

(i) *Bamboo.* It is a common plant in Latin America. *Bambusa vulgaris* is the best known and most widely used species, although for building, *B. guada* is also applied. The culm consists of longitudinal strands of fibres strongly linked with lignin. The fibres, which represent 60–70% of the bamboo weight, are concentrated in the external part of the culm.

It is very difficult to separate the fibres: therefore in building, bamboo is used as culms or as sticks. It is necessary to point out that there is no tradition of bamboo processing in this continent. The pulp and paper industry also consume bamboo because its fibrils are relatively long, from 2.5 to 3.5 mm. The properties of bamboo sticks are presented in Table 6.2. As observed in microscopic analysis, bamboo behaviour in an alkaline medium is good[17].

(ii) *Jute.* This is not a typical American plant, but Brazil is already producing it on a large scale, mainly in the Amazon region. The plant grows quite easily in wet and warm areas, and is harvested 130 days after planting. The average production varies from 1500 to 2000 kg/ha.

The fibres are separated by maceration or decortication. However, for large-scale production only mechanical decortication is suitable, and this kind of equipment is not available in Latin America. The present production relies on low-cost labour in the north of Brazil. The diameter of these fibres varies from 20 to 100 μm and their length can be several metres.

There are few data from Latin American fibres. Cook[11], quoting from studies by Sridhara *et al.*, states that the fibres presented low performance in durability evaluation. Ramaswamy *et al.*[18] state that the tensile strength of dry fibres is 226 MPa and the elongation at break is 1.28%. Table 6.3 presents average properties of these fibres used in Brazil for textiles[19].

Table 6.3 Properties of jute fibres (19)

Property	Value
Strand diameter	4.8 mm
Water absorption	214.1%
Bulk density	301 kg/m^3
Tenacity*	9.48 to 20.30 gf/tex
Elongation at break*	3.7 to 6.5%

*ASTM D 3822–82 test method (20)

(iii) *Piassava. Attalea* is a typical tropical American genus. *Attalea funifera* (piassava) is probably the only one with commercial value, and is produced mainly in the northeast of Brazil. After being planted it takes a long time for the leaves to produce fibres, a minimum of 4 years. Each leaf has 500–800 g of fibres; these are 50 to 60 cm long. The fibres are fixed on the petioles of the leaves and are easily pulled off.

The diameter of the Brazilian fibres is about 5 mm at the junction of petioles, decreasing at the other end. Its elongation at break varies from 1.75 to 9%[21]. Production data for these fibres are not available, but may be about 10 000 metric tonnes. Table 6.4 presents properties of Brazilian piassava fibres. No durability data are available.

(iv) *Sisal.* The most common sisal plant produced in Brazil is *Agave sisalana* P. It is common because in this species it is easy to get fibres from the leaves, although these fibres are not the best ones from *Agave* plants. Sisal is also produced in Haiti, Venezuela and Mexico, where it originated. Other Central American countries produce small amounts.

The annual production of fibre in Brazil is about 250 000 metric tonnes, making it probably the largest producer of sisal in the world. However, this production has not increased since 1975, due to drought which affected the producing region, and also due to a reduction in consumption. Sixty percent of the Brazilian production is concentrated in the state of Bahia and 30% in the

Table 6.4 Properties of piassava fibres (14, 19)

Property	Value
Water absorption	34.4 – 108%
Bulk density	470 – 861 kg/m^3
Density	1054 kg/m^3
Tensile strength*	143 MPa
Young's modulus*	5.6×10^3 MPa
Tenacity**	4.00 – 15.85 gf/tex
Elongation at break*	5.99%

*ASTM D 3379-75 test method (16)
**ASTM D 3822-82 test method (20)

Table 6.5 Properties of sisal fibres (14)

Property	Value
Water absorption	239%
Density	1270 kg/cm³
Tensile strength*	458 MPa
Young's modulus*	15.2×10^3 MPa
Elongation at break*	4.3%

*ASTM D 3379-75 test method (16)

state of Paraiba, both located in the northeast of the country. The main use for these fibres up to now has been in the carpet and baler twine industries. Nowadays sisal leaves are also being used by the pulp and paper industry.

The sisal fibre decortication procedure is very crude and dangerous for the workers. Thousands of simple machines powered with diesel engines are spread out through the sisal plantations. These machines mechanically separate the fibres from the mucilage, but about 40% of the fibres, the short ones, are lost in the mucilage residues. The acidity of the fibres is neutralized simply by washing in water. The fibres are then bleached in the sun. During the processing a further 10% of fibres is lost as residues. Therefore only 3% of the weight of the leaves is recovered as fibres. The short fibres cost about US$ 0.20/kg, and the processing residues ('wads') cost US$ 0.10–0.15 per kg.

The main properties of Brazilian sisal from Bahia are presented in Table 6.5. Durability tests show a sharp decrease in the tensile strength when these fibres are immersed in a solution of calcium hydroxide (pH 12). After 119 days of immersion, the strength is reduced to about 10% of the initial value.

6.2.3 *Coir fibres*

Coconut is a tall cylindrical-stalked palm tree, reaching 30 m in height and 60–70 cm in diameter[21]. It is a tropical plant for low altitudes. It needs sunshine and a soil rich in calcium and phosphorus, and is thus generally suitable for cultivation in sandy sea-shore. Figure 6.1 shows a coconut plantation close to the sea.

Although coconut cultivation is concentrated in tropical belts of Asia and East Africa[5], it is also found in Latin America on a smaller scale. Coconut is cultivated on a commercial basis in Brazil. The most important part of the tree is its fruits, which are egg-shaped, about 30 cm long, and 25 cm in diameter[22]. The more external layer of the fruit is thin and smooth: its fibrous mesocarp is 3–5 cm thick, and the endocarp is very hard[21]. The fruit has a large central cavity which contains a sweet liquid (coconut water). The number of fruits per tree varies, depending entirely on soil conditions; it is possible to get from 20 up to 600 fruits per year. Figure 6.2 shows a tree in a soil without fertilizer.

The main products come from the fruit itself: dehydrated coconut, coconut water, copra, oil, etc. The external part of the fruit, the husk, is a residue.

Figure 6.1 Coconut plantation in Brazil.

Figure 6.2 Coconut tree and fruits.

Figure 6.3 Simple equipment used in coir fibre production.

Therefore the coir fibres, which are obtained from the husk, basically from the mesocarp, are a by-product of the coconut processing industry. Usually, in Brazilian conditions, only 12% of the income of the coconut industry comes from the fibres[23].

The coir fibre production in Brazil is also rudimentary. Old fashioned equipment crushes the husk and separates the fibres (Figure 6.3). Only two industries located in the state of Sergipe have modern equipment, and can produce long fibres (more than 110 mm) suitable for brushes and threads. The main production of the other industries is fibre-wad for upholstery. The shortest fibres and the residues from the two industries previously mentioned have a low value, about US$ 0.15/kg.

The Brazilian annual production of coir fibres is higher than 2 million metric tonnes; however, only a small amount of the five hundred million fruits annually cultivated in the country, are used for fibre production. This production is concentrated in the states of Bahia, Pernambuco and Sergipe, in the northeast of the country.

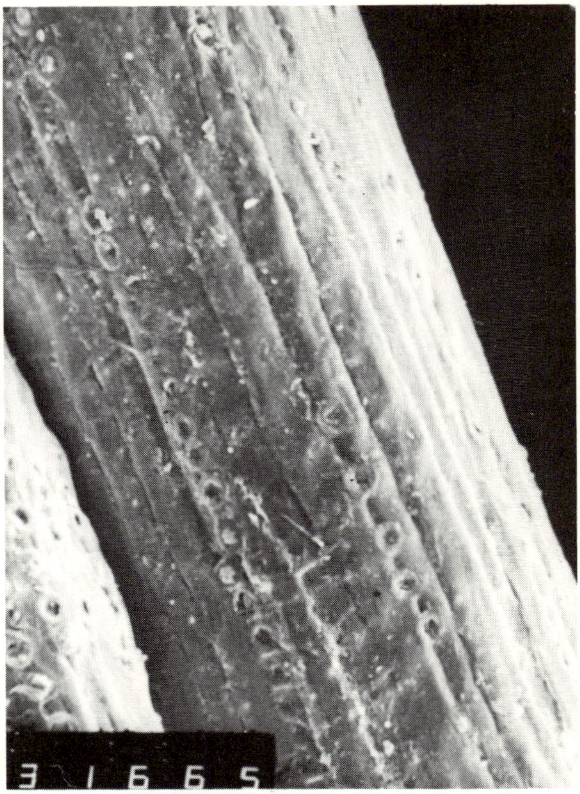

Figure 6.4 Scanning electron micrograph showing coir fibre surface (× 255).

Fibre length varies from 10 to 200 mm, and the diameter is about 0.3 mm. The fibrils are very short; their length varies from 0.4 to 1 mm, and their diameters are within the range 0.016–0.020 mm: for this reason the coir fibres are not suitable for pulp production.

Coir fibres also have a low cellulose content when compared with other fibres used in fibre-cement. The total cellulose, including hemicellulose, is about 50%[24], which is still lower than the international value of 53% quoted for these fibres[21]; finally, the amount of lignin is about 48%. In spite of the low cellulose content, the fibre structure is very close, as can be seen in Figure 6.4. This must be the reason for its better durability performance when compared with fibres with higher cellulose content, such as sisal, which has 72–77% total cellulose[24].

Table 6.6 presents the properties of seven different types of coir fibre from Brazil. The coefficient of variation for some values is higher than 40%. The specific gravity of these fibres varies from 0.89 to 1.11. The results presented in this table are quite similar to those presented by Dantas[19], which refer to two

Table 6.6 Properties of coir fibres

Type	Humidity* (%)	Water absorption (%)	Diameter** (mm)	Elongation at break** (%)	Tensile strength** (MPa)
Long A	10.9	120	0.259	44.7	118
Long B	12.9	148	0.366	51.4	103
Short	11.5	153	0.262	29.4	95
'Fibril'	12.6	117	0.212	23.9	99
Residue A	11.8	171	0.274	36.9	116
Residue B	10.5	161	0.230	36.0	101
Mat shreds	10.8	159	0.366	25.0	114

*Percentage of the weight of dry fibre
**ASTM D 3822-79 test method (20)
Note: Each figure represents the average for a series of 30 trials.

other types of coir fibres, and the mechanical properties are slightly lower than those presented by Guimarães[14], probably because the latter used another test method. However, it is very interesting to note that the properties are of the same magnitude as those presented by Aziz *et al.*[5] for African and Asian fibres.

Cook[11] states that the chemical composition and morphological properties of coir fibre clearly provide a better protection against decomposition than is the case for sisal fibre. He presents data where coir fibres immersed for 6

Table 6.7 Durability evaluation of coir fibres

Conditioning		Long B		'Fibril'	
Type	Duration (days)	Load[1] (N)	Elongation[2] (%)	Load[1] (N)	Elongation[2] (%)
Control group	0	8.6	51.4	7.0	23.9
Treatment 1:					
24 h immersed in water	28	6.4	33.1	5.3	29.9
24 h air drying	91	4.9	6.8	3.1	5.5
Treatment 2:					
24 h immersed in water	28	6.1	27.4	4.5	23.2
24 h oven drying	91	5.0	10.0	4.2	5.4
Treatment 3:					
Immersion in calcium hydroxide	7	8.9	40.6	7.2	26.3
solution (pH 12)	28	11.5	42.7	3.8	35.7
	63	7.7	36.8	4.6	28.0
	119	4.7	29.3	1.7	25.0
	182	3.3	20.2	3.0	9.9

Key:
(1) Load: rupture load as ASTM D 3822-82 (20).
(2) Elongation: elongation at rupture as ASTM D 3822-82 (20)
Note: Each figure represents the average for a series of 30 trials.
Coefficients of variation are very high, in most cases above 40%.

months in lime solution (pH 12.6) lost only 35% of their initial strength. Nevertheless, the Brazilian coir fibre performance is not as good. This can be seen in Table 6.7; in similar conditions of immersion, the fibres lost their strength and also their ductility very sharply. These results are still far better than those obtained with Brazilian sisal fibres (see section 6.2.2). However, the results point to the need for comprehensive durability studies before fibre-cement components with coir fibres can be put into use. Figures 6.5–6.7 show micrographs of coir fibres conditioned in lime solution; the surfaces of fibres change because of the reaction of calcium hydroxide with lignin. After 182 days of conditioning, the surface seems to have less adherent particles, but it is very different from that of the reference surface (Figure 6.4).

6.2.4 *Henequen or 'maguey' fibres*

'Maguey' is the usual term for various species of *Agave* in Latin America: for this reason, some confusion in the analysis of the available data occurs. This term is mainly used for *Agave cantala*[10,21,25]; however, in Cuba it is used for *Furcraea hexapetala* and in Peru for *F. macrophylla*[10]. In Mexico, this term also refers to *Agave atrovirens*, *A. kirchneriana*, *A. weberi*, *A. kerchovei*, etc.[21].

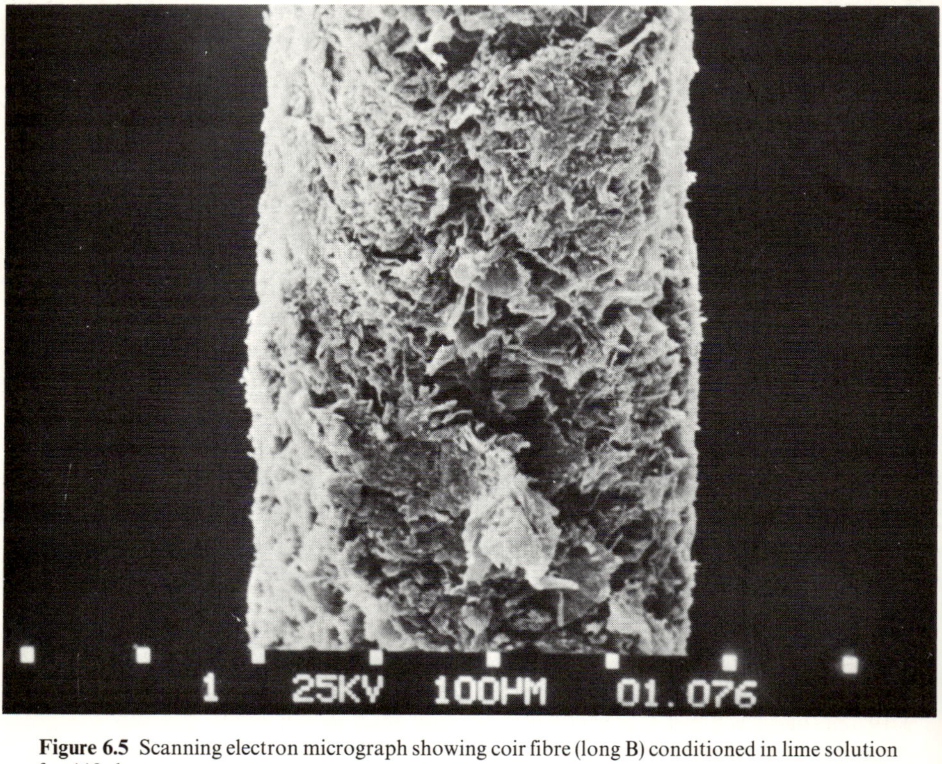

Figure 6.5 Scanning electron micrograph showing coir fibre (long B) conditioned in lime solution for 119 days.

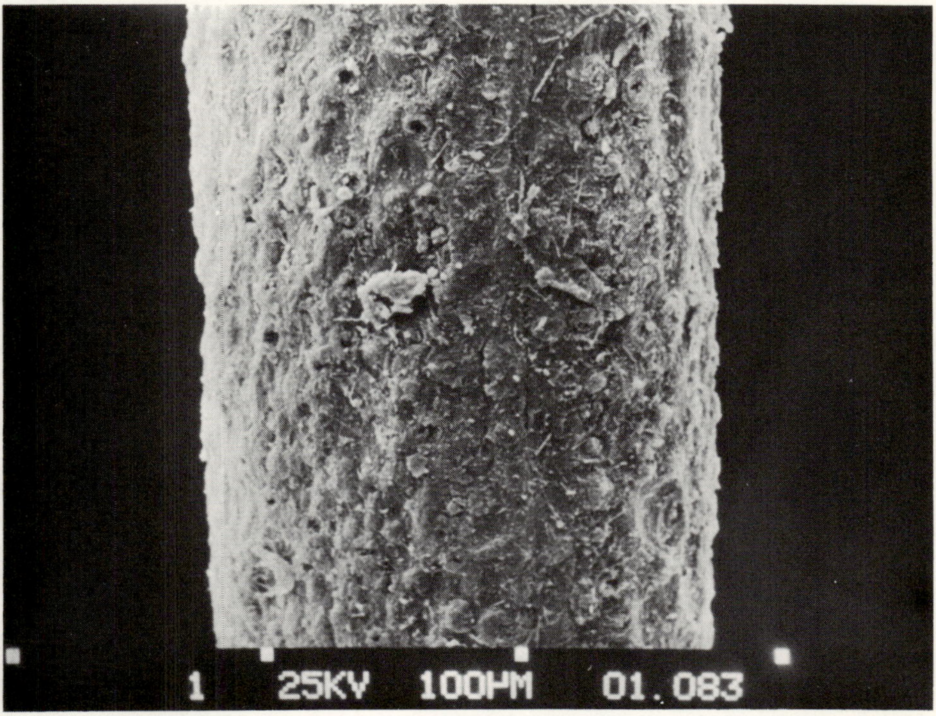

Figure 6.6 Scanning electron micrograph showing coir fibre ('fibril') conditioned in lime solution for 119 days.

In addition, in Central America and Mexico, maguey refers to *A. fourcroydes* Lemaire, also known as henequen, white henequen, Mexican sisal or Cuban sisal. In this section, the properties of *A. fourcroydes*, which is used in some countries as cement matrix reinforcement, are presented.

Henequen is very similar to sisal, and is therefore called false sisal. It has bluish grey leaves with strongly spiny margins and stout tip spines; the leaves are rigid, 100 to 130 cm long and less closely packed at the bole than the sisal[25] (see Figure 6.8).

The Yucatan peninsula, Mexico, is the main producer of henequen, probably because this species thrives better on rocky ground. Cuba is also an important producer. Central America, as a whole, produces henequen, but little on a commercial basis. Besides Mexico and Cuba, the only moderate production is in the Dominican Republic and in Guatemala. The total world production must be around 200 000 metric tonnes a year, and most of this is produced in Yucatan.

The fibre decortication is more efficient in Mexico and Cuba than that in Brazil for sisal fibres (see section 6.2.2). Automatic equipment produces large amounts of fibres. Also, the fibre residues in the mucilage are recovered: therefore, it is possible to obtain more than 3% of the leaf weight in fibres. The fibre length varies from 60 to 160 cm, and they are usually finer than those of

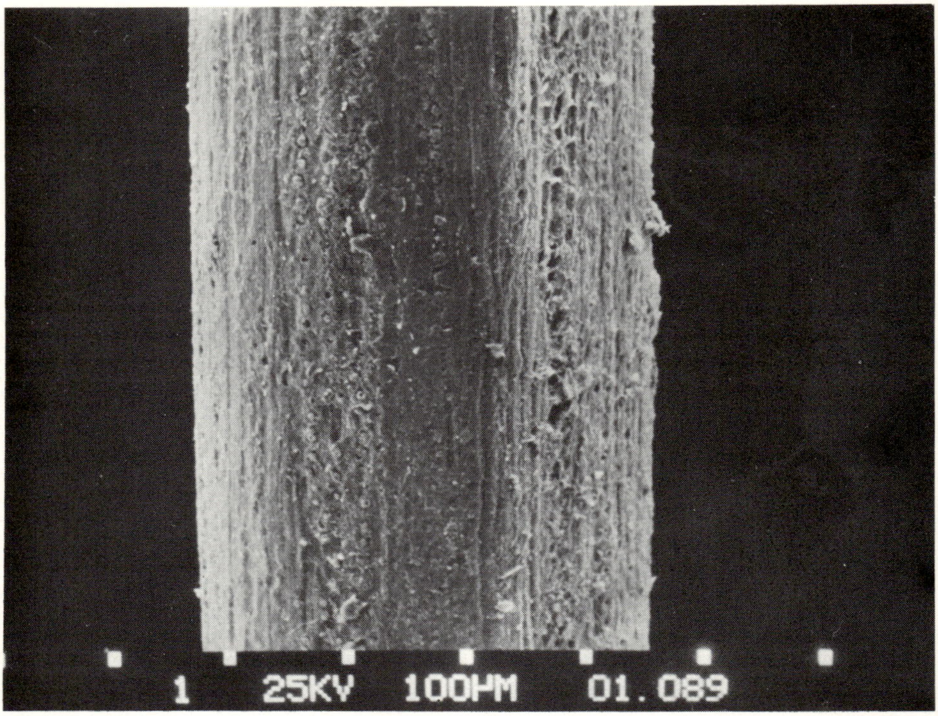

Figure 6.7 Scanning electron micrograph showing coir fibre (long B) conditioned in lime solution for 182 days.

Figure 6.8 Wild henequen, Guatemala.

Table 6.8 Properties of henequen ('Maguey') fibres (26)

Property	Value
Density	1395 kg/m^3
Water absorption	163.1%
Tensile strength*	91.0–307.0 MPa
Elongation at break*	2.3–7.6%

*Based on ASTM test methods, mainly ASTM D 3822 (20)

sisal – the diameter varies from 0.125 to 0.500 mm[21]. The fibrils are 1.5–4 mm long and their diameter varies from 0.02 to 0.03 mm. These fibres are dried and bleached in the sun, as it is done for sisal, resulting in a fibre with a whitish colour and good appearance. Nevertheless, this is not much appreciated by the textile industry, which prefers sisal fibre, due to its better workability in machinery. The main use of henequen is for twine production.

The amount of cellulose in henequen fibres is about 72%[21] which is about 5% less than that in sisal. These fibres are therefore less strong, and probably less durable, than sisal.

Table 6.8 presents properties of henequen fibres from Guatemala. There are no durability studies; however, Quiñonez[26] has experimented with these fibres in different environments and observed their loss of strength and ductility. The results of this study are presented in Table 6.9.

6.2.5 *Malva fibres*

Malva is a general name for various species of the Malvaceae family, which has more than a thousand types of herbaceous plants widespread in the world[27],

Table 6.9 Properties of henequen ('Maguey') fibres conditioned in different environments for 10 days (26)

Type of conditioning	Rupture Load (N)	Elongation at break (%)	Visual appearance
Control group (normal environment)	6.5–27.2	2.3–7.6	White-yellow colour, clean
Treatment 1: immersion in water	1.0–4.0	0.8–1.5	Brown-black colour mostly on surface, evidence of fungus
Treatment 2: in wet-chamber	2.7–4.9	0.8–2.2	White-yellow colour, but darker than control, no evidence of fungus
Treatment 3: dry surface saturated	1.5–3.3	0.8–1.1	White-yellow colour, but clearer than the control, white fungus

mainly in the tropical and subtropical regions. This section relates to *Urena lobata* Linn., a perennial shrubby plant 1–3 m high, with many branches.

Urena lobata is known under different common names. Cook[10] lists the following names: ake-ire, caesar weed, cadillo, canhamo, carrapico, Congo jute, cousin rouge, culut culutan, grand cousin, grand mahot cousin, guaxima, guizazo, malva branca, malva roxa, malvas, otote grande, paka, toja and uacima. In Brazil it is more commonly known as malva roxa, in Cuba as malva branca or guizaso, and in Venezuela, Ecuador and Puerto Rico as cadillo. The former Congo (Zaire) is the main producer, and for this reason it is internationally known as Congo jute.

Malva grows wild in many Latin American countries, but it is cultivated only in Brazil as a fibre plant on a commercial basis. The world commercial production must be around 10 000 metric tonnes a year, not considering the consumption of wild *Urena* which peasants use; most of the commercial production is from Zaire. In Brazil, malva is cultivated in the Amazon region, mainly in the northeast of the state of Pará, and also in the state of Minas Gerais in Central Brazil[7,12,27].

The fibres are removed from the bark of the trees. The stalks or bark, which is mechanically removed from the stalks, are immersed in water for 8–20 days. It is preferred to do the maceration in slow streams, but when these are not available, it is possible to use ponds for this purpose. The fibres are easily stripped from the retted bark by hand, washed in clean water and dried in the open air. About 5–6% of the stalk weight consists of fibres. It is possible to obtain, under normal conditions, 1300–1500 kg of fibres per hectare of plantation. These are cropped annually.

The commercial fibre length varies from 1.0 to 2.5 m and their mechanical properties are very similar to those of jute (see Table 6.3). The fibres can be near-white when carefully retted[10] and their fibrils are 1.5–5 mm long with a diameter of 0.009–0.012 mm. The amount of cellulose of these fibres is about 76%[21], which is very similar to the jute (76–77%).

The use of malva as brittle matrix reinforcement was suggested only recently; there are therefore, at present no published experimental results for the use of these fibres for building-material purposes. The Agricultural Institute of Campinas (state of São Paulo) has been carrying out research on *Urena* plants and fibres, and so far, has improved plantation processes and fibre properties, but studies on fibre uses have not started yet. The main industrial use of malva fibres is for sacking.

6.2.6 *Disintegrated newsprint*

When vegetable fibres are analysed for cement matrix reinforcement, cellulose fibres are considered to have high potential. In spite of the fact that they are the cheapest fibres for asbestos replacement[3], they are still expensive when compared to natural vegetable fibres such as sisal and coir, mainly because the latter fibres are actually by-products of fibre processing industries. Fibre-cement does not demand fibres at their natural length, but short ones which are more easily mixed with the matrices; therefore, the residues from vegetable fibres would be adequate.

The price of cellulose fibre varies from US$ 300 to US$ 400 per metric tonne, which is more than double that of available residual vegetable fibres. In order to make these fibres appropriate for building, it has been suggested that they should be obtained from disintegration of waste papers[28,29]. Printing papers are the most commonly available papers, and among them, newsprint seems to be the most promising for building purposes. Newsprint, a type of paper generally used for printing newspapers, furnishes largely mechanical wood pulp, with some coniferous chemical wood pulp[30]. The usual woods used are from *Pinus elliottii* and *Eucalyptus citriodora.* In general, the quality of this paper is poor, with lower strength values and durability[31] than those of other types of printing papers. The fibres are short, 1–2 mm long, and twisted, as can be seen in Figure 6.9. However, this type of paper presents an important advantage in relation to other papers: it has not been glued, and therefore is the easiest type of paper to disintegrate in water. Glues used in paper making are usually slightly acidic which can be harmful to the cement – for instance, ordinary printing paper paste sometimes has a pH value as low as 4; the pH values of newsprint pastes, however, are usually above 6, almost neutral.

Latin America is a newsprint producer but also an importer. The Food and Agriculture Organization of the United Nations[32] predicted that in 1986, the total capacity of newsprint production in Latin America would be about

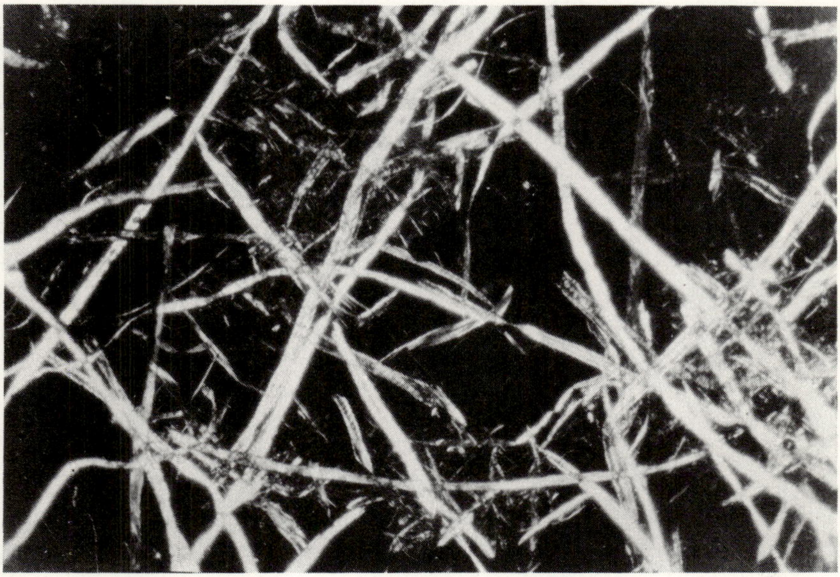

Figure 6.9 Disintegrated newsprint fibres (×27).

Table 6.10 Properties of cellulose fibres (5, 29, 33)

Property	Value
Density	1200–1500 kg/m³
Water absorption	about 400%
Young's modulus	10–40 GPa
Tensile strength	300–500 MPa

1 157 000 metric tonnes per year. The main producers are, respectively, Mexico, Brazil, Argentina, Chile and Peru. Cuba and Uruguay also produce small amounts of this type of paper. For economic reasons the total capacity of the local industry is not used; for instance, nowadays the Brazilian actual production is 210 000 metric tonnes per year, in two industries located in the state of Paraná; however, these industries are sure that they can expand their production to up to four times the present amount.

As there are newspapers all over the continent, it is possible to assume that waste newsprint is a ubiquitous product. Usually newsprint is recycled for the production of other types of paper, but possible consumption by the building industry is not significant when compared to the total production.

The cost of waste newsprint (newspaper) is very low, varying from US$ 0.10 to US$ 0.15 per kg. The building industry can buy the paper from 'junk deposits', as the amounts are relatively small.

The disintegration process is quite simple[29]: pieces of newsprint are dispersed in water with the help of a high-speed mixer for about 30 minutes. The solution is then kept for 24 h in a wire screen container to remove excess water. This procedure allows the water content of the paper pulp to be 78.3% with a coefficient of variation of 6.5%. This pulp is then used as reinforcement.

Disintegrated newsprint pulp actually contains cellulose fibres. Their properties are well known and consistent (Table 6.10). Although they are known as durable fibres, their durability in alkaline environments has been studied by a special method. This method consists of immersing the dissolved paper in calcium hydroxide solution (pH 12); later the paper is remade in the laboratory with a base-weight of about 60 g/m^2. Its rupture load is measured with zero-span. After 182 days of immersion, the tensile index was 70.3 Nm/g, similar to the original value of 66.1 Nm/g.

6.3 Fibre-cements

Vegetable fibres as reinforcement for building materials have been used for generations, but apparently without extensive research studies. Mud overlying wooden or bamboo skeletons, and grasslike plants reinforcing mud for walls or bricks, were usually used in construction till last century in various Latin

American countries, and even nowadays they are sometimes used in rural areas.

Only very recently have researchers attempted to improve these materials and to develop new ones reinforced with vegetable fibres. The first published paper on this subject probably dates from 1976[34]. Before this date, the only available information refers to bamboo or culms used as reinforcement. The main technological knowledge has therefore been accumulated during this decade and the resulting products have been in use for only a few years. Time has not permitted extensive and thorough research to be produced for each of the materials considered. Nevertheless, some of this research has been thorough and can be translated into practice. This is presented in this section, with the main properties of the resulting products.

6.3.1 *Bamboo and cane culms or sticks as reinforcement*

The main purpose of this chapter is to study fibre reinforcement, but as culm is probably the most common vegetable reinforcement for building in this subcontinent, it must also be considered.

In section 6.2.1 it was mentioned that Latin America has not succeeded in obtaining bamboo fibres through a simple method, such as the one presented by Pakotiprapha *et al.*[35]. Solutions such as hammering the stems or even immersing them in sodium hydroxide solution for 24 hours were found to be ineffective[7].

Nevertheless, the use of culms or sticks as reinforcement, which is the usual process in Asia, as described by Subrahmanyam[36], is also the usual building process in Latin America. The traditional 'quincha', 'taipa' or 'pau-a-pique' are building systems where the skeleton consists of an arrangement of bamboo or cane culms or sticks.

Hidalgo Lopez[37,38] was probably the first Latin American researcher to study the improvement of this traditional process. The same approach has been tried in Peru by Díaz Gutiérrez[39] and Gallegos Vargas[40]. In Brazil there have been some attempts to use the fibres of these plants[7] and today the emphasis is on bamboo reinforced concrete[7,14,15,17,41].

Hidalgo's expertise was utilized for planning and building an experimental housing program with 12 units, in Guayaquil, Ecuador[38]. The houses have one floor and 42 m^2 area. They were produced with intensive use of bamboo. The flooring consists of an 8-cm thick concrete slab reinforced with bamboo mesh under the wall regions; the slab is surrounded by bamboo reinforced beams. The walls are prefabricated by a technique very similar to that of the platform frame used in the United States. The bamboo frame is covered on both sides with culms which produces an internal air chamber very convenient for thermal and acoustic insulation. Cement mortar is used as coating. Finally, the structure of the roof is also made with bamboo culm. The author states that the costs of producing this experimental housing are about 60% of those of similar-sized houses built by the Government.

A similar approach was taken by Gallegos[40] and Díaz[39], but the latter adapted a similar process to make it more flexible to suit local conditions. Díaz calls his process 'pre-fabricated cane lathing'. This consists of panels 2.4 m high and 0.6–1.2 m wide. The frame is made with sawn timber with a section of 30 mm × 65 mm, and includes diagonal crossings. The frames are filled with round wattle (diameter 10–20 mm), cane and strips of bamboo (25 mm wide), all plaited together so that they can be placed and fixed in the frame without nails. Once these panels have been assembled and fixed in place to form the walls, they are covered with a mixture of straw and mud (proportion of 1:50 by weight) forming a first layer, and finally, they are given a top layer which can be a mud layer of 1.5–2.0 cm thick; two coats of cement paint; or plastering with mortar, plaster of Paris or soil-cement. The top coat of the panels depends on preferences, costs and climatic or other conditions to which the wall is to be exposed. Díaz considers his system as a timber one. Therefore the behaviour of the system relies on the performance of the timber of the frame and also on that of the cane and the bamboo sticks. There are no figures for these panels; however, the author stresses the good performance of the system, and also good resistance to earthquakes. This was proved in experimental conditions. Gallegos[40] also states that a similar module, when tested to acceleration equal to gravity, suffered damage only on the top mud plastering.

Guimarães[7,14] has produced concrete beams reinforced with bamboo cables; these cables consist of bamboo sticks without the inner part of the culm, of 2–5 mm thickness. The sticks are twisted to form cables. The results were not very encouraging: the cement hydration close to bamboo is affected, the concrete deformations are not altered, and it is necessary to use large amounts of bamboo to achieve a similar strength in the beams to those reinforced with steel bars. For instance, the section of bamboo should be twelve times the section of the steel bars. Ghavami[15] also did not obtain good results in beams in bending tests, but he has stressed that the process of bamboo reinforcement is a feasible one and that this kind of reinforcement can be used in rural areas[41], located far away from the steel industry.

6.3.2 *Coir-cement*

Due to its availability, as shown in section 6.2.3, coir fibre reinforcement has been studied widely in Brazil. In São Paulo[19,42,43] researchers have been studying mainly the composite. No building components have yet been produced. Savastano Jr.[42], using mat processing residues, produced a less brittle material than the mortars, suitable for producing partitions. The IPT[19,43] has worked with concrete with low cement content (175 kg/m^3) and a high percentage of mortar (60%), reinforced with fibres with a constant length of 50 mm, and where the fibre volume fraction varied from 1 to 6%. The specific gravity was varied from 2.12 (with 1.4% of fibre volume) to 1.90 (with 5.1% of fibre volume). The Young's modulus and the compressive strength were reduced with the increase of fibre content, and the best tensile strength in bending tests (1.25 MPa) was obtained with about 4% fibre volume.

Ghavami and Hombeeck[44] have tried to produce tiles using this material. Their best results were achieved with a cement–sand ratio of 4:9, a fibre–cement ratio of 1:6 and a water–cement ratio of 19:24, all by weight. The tiles are pressed under 2 MPa pressure for about 5 minutes. The authors point out that the thermal insulating property of this product is 20% higher than that of asbestos-cement tile. This component is presented as a low-cost lightweight product for non-load bearing walls, ceilings, roofing, etc.

Researchers from the Centro de Pesquisas e Desenvolvimento, CEPED-Bahia, have been able to transfer their studies into practice[7,14,28,45,46]. Their main product is a roofing tile of gutter-shape, 1.80 m long, about 41 cm wide and 1.5 cm thick. This shape is found easier to be produced than corrugated sheets. In a comprehensive study, this Centre has analysed the effects of fibre length and volume; matrix composition; manufacturing method; and

Figure 6.10 Roofing tile with coir fibre reinforced cement produced by CEPED's process: fibre distribution over the first layer of mortar.

workability. The methodology is presented in a booklet in non-technical terms[46] directed to the final users.

The tile matrix is a cement–sand mortar in the proportion of 1:3 by volume; the fibres are chopped to 4 cm long; the water–cement ratio is 0.7 and the fibre volume is about 2%. If the length of the tile is shorter, this volume of fibre can be reduced; for instance, for 1 m long tiles 1% of fibres have been used.

Initially, a flat plate is cast on a plastic sheet which is fixed to a wooden frame. The fibres are put manually between two layers of the mortar (Figure 6.10); the top surface is trowelled because this will be the external face. After casting the tile, the wooden frame is removed and the flat plate is suspended with the plastic base and fixed on a hollowed bench to take its final shape (Figure 6.11). It is a very simple process which can be performed by unskilled workers and without equipment. For this reason it has been used in rural areas of the state of Bahia. The only known problems occur when a

Figure 6.11 Roofing tile with coir fibre reinforced cement produced by CEPED's process: corrugated shaping/modelling.

Table 6.11 Results of bending test applied to fibre reinforced mortars (7, 45) (cement–sand 1:1 by volume, water–cement ratio 0.43)

Fibre		Composite		
Type	Volume ratio (%)	Length (mm)	Modulus of rupture (MPa)	Strength ratio of composite to control
—	Control group	—	5.62	1
Coir	5	5	5.74	1.02
	10	5	5.43	0.97
	5	30	5.93	1.06
	5	60	5.85	1.04
	4	270*	4.47	0.80
Sisal	5	5	5.45	0.97
	10	5	3.61	0.64
	3	30	5.51	0.98
	5	30	6.56	1.17
	5	60	5.71	1.02
	4	270*	11.97	2.13
Piassava	5	5	4.17	0.74
	10**	5	5.64	0.75
	5	60	4.59	0.82
	4	270*	8.46	1.51

*Aligned fibres
**Water–cement ratio 0.38

smaller percentage of fibres than that advised (2%) is used. Tiles with a length of 1 m, 1% fibre volume and water–cement ratio of 0.85 were tested in flexure. Those tiles with 4 cm long fibres presented 83% more strength than the unreinforced tiles; even when shorter fibres (length 1 cm) were used, the strength was still 46% higher than that of the matrix.

Another study has been carried out by CEPED with small beams 300 mm long and with 50 mm square cross-section. The matrix is cement–sand mortar in a proportion of 1:1 by volume, and a water–cement ratio of 0.43. The fibres, in different volumes and lengths, were added to the mortar. The mixing and casting were performed by hand. The bending test results at the age of 7 days are presented in Table 6.11.

6.3.3 *Henequen-cement*

Studies of henequen fibre reinforced cement are mainly reported by Quiñonez[26] from Guatemala, and Alva Mejía[47] from Peru. Both authors mention corrugated roofing sheets resembling the asbestos-cement ones which are very popular in Latin America. Quiñonez has produced a detailed study of the composites. He has analysed the effect of matrix, fibre length and fibre volume on the composite properties. He tested cement–sand ratios varying from 1:05 to 1:2 in weight; fibre volumes up to 2%; and fibre lengths of 4, 15, and 100 cm.

The developed component is a standard corrugated sheet[48] 90.0 cm wide and 0.61 m or 1.00 m long; the average thickness is 0.8 cm or 1.0 cm.

The specimens were prepared by hand. They consist of a dry mixture of cement, sand and fibre, to which water is added; finally the sheet is cast on a flat surface and put on the mould in the final shape. This process is similar to that developed by CEPED for coir-cement, with a more refined mould (section 6.3.2). The products are kept wet for 14 days. The water permeability and absorption, impact resistance and bending strength were determined.

In a 24-h test, no water under pressure of 0.025 m penetrated the thickness of the sheet. Water absorption varied from 16.46% to 20.83%. The bending test results varied significantly from 15.00 MPa up to 56.16 MPa; these values are also obtained with unreinforced sheets.

Some of the conclusions stated by the author are:

(i) The best proportions of materials for the composite are 1:1 (cement:sand) by weight, and 0.5% by volume of 4 cm long fibres
(ii) Short sheets (0.61 m long) are most easily manufactured
(iii) There is no increase in bending strength due to fibre reinforcement, but the performance of the sheet is quite different
(iv) There is a substantial increase in impact strength of the sheets due to fibre reinforcement (Figure 6.12).

No durability evaluations of these components are available. However, some experimental roofs have already been produced to test the material under natural ageing (Figure 6.13).

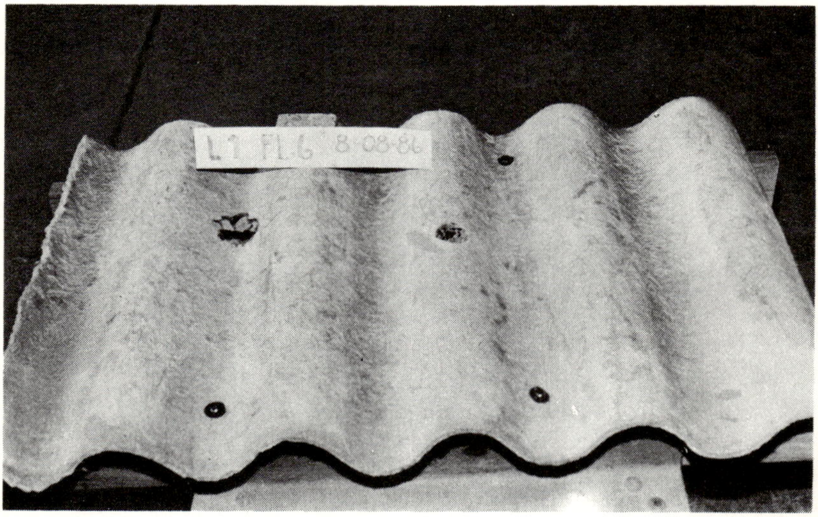

Figure 6.12 Corrugated roofing sheet made with henequen fibre reinforced cement, submitted to impact test (dropping ball) – there are local faults but no cracks.

Figure 6.13 Roof made with corrugated sheets of henequen fibre reinforced cement.

Quiñonez draws attention to the low cost of these components, which, in Guatemala, is 70% less than that of asbestos-cement products and 40% less than that of zinc sheets.

Alva Mejia's experimental work refers to jute fibre, which is discussed later (section 6.3.5.1).

6.3.4 *Sisal-cement*

Agave sisalana fibres offer a very interesting fibre for reinforcement because of its high strength and Young's modulus; nevertheless, its somewhat low durability must be considered, as this can be a restricting factor to its use. All research done by CEPED on coir fibres has been repeated with sisal fibres including the gutter-shape roofing tile[7,14,28,45,46]. The proportions of the materials and the production method are the same as described in section 6.3.2. In these studies, the mechanical behaviour of the roofing sheets reinforced with sisal are very similar to those reinforced with coir fibres. The properties of fibre-cement beams are presented in Table 6.11. Another analysis has been made for this type of fibre[14,28] by CEPED. The effects of casting and compacting procedures have been studied. The results of this evaluation are presented in Table 6.12. This table shows the importance of compacting the material during production.

Despite these results, there are still doubts about the durability performance of these fibres in cementitious matrices. CEPED has therefore encouraged the use of coir-fibre reinforcement. The durability aspects are very worrying. Silva and Campolina[49] also report that the mechanical strengths of different types of specimens (cylinders, prisms and slabs) have not increased significantly with

Table 6.12 Effect of casting and compacting procedures on composites reinforced with sisal fibres (14)

Casting compacting procedures	Modulus of rupture (MPa)
Hand pressed (2 layers)	4.07
Mechanical pressure (MPa)	
1.6	3.84
1.9	4.79
2.1	5.39
2.2	6.17
2.3	5.63
2.5	4.46
3.0	6.02
Vibrating table amplitude (mm)	
1	5.74
2	5.38
3	5.56

Note: Cement–sand ratio 1:1 by volume
Initial water–cement ratio 0.656
Initial fibre volume 2.63%
Fibre length 5.0 cm

age as with the portland cement mortar matrix. The results for 7-day specimens are very similar to those at 28 days.

The appropriate technology approach is also widespread in Latin America; the ideas of the Intermediate Technology Building Materials Workshops, UK, have been tried in several countries such as Colombia, the Dominican Republic, Guatemala and Nicaragua. Sisal-cement roofing sheets using ITBMW methodology or an adapted form[49] have been produced in these countries, mainly through self-help production. The process basically consists of preparing a flat sheet with premixed composite and setting it on the top of a mould with the final shape. This approach to building materials production has usually been promoted by international aid agencies. Lola[50] describes a detailed appraisal of roofing sheet technology as applied to production/dissemination and adaptive research projects independently financed by Appropriate Technology International (USA) and the Save the Children Fund. This appraisal includes roofing sheets produced in the Dominican Republic and Guatemala. In the first country, 19–47% of the sheets analysed had cracks; in the latter, about 20% of the sample sheets showed cracks. In the analysis of these results, which also include data from Colombia and Sri Lanka, Lola[50] presents the main conclusions of the appraisal, which can be summed up as:

(i) The products have a limited useful life of 2–4 years
(ii) The technology packages, even those locally adapted, demand greater levels of production skill, management and quality control than those available at community level

Table 6.13 Characteristics of jute-cement roofing sheet (47)

Characteristics	Corrugated sheet	Arched sheet
Dimensions:		
Length	100 cm	100 cm
Width	85 cm	86 cm
Thickness	1.0 cm	1.2 cm
Weight	32 kg	32 kg
Materials/quantity		
Cement	10 kg	14 kg
Sand	14 kg	14 kg
Chopped jute	0.20 kg	0.25 kg
Water	~ 5l	~ 5l
Costs:		
Materials only	US$ 1.00	US$ 1.03
Total	1.71	—

(iii) The initial cost of the roofing system is reduced, but it can be considered expensive due to its limited service life.

Lola's report causes concern, as his conclusions are based on thousands of sheets produced in the above-mentioned countries. Nevertheless, he also indicates a successful experiment in Asia, probably also with sisal fibres.

6.3.5 *Cement reinforced with fibres other than bamboo, coir, henequen, and sisal*

As mentioned previously, Latin America has a large diversity of types of vegetable fibres. Therefore researchers, in order to use their local resources, have been engaged in developing reinforced materials with several types of vegetable fibres. In this section, these main research works are presented. The depth of these studies differs, and some of the materials studied are already produced on a commercial basis by local industries, whereas some composites, which seem promising, have only been produced on a laboratory scale.

6.3.5.1 *Jute-cement.* In section 6.3.3, the research of Alva Mejía[47] on roofing sheets was reported. This author reported the development of a low-cost building system where the roofing consisted of fibre-cement sheets. In this case, it was possible to use corrugated or arched sheets. The latter are used as a permanent mould for concrete slabs. The composition and characteristics of these products are presented in Table 6.13. Figure 6.14 shows experimental houses covered with the corrugated roofing sheets. The author stresses that the plates can be reinforced with other types of fibres, such as maguey, plantain, henequen or even man-made fibres. However, the experiment has been carried out with chopped jute fibres about 10 mm long.

Figure 6.14 Roof made with corrugated sheets of jute fibre reinforced cement.

These sheets present a good mechanical strength. The corrugated sheet can support a load of 480 kgf and the arched sheet supports a load of 250 kgf. However, if a slab is built on top of the roof, each sheet can support a load greater than 800 kgf. The production process is very similar to that described in section 6.3.4. The product is taken out of the mould 24 h after preparation, and curing takes a further two weeks. Each sheet is tested before use.

This paper also mentions, without presenting experimental test results, that these sheets have better impact strength, similar durability and better workability on site than those of asbestos-cement. They are also cheaper than commercial products.

6.3.5.2 *Piassava-cement.* Table 6.11 shows results from CEPED tests on piassava-cement, which were performed similarly to those mentioned in section 6.3.2[7,45]. It was also attempted to produce concrete beams reinforced with piassava cables of three fibres, but the fibre cable–concrete bond is very poor, and therefore this process is not recommended[7].

6.3.5.3 *Sugar-cane husk and bagasse.* In Cuba[51], the Technical Centre for Construction and Material, with the help of the Cuban Institute for the Research of Sugar Cane By-Products, has developed sugar-cane husk-cement panels, similar to those produced from chipwood bonded with cement. Panels of up to 1.20 m^2 have been produced and found useful for construction purposes. The results of the evaluation of these panels have been very favourable, and it has been suggested that the Institute develops a plant for panel production. Sosa, in several papers[52–54], has proposed the use of pressed sugar-cane bagasse for the production of panels and sheets. This bagasse has 48.5% fibre[52]. This author also states that in Venezuela, one industry is already producing building materials using the bagasse bonded with resin. In Brazil, bagasse is used as a fuel.

6.3.5.4 *Mineralized wood shaving-cement.* In Brazil, mineralized wood shaving-cement panels for walls and partitions are already commercially produced. The actual industrial process is not publicly known. The mineralization process should consist of drying the wood and injecting cement paste into the shavings, which come from different wood species. The panels are usually 5 cm thick, their bulk density is 440 kg/m^3, and thermal conductivity is 0.037 W/mK, according to the manufacturer.

When used for wall purposes, the panels are covered with cement-lime mortar, and when used for partitions, they can be plastered over or simply painted. Nowadays, their main use is as prefabricated housing systems or for ceilings. Panels with 2.5 cm mortar on each face have shown good mechanical strength, waterproofing and thermal comfort performance.

6.3.5.5 *Other experiments.* Several other fibres, not mentioned in section 6.3, are also in use as reinforcement. The research data on such fibres are scarce. Here, results for the most widely used fibres are presented briefly.

Hernández and Odone[55] have studied flax (*Linum usitatissimum* Linn.), hemp (*Cannabis sativa* Linn.) and phormium (*Phormium tenax* Forster) and analysed their durability. They have not found adequate improvement of composites when the fibres are treated with sodium silicate. However, when the fibres are treated with white benzine, sodium hydroxide or ordinary household washing-up liquid, the strengths of the composites increased by about 25%. All of the later treatments refer to a cleaning-up of the fibres, and consequently improving the specific surface area of these fibres, and also probably the fibre–matrix bond. The best results have been achieved with sodium hydroxide.

Castro and Naaman[56] have performed a comprehensive study of composites with lechuguilla (*Agave lophanta* Schiede) and maguey (*A. atrovirens* Karw.) fibres, available in Mexico. They prepared beams of cement mortar of 300 mm long, 75 mm wide and 12.5 mm thick, with a cement ratio of 1:1, water–cement ratios of 0.5 and 0.6, and fibre volume fractions of up to 9%. For the series with water–cement ratio 0.5, they added a superplasticizer, 2.3% by weight of the cement. The fibre length used was between 50 and 75 mm. Elasto-plastic behaviour in flexure and multiple matrix cracking were achieved for fibre volume fractions above 7%. The long-term behaviour of the composites has been analysed but no definitive conclusions were reached.

In Cuba[34,57] rice husk used as reinforcement has been tested. The husk also reduces the weight of the composites, which makes the panels more convenient for building. With husk, cement and sand ratios of 1:2.5:1.4, it was possible to build panels of 50 cm × 240 cm, with a thickness of 10 cm, and a weight of 84 kg. The cement consumption is about 45.6 kg per panel or 380 kg/m^3.

6.4 Final remarks

In the previous section it was pointed out that research on this subject is still in progress in Latin America. In addition, it is necessary to remember that these fibres are also suitable as reinforcement for other matrices. Here new research in progress and other uses for vegetable fibres, as well as a final discussion, are presented.

6.4.1 *New research in progress*

Certainly all of the main current research works on this subject refer to the durability aspects of the fibres in alkaline media. Guimarães, from CEPED-Bahia (Brazil), is trying to improve the durability of coir fibres in cement matrices by coating the fibres with alkaline products or resins. Quiñonez, from the Centro de Investigaciones de Ingenieria (Guatemala), is doing the same to henequen fibres. In this case he is testing water-repellent agents to impregnate the fibres, and also adding natural pozzolana (pumice sand) to the mortar to reduce its alkalinity. The IPT–São Paulo (Brazil) has failed to achieve good results by adding pozzolanic materials to the mortar[58]. In this case tests were performed with two blended binders: cement–rice husk ash and hydrated lime–blastfurnace slag. At present, the IPT is testing ternary binders to find the correct proportions to eliminate free lime from the composite. The fibres being analysed at the IPT are coir, sisal and disintegrated newsprint.

Kenaf and malva fibres, not previously analysed in Latin America, are now under study. Kenaf is being investigated in Guatemala and in Brazil, and malva fibres at the University of São Paulo.

Salas Serrano, in his appraisal of the state-of-art of research on housing in Latin America[59], mentions additional research works on fibres currently in progress. Unfortunately the published results are not known. The themes are: wool–cement development, being pursued at the University of Biobio (Chile); concrete reinforced with Manila hemp and cabuya (*Furcraea cabuya* Trelease, from the same family as the agaves, the Amaryllidaceae), being studied at the Escuela Politecnica Nacional, Quito (Ecuador); and adaptations of ITBMW's process (section 6.3.4), being applied in Nicaragua and in the Dominican Republic.

6.4.2 *Vegetable reinforcement for non-cementitious matrices*

The mechanical behaviour of vegetable fibres produced in Latin America, discussed in section 6.2, makes them a suitable reinforcement for other matrices. Moreover, if the matrix is a non-alkaline one, the composite may profit from the advantage of the high strength of some of the fibres without problems of durability.

Belmares *et al.*[60] have demonstrated the feasibility of vegetable fibre reinforced polyester. Systematic studies have been carried out with materials based on palm fibres. They have used fibres from *Yucca carnerosana* Mckerney, which is common in the authors' region of Coahuila, Mexico[21]. In addition, they assume that their results can be extended to many other natural hard fibres, such as henequen and sisal. In this case, the vegetable fibres are used, at least partially, as a replacement for glass fibres.

In order to avoid the harmful alkaline medium used for processing vegetable fibres, the IPT has been studying fibre reinforced gypsum[8,29,58] based on sisal, coir and disintegrated paper fibres. The performance of the composites has not yet been determined, but as far as is known, they seem to be suitable building materials for indoor purposes.

6.4.3 *Concluding comments*

Twenty types of fibres have been tested in Latin America and at least seven have produced promising results for composite materials for building purposes. The diversity of fibres is due to the great number of available fibres, and also to the attempts of researchers to use materials readily available within their own regions.

This diversity has two main drawbacks:

(i) There are no extensive studies concerning a wide variety of species and providing basic data for each one
(ii) The scarcity of these studies means that there has been no accumulation of data concerning a single species.

As a result, every time a new research project starts on a new type of fibre, or on a fibre from a region not studied previously, the research must start from scratch, first determining all the basic characteristics.

Unfortunately, several research projects stopped short when promising results were becoming available. Some of the interruptions are due to cuts in funding, or to changes in the research policies of institutions. Some research projects have been re-started after long breaks, when the researchers managed to get resources, but this has delayed development.

Lack of exchange of information is another problem. Not only are resources for proper reviews scarce, but there is also no dissemination of the work carried out within the subcontinent. Even when results are published, this takes place mainly in local journals or as internal reports. When presented at local meetings, proceedings are not available to the majority of researchers on this subject, and are not retrieved through international data systems. One of the consequences is the duplication of research with unnecessary wastage of limited resources. For instance, some of the studies mentioned in this chapter have repeated work previously done by another research team, which already had achieved negative results.

It is necessary to stress that the Latin American researchers, in spite of obstacles, have managed to develop very useful and appropriate products for low-cost buildings, and nowadays they are advancing their achievements, trying to improve the materials already developed for better use by the community.

Statistics on building shortages and the tendency towards rapid urbanization in Latin America were presented at the beginning of this chapter. Such figures indicate the large numbers of buildings urgently needed, but contrast with the main approach detected in the research reviewed. Roofing components, instead of panels, are the subject of most of the research work in this field, and represent most of the resulting products. This could be interpreted as a tendency to follow ITBMW's proposals for fibre-cement roofing sheets[49]. Certainly, a shelter is a basic need for human beings; and as, in most cases, the surface area of vertical components is far greater than that represented by the roof, the relative costs of vertical components are correspondingly higher. Another link can be traced between the ideas disseminated by intermediate or appropriate technologies and the research reviewed here: that is, the authors tend to avoid the use of heavy equipment and to produce their materials and components mostly by hand, when possible, through self-help. Nevertheless, if the building, mainly housing, deficit is to be diminished, large-scale production is necessary. As vegetable fibres present such promise in cost reduction while maintaining adequate performance levels, it seems necessary, if not essential, to develop research programs aimed at large-scale production of composites reinforced with the vegetable fibres.

Acknowledgements

The author would like to thank his colleagues, Dr M.A. Cincotto and Dr N.G. Cardia, who undertook the task of proofreading and commenting on this chapter. He would also like to extend his gratitude to all the researchers who answered his request for further information for this text.

References

1. Bocalandro Montoro, M. (1987) Propuesta de proyecto de investigacion/desarrollo—*Tecnologia para viviendas de interes social.* Centro Técnico de la Construcción y los Materiales, Havana, 4–6.
2. Sosa, M. (1987) *Rapport concernant le colloque, 'Building Materials for Low Income Housing'*, Centre Scientifique et Technique du Bâtiment, Paris, 6 pp.
3. Keer, J.G. (1984) Fibre reinforced concrete. *New Reinforced Concretes,* ed. R.N. Swamy, Surrey University Press, [Blackie], Glasgow and London, 52–105.
4. Agopyan, V. (1986) GRG prepared by premixing for wall panels in developing countries. *Proc. 10th Trienn. Congr. CIB,* Washington DC, September 1986, 1197–1203.
5. Aziz, M.A., Paramasivam, P., and Lee, S.L. (1984) Concrete reinforced with natural fibres. *New Reinforced Concretes,* ed. R.N. Swamy, Surrey University Press, [Blackie], Glasgow and London, 106–140.
6. Fernandes, V.C., Silva, M.M.A., and Wolf, A. (1987) *Fibras disponíveis no país.* EPUSP, São Paulo, 75 pp.
7. Centro de Pesquisas e Desenvolvimento (1982) *Utilizacão de fibras vegetais no fibrocimento e no concreto.* Banco Nacional da Habitação, Rio de Janeiro, 120 pp.

8. Agopyan, V., Terzian, P.R., and Derolle, A.P. (1987) Pasta de papel-imprensa reutilizado como reforço de materiais de construcão. *Proc. 1st Int. Symp. Housing Tech. Production and Transfer*, São Paulo, April 1987, 111–119.
9. Ayyar, T.S.R. and Mirihagalla, P.K. (1980) Elephant grass fibres as reinforcement in roofing sheets. *Appropriate Technology* **6,** 8–10.
10. Cook, J.G. (1968) *Handbook of Textile Fibres—1. Natural Fibres.* Merrow, Watford, 208 pp.
11. Cook, D.J. (1980) Concrete and cement composites reinforced with natural fibres. *Proc. Symp. Concrete Int.-Fibrous Concrete*, Lancaster, 1980 99–114.
12. Salgado, A.L.B. (1985) *Plantas fibrosas.* Instituto Agronômico, Campinas-SP.
13. Gram, H.-E. (1983) *Durability of Natural Fibres in Concrete*, Swedish Cement and Concrete Research Institute, Stockholm. 79–80.
14. Guimarães, S.S. (1984) Experimental mixing and moulding with vegetable fibre reinforced cement composites, *Proc. Int. Conf. Low-cost & Energy Saving Constr. Mat.*, Rio de Janeiro, July 1984, 37–51.
15. Ghavami, K., and Hombeeck, R. (1981) Application of bamboo as a construction material. *Proc. Lat. Amer. Symp. Rat. Org. of Building Appl. Low-cost Housing*, São Paulo, October 1981, 49–66.
16. ASTM D 3379-75 (82), Standard test-method for tensile strength and Young's modulus for high modulus single-filament materials. American Society for Testing and Materials, Philadelphia, 1985, 6 pp.
17. Ghavami, K., and Veloso, R.F.R. (1985) *Análise microscópica de bambú e comportamento mecânico de fibras de coco.* PUC-RJ, Rio de Janeiro, 28 pp.
18. Ramaswamy, H.S., Ahuja, B.M., and Krishnamoorthy, S. (1983), Behaviour of concrete reinforced with jute, coir and bamboo fibres. *Int. J. Cement Comp. and Lightweight Concr.* **5**, 3–13.
19. Dantas, F.A.S. (1987) Concretos de baixo consumo de cimento reforcados com fibras—Propriedades e análise de fissuracão devida à retracão. Thesis, EPUSP, 214 pp.
20. ASTM D 3822-82, Standard test method for tensile properties of single textile fibres, American Society for Testing and Materials, Philadelphia, 1985, 7 pp.
21. Medina, J.C. (1959) *Plantas fibrosas da flora mundial.* Instituto Agronômico, Campinas-SP, 913 pp.
22. Instituto de Pesquisas Tecnológicas do Estado de São Paulo (1983) Fichas de caracterizacão de materiais têxteis. *Tópicos Têxteis* **5**, 7–12.
23. Centro de Apoio à Pequena e Média Empresa do Estado de Sergipe (1980) Diagnóstico das empresas de beneficiamento do coco do Estado de Sergipe, CEAG-SE, Aracajú, 115–149.
24. Belém Jr., T.F., and Uemoto, K.L. (1987) *Durabilidade das fibras vegetais no meio alcalino*, EPUSP, São Paulo, 32 pp.
25. Lock, G.W. (1969) *Sisal.* Longman, London, 17–33.
26. Quiñonez, F.J. (1987) *Laminas onduladas de mortero reforzado con fibras de maguey*, Centro de Investigaciones de Ingenieria, Guatemala, 80 pp.
27. Ferri, M.G. (1976) *Plantas produttoras de fibras.* EPU, São Paulo.
28. Centro de Pesquisas e Desenvolvimento (1984) *Utilizacão de fibras vegetais na construcão civil.* CEFED, Camaçari—Bahia, 61 pp.
29. Agopyan V. (1987) *Desenvolvimento de materials de construcão reforços com fibras vegetais—Relatório final.* IPT Secr. Ciência e Techn. do Est. São Paulo, São Paulo, 28 pp.
30. ASTM (1963) *Paper and paperboard—Characteristics, nomenclature and significance of tests.* American Society for Testing and Materials, Philadelphia, 40–41 (ASTM STP n. 60-B).
31. Higham, R.R.A. (1968) *A Handbook of Papermaking.* Business Books, London, 346 pp.
32. FAO (1982) *Pulp and paper capacities: 1981–1986.* Food and Agriculture Organization of the United Nations, s.l.p., 15, 105–124.
33. Hannant, D.J. (1978) *Fibre Cements and Fibre Concretes.* John Wiley, New York, 3.
34. Bruthans, Z., Collazo, R., and Romero, E. (1976) Materiales ligeros de fibras vegetales aglomerados con cemento. Càscara de arroz-cemento. *Boletin Técnico CIEC-Cuba* **2**, 1–11.
35. Pakotiprapha, B., Pama, R.P., and Lee, S.L. (1983) Behaviour of a bamboo fibre-cement paste composite. *J. Ferrocement* **13**, 235–248.
36. Subrahmanyan, B.V. (1984) Bamboo reinforcement for cement matrices. *New Reinforced Concretes*, ed. R.N. Swamy, Surrey University Press, [Blackie], Glasgow and London, 141–194.

37. Hidalgo Lopez, O. (1974) *Bambú: su cultivo y aplicaciones en fabricacion de papel, construcción, arquitetura, ingenieria, artesania*, Estudios Tecnicos Colombianos, Cali, 318 pp.
38. Hidalgo Lopez, O. (1986) Programa de autoconstrucción de vivienda prefabricada de bambú 1983. *Proc. Int. Conf. Mat., Construction Techniques and Constr. Econ. in Developing Countries*, Paris, December 1986, 231–237.
39. Díaz Gutiérrez, A. (1984) Sistema constructivo Quincha prefabricada. *Informes de la Construccion* **36** (361), 25–34.
40. Gallegos Vargas, H. (1986) Use of vegetable fibers as building materials in Peru. *Proc. Symp. Use of Veg. Plants and Fibres as Build. Mat.*, Baghdad, October 1986, A25–A34.
41. Ghavami, K. and Martinesi, R.A. (1987) Utilização de bambú como material em habitação de baixo custo. *Proc. 1st Int. Sym. Housing Tech. Production & Transfer*, São Paulo, April 1987, 181–188.
42. Savastano Jr., H. (1987) Fibras de coco em argamassa de cimento portland para produção de componentes de construção civil, Dissertation, EPUSP, 201 pp.
43. Tezuka, Y., Marques, J.C., Crepaldi, A. and Dantas, F.A.S. (1984) Behaviour of natural fibers reinforced concrete. *Proc. Int. Conf. Low-cost & Energy Saving Constr. Mat.*, Rio de Janeiro, July 1984, 61–69.
44. Ghavami, K., and Hombeeck, R. (1984) Application of coconut husk as low cost construction material, *Proc. Int. Conf. Low-cost & Energy Saving Constr. Mat.* Rio de Janeiro, July 1984, 53–60.
45. Guimarães, S.S. (1987) Fibra vegetal-cimento: resultados de algumas experiências realizadas no THABA-CEPED. *Proc. 1st. Int. Symp. Housing Tech. Production & Transfer*, São Paulo, April 1987, 103–109.
46. Centro de Pesquisas e Desenvolvimento (1985) *Cartilha para fabricação de telhão em argamassa armada*. BNH, Rio de Janeiro, 25 pp.
47. Alva Mejía, A. (1984) Sistema constructivo de tecnologia intermedio como solucion al problema de vivienda en las areas marginales e socioeconomicamente deprimidas de una ciudad. *Informes de la Construcción* **36**(362), 73–80.
48. ISO R 393, Asbestos-cement corrugated sheets for roofing and cladding, International Organization for Standardization, Geneva, 1964, 15 pp.
49. Parry, J.P.M. (1979) Low-cost handmade roof sheets of fibre-reinforced cement. *Appropriate Technology* **5**, 6–7.
50. Lola, C.R. (1986) Fibre reinforced concrete roofing sheets: a technology appraisal. *Proc. 10th Trienn. Congr. CIB*, Washington DC, September 1986, 2131–2139.
51. Rivera Alés, N., Mata de la Cruz, A., and Rojas Garcia, R., (1986) *Use of Vegetable Plants and their Fibres as Building Materials*. Centro Técnico de la Construcción y los Materiales, Havana, 9 pp.
52. Sosa, M., and Chercheur, E. (1984) *Utilisation des végétaux dans l'habitat—à faible coût*. Ecole Nationale des Ponts et Chaussées, Paris, 75–80.
53. Sosa, M. (1985) El vegetal: material de construccion, *IV Jornadas de Investigacion del IDEC*, Caracas, December 1985, 10.
54. Sosa, M., and Trebbi, J.C. (1986) Building with plants. *Proc. Symp. Use of Veg. Plants & Fibres as Build. Mat.*, Baghdad, October 1986, A47–A50.
55. Hernández, A., and Odone, C. (n.d.) *Fabricación de paneleria y techumbre a partir de fibras vegetales*, Universidad de Santiago de Chile, 96 pp.
56. Castro, J., and Naaman, A.E. (1981) Cement mortar reinforced with natural fibres. *ACI Journal* **78**, 69–78.
57. Bruthans, Z., and Romero, E. (1976) Analisis de la possibilidad de fabricar paneles de cascara de arroz-cemento. *Boletim Técnico CIEC-Cuba* **2**, 12–22.
58. Agopyan, V. (1987) *Fibre/agro-industrial by-products bearing walls—1st progress report*. IPT-IDRC, São Paulo, 35 pp.
59. Salas Serrano, J. (1987) *Programa de ciencia tecnologia para el desarrollo—V Centenario (CYTED-D)—Tecnologia para viviendas de interes social*. Comision Nacional Quinto Centenario, Madrid, 565–781.
60. Belmares *et al.* (1981) New composite materials from natural hard fibres, *Ind. Eng. Chem. Prod. Res. Dev.*, **20**, 555–561.

7 Use of natural fibre concrete in India

S.S. REHSI

Abstract

The use of natural fibres as reinforcement in concrete (cement–sand matrix) has been extensively investigated in many countries. An account of the findings with particular reference to the fibres available in India and neighbouring countries is given here. Physicochemical testing of fibres such as those obtained from coconut coir, sisal, jute, bamboo, banana, pineapple leaf, palmyrah, etc., shows that (i) tensile strength, modulus of elasticity and elongation at break range from 9–740 kg/mm^2, 4–510 GN/m^2 and 1.1–40 percent, respectively, and (ii) chemical constituents consist of cellulose, hemicellulose, lignin and pectin in varying amounts. X-ray examination of the fibres shows moderate to high degree of crystallinity. Water-soluble constituents in the fibres cause retardation in setting and strength development of the composite. Among different fibres, only coir fibre is found to resist deterioration when subjected to alternate cycles of wetting in saturated lime solution and decinormal solution of sodium hydroxide and drying. Different measures to render natural fibres resistant to alkali attack are mentioned.

Production and properties of fibrous gypsum plaster boards and coir fibre corrugated roofing sheets are described. Manufacture of such boards and roofing sheets, on cottage industry scale, in developing countries in tropical and subtropical regions, where natural fibres are abundantly available, is suggested.

7.1 Introduction

Fibre reinforced concrete has been the subject of extensive investigations in various countries for many years[1–4]. Different types of mineral, metallic and organic fibres have been used. Among the mineral fibres the use of asbestos in the production of asbestos cement products is well known. Water absorption of asbestos fibre is high; it is also a health hazard. Alkali-resistant glass fibre is considered a good reinforcement[5–7]. With metallic fibres, the use of steel fibres in cement mortar and concrete has received wide attention[8–12].

Organic fibres consist of natural as well as man-made fibres, as shown in Figure 7.1. Synthetic fibres, such as nylon, polypropylene and polyethylene,

Table 7.1 Physical properties of some natural fibres[16–19]

Fibre	Colour	Texture	Specific gravity	Water absorption (24 h), %	Tensile strength (kg/mm^2)	Modulus of elasticity (GN/m^2)	Elongation at break (%)	Electrical resistivity ($\times 10^5\ \Omega$ cm)
Coir	Golden brown	Smooth and tough cylindrical and twisted fibre	1.31	80	9–14	4–6	15–40	9–14
Sisal	Dirty white	Long, rough and twisted fibre	1.35	150	100–200	34–62	3–7	4.7–5.0
Jute	Light yellow	Soft and resilient fibre	1.36	250	400–500	17.4	1.1	—
Bamboo	White to light yellow	Smooth to rough fibre	1.50	—	—	—	—	—
Banana	Light yellow to light brown	Smooth to rough, long cylindrical fibre	1.30	400	110–130	200–510	1.8–3.5	6.5–7.0
Aak	Dirty white	Soft, cotton-like	1.17	350	74–87	—	—	—
Bhabar	Dirty green	Rough and brittle fibre	1.30	185	5–70	—	—	—
Castor	Slaty white	Short crumbled soft fibre	1.01	235	25–40	—	—	—
Dhakala	Yellowish white	Bundle of soft fibres	1.06	120	40–85	—	—	—
Hemp	Light yellow	Soft and resilient fibre	1.36	140	40–200	—	—	—
Munja	Creamy yellow	Hard and brittle fibre	1.29	160	20–75	—	—	—
Shinio	Greyish white	Short, rough crumbled and twisted fibre	0.94	230	30–55	—	—	—
Khimp	Dirty yellow	Short bundle like cotton lint	1.33	450	60–65	—	—	—
Pineapple leaf	Creamy white	Soft with pointed ends	1.44	—	360–740	24.3–35.1	2.0–2.8	—
Palmyrah	Yellowish	Stiff, thick, strong	—	—	180–215	4.4–6.1	2.0–2.8	—

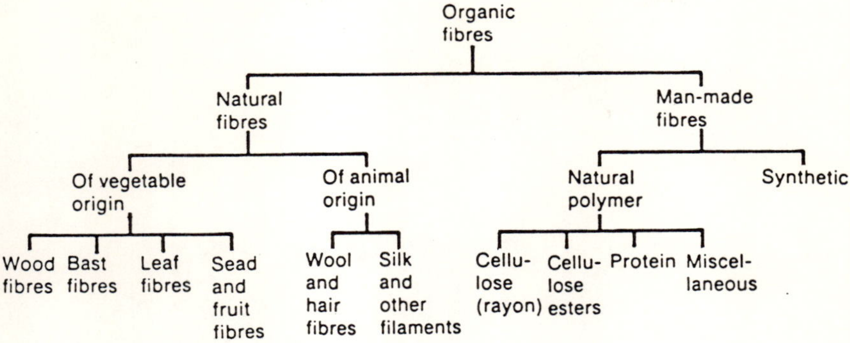

Figure 7.1 Fibre classification chart[13].

are highly resistant to the alkaline environment in concrete. But, due to their low elastic, modulus, the incorporation of these fibres in the matrix does not generally increase strength. In fact, the strength of the composite is often reduced[14]. However, concrete containing nylon or polypropylené fibres is reported to possess 10 to 25 times higher impact resistance[15].

Natural fibres such as those obtained from coconut coir, sisal, jute, sugarcane bagasse, banana, wood, palms, flax, akwara, elephant grass, etc., have been examined in various countries, including India, for their structure, physical properties and chemical composition[16–20]. The physical properties and chemical composition of some Indian natural fibres are listed in Tables 7.1 and 7.2. Stress–strain diagrams for retted coir fibre, sisal fibre, banana fibre, pineapple fibre and palmyrah fibre are shown in Figures 7.2–7.6. Moisture absorption characteristics of jute, coir and bamboo fibres are shown in Figure 7.7. X-ray analysis[16] has shown that:

(i) Coir fibre has a high degree of crystallinity, the spiral angle of the microfibrils ranging between 30 and 40°. This imparts to coir fibre greater extensibility compared to other natural fibres.

Table 7.2 Chemical composition of some natural fibres[16–18]

Fibre	Ash (%)	Cellulose (%)	Hemicellulose (%)	Lignin (%)	Pectin (%)
Coir	1.44	32–43	0.15–0.25	40–45	2.75–4.0
Sisal	1.13	66–72	12.0	10–14	—
Jute	7.90	63.0	—	—	—
Banana	1.02	63–64	19.0	5.0	—
Pineapple leaf	1.10	81.5–85	—	12.7	—
Palmyrah	—	40–52	—	42–43	—

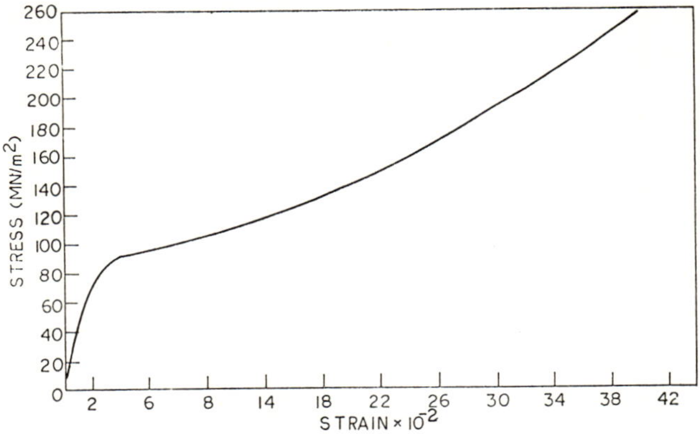

Figure 7.2 Stress–strain diagram for retted coir fibre[16].

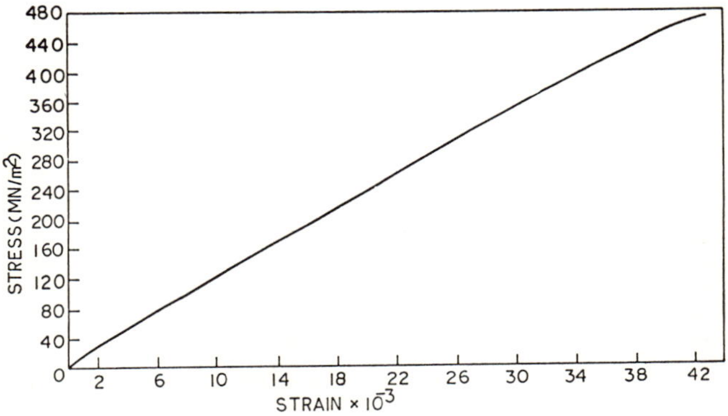

Figure 7.3 Stress–strain diagram of banana fibre[16].

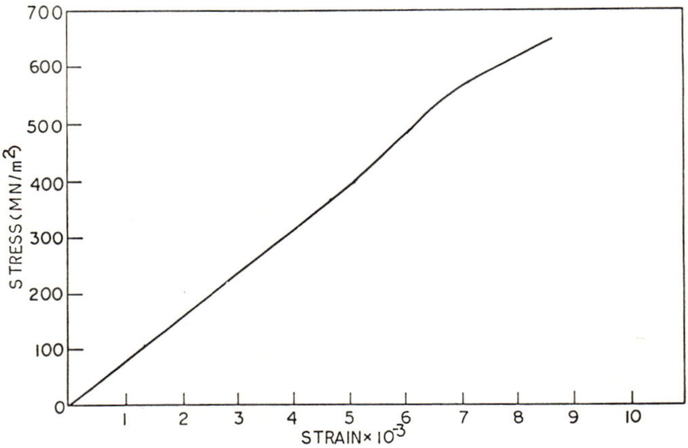

Figure 7.4 Stress–strain diagram of sisal fibre[16].

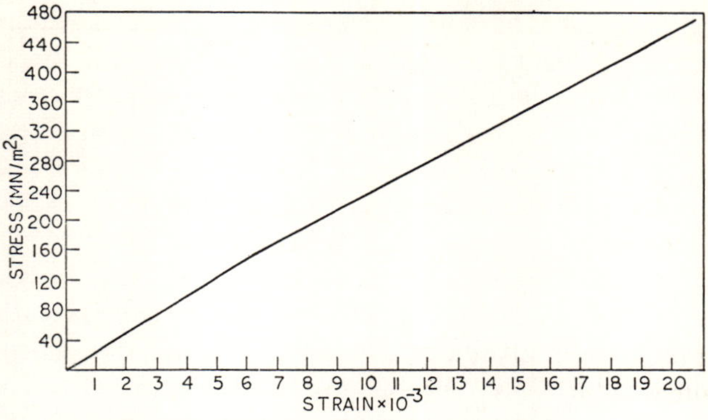

Figure 7.5 Stress–strain diagram of pineapple fibre[16].

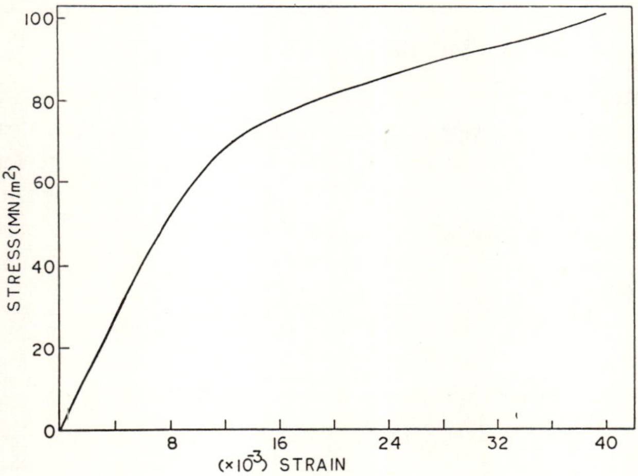

Figure 7.6 Stress–strain diagram of palmyrah fibre[16].

(ii) Sisal fibre is moderately crystalline, corresponding to cellulose in spiral orientation, with a disordered alignment of crystallites. The spiral angle is 20–25° around the fibre axis.

(iii) Banana fibre has a high degree of crystallinity with good alignment of crystallites parallel to the fibre axis. The fibre has a spiral angle of 30°.

(iv) Pineapple leaf fibre has a high degree of crystallinity along with the orientation of spiral angle of 40° with the fibre axis.

(v) Palmyrah fibre has crystallites oriented in a somewhat disorderly fashion at an angle to the fibre axis.

As they are available in most developing countries and require not much skill, plant or equipment for processing, the use of natural fibres to produce low-cost alternative building materials is an attractive proposition. Research and development work carried out in India in this direction is described here.

The term 'concrete' used here actually refers to a cement–sand matrix. Normally, ordinary portland cement is used. But considering the deterioration of the natural fibres in the highly alkaline environment (pH 12.5 or more) in the matrix prepared with ordinary portland cement, the use of blended cements, namely portland pozzolana cement, portland slag cement and supersulphated cement having lower alkalinity compared to ordinary cement, is considered more suitable. The type of sand used in normal concrete is also appropriate for use in fibre reinforced concrete. Fibres in the form of chopped short lengths or in continuous lengths can be used. The use of fibres pre-twisted to twine has an adverse effect on the strength and other properties of fibre reinforced concrete, and should be avoided.

7.2 Properties of fresh fibre reinforced concrete

The fresh concrete matrix shows a certain amount of stiffening and reduction in workability when chopped fibres are added to it. Additional water is thus required to maintain the same workability. The fibres may interlock and form balls: this is particularly true if too long a fibre is used. Fibres having aspect ratios (length–diameter ratios) ranging from 75 to 250 do not cause any

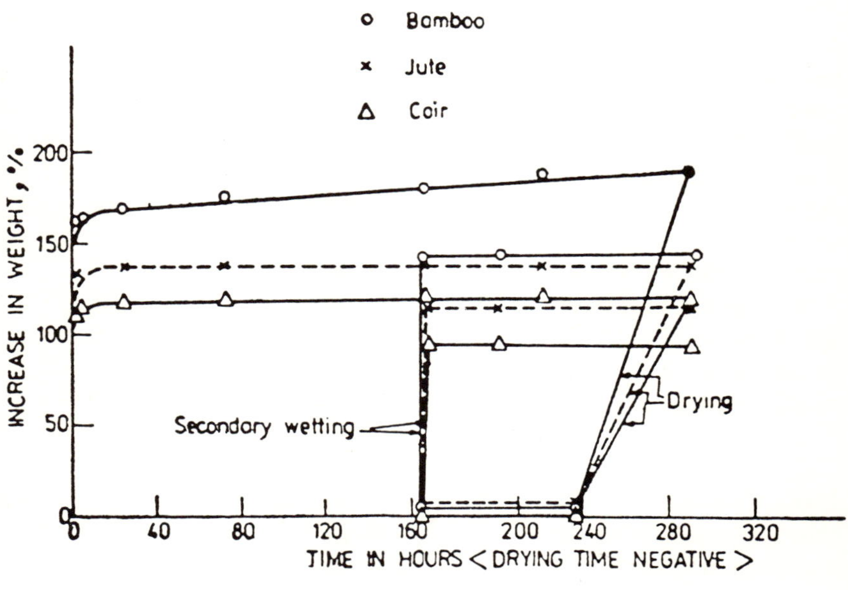

Figure 7.7 Moisture absorption of fibres.

problem of balling. Unlike metallic fibres, the length of the chopped natural fibres added does not influence the additional water requirement to any significant extent. This can be attributed to the ability of the natural fibres to bend easily during mixing. Long continuous fibres are usually sandwiched between layers of the matrix laid in the casting mould, and do not as such play any role in influencing the properties of the fresh matrix.

Water-soluble constituents in the fibres cause retardation in the setting and hydration of the cement. Consequently, the strength of the fibre reinforced concrete is likely to be adversely affected.

7.3 Properties of hardened fibre reinforced concrete

In hardened fibre reinforced concrete, the efficacy of the reinforcement in the matrix and the transfer of stress between the matrix and the fibre depend on relative fibre–matrix stiffness, fibre–matrix interfacial bond and strain compatibility between the fibre and the matrix. For reinforcement to be effective in the precracking stage of the composites, the elastic modulus of the fibre has to be much higher than that of the matrix. This is evident from the law of mixtures. There is a minimum modular ratio (E_{fibre}/E_{matrix}) below which improvement in the mechanical properties of the composites, particularly in the precracking stage, cannot be achieved. While natural fibres, due to their low modulus and high elongation characteristics, do not contribute towards strength improvement, they are capable of absorbing large strain energy and thus impart a greater degree of toughness and resistance to impact and explosion as compared to high-modulus rigid fibres.[21]

The strength of interfacial bond between the matrix and the fibre determines the effectiveness of stress transfer from the matrix to the fibres when the matrix tends to crack. A strong bond is essential for improved tensile strength of the composite. It is a common observation that, at the point of failure in a fibre reinforced concrete, fibres invariably are pulled out but do not fail in tension. This shows that it is the interfacial bond strength which is the limiting factor rather than the high tensile strength of the fibre. In the post-cracking stage, natural fibres with relatively low tensile strength perform equally well provided that the interfacial bond is adequate.

With natural fibres such as coir, the fibre–matrix bond is mostly mechanical and the bond strength depends upon the keying action between the matrix and the individual filaments in the fibre. The length of the fibre has to be sufficient in order to mobilize the interfacial bond force.

7.3.1 *Mechanical properties*

Table 7.3 shows some of the mechanical properties of concrete reinforced with coir, jute and bamboo fibres[21]. It was observed that while the unreinforced specimens broke suddenly into two, the fibre reinforced specimens remained in one piece even when the maximum load was applied. The fibre reinforced composite thus behaves like a ductile material. Shrinkage of the fibre reinforced concrete was found to be about half of that of the unreinforced concrete. Creep of the concrete with and without fibre reinforcement was about the same. The impact resistance of concrete was found to increase when reinforced with coir and jute fibres[22,23].

7.3.2 *Alkali resistance*

Vegetable fibres are known to possess poor resistance to alkalis. They get damaged in the alkaline environment of the cement matrix. The deterioration of the fibres is considered to be due to the decomposition of cellulose, hemicellulose and lignin by reaction with alkalis in the pore water in the concrete matrix[13,17,24]. Data given in Table 7.4 show that, with the exception

Table 7.3 Mechanical properties of concrete reinforced with coir, jute and bamboo fibres[21]

Fibre	Aspect ratio	Fibre content (vol. %)	Strength (% of control) Compression	Tension	Impact	Failure strain of the composite (%)
Coir	75	1.0	0.7–1.1	0.9	1.20	1.5
Jute	250	1.0	0.7–0.9	0.9	1.21	1.5
Bamboo	200	1.0	0.7–0.95	0.9	1.21	1.0

Table 7.4 Tensile strength of fibres subjected to alternate wetting and drying cycles in saturated lime solution and decinormal solution of sodium hydroxide[17]

Fibre	Tensile strength (kg/mm^2)	Tensile strength after 60 cycles of wetting and drying: In saturated lime solution	In 0.1 N sodium hydroxide solution
Coir	9–14	7–20	10–17
Jute	400–450	broken to pieces	42–110
Sisal	100–200	10–20	23–82
Banana	110–130	60–120	21–46
Bhabar	5–70	5–20	8–65
Hemp	40–200	12–28	30–55
Munja	20–150	15–40	22–58

of coir fibre, other vegetable fibres, namely banana, bhabar, hemp, jute, munja and sisal, deteriorate and lose strength when subjected to 60 cycles of alternate wetting in saturated lime solution and 0.1 N solution of sodium hydroxide for 24 h, washing with water and drying at a temperature of 20–23°C. It is seen that the damage in lime solution is more severe than in sodium hydroxide solution.

Some of the measures suggested to render vegetable fibres resistant to alkali attack are[13]:

(i) Reducing alkalinity in the concrete matrix by replacing up to 40–50% of the portland cement with ultra-fine reactive silica fume (replacement of up to 67% portland cement with fly ash, in place of silica fume, is also found effective)
(ii) Using portland slag cement in place of ordinary portland cement
(iii) Carbonation
(iv) Sandwiching fibres between layers of the concrete matrix
(v) Sealing the pores in the concrete matrix by admixing small beads of wax in the fresh concrete; when the concrete has hardened and dried, the composite is heated so that the wax melts and flows out into the pores
(vi) Impregnating the fibres with both blocking and water-repellent agents.

7.4 Production of fibre reinforced building boards and roofing sheets

Extensive work has been carried out in India on the use of bamboo as reinforcement in concrete[25–35], and coconut husk and coconut pith in the production of building boards and insulating concrete[36–40]. Production of building boards and corrugated roofing sheets using sisal and coir fibres, and their properties, are described below.

7.4.1 *Fibre gypsum plaster boards*

Gypsum plaster obtained from mineral gypsum or by-product phosphogypsum is used to produce building boards of size 120 × 60 × 1.2 cm[41,42].

For casting the boards, a thin slurry of gypsum plaster containing about 80% water is poured and evenly spread up to 6 mm thickness in a steel mould placed over a concrete casting platform. Treated sisal fibre, about 250 g per square metre, is then spread over the surface. The fibre layer should extend 5 cm beyond the edge of the mould. The overhanging part of the fibre is turned back into the sheet to give extra reinforcement. The remaining portion of the slurry is then poured, levelled and allowed to set. The set board is demoulded after about 2 h and dried on wooden racks in air. Other fibres, such as coir, bamboo and bhabar, can be used in place of sisal fibres.

The fibre gypsum plaster board is light and possesses good thermal insulating and fire-resisting properties. At a bulk density of 1150 kg/m^3, its thermal conductivity was found to be 0.16 kcal/m/h/°C as compared to 0.25 kcal/m/h/°C for 6 mm thick asbestos cement sheet of density 1700–1800 kg/m^3. Data on its fire-resistance are shown in Table 7.5. It can be seen that the grading period of fire resistance of 12 mm thick fibre gypsum plaster board is 30 minutes and is far superior to that of other commercial building boards.

The fibre gypsum plaster boards are usable for false ceilings, lightweight partitions and as a dry surfacing material for walls. The boards should not be used in persistently damp conditions, as their strength and structural properties are adversely affected.

Table 7.5 Fire resistance of fibrous phosphogypsum plasterboard[42]

Time (minutes)	Furnace temp. (°C) Standard ($T-T_0$)	Average ($T-T_0$)	Temperature of unexposed surface (°C)	Remarks
0	0	0	20	1. Appearance of smoke from side joints after 5 minutes
5	556	555	60	
10	659	670	90	2. Hair cracks on outer face after 13 minutes
15	718	720	90	
20	760	770	120	3. Cracks became wider after 29.5 minutes
25	780	790	160	
30	—	800	180	4. Failed

Manufacture of fibre gypsum plaster boards can be taken up on a cottage industry scale in developing countries.

7.4.2 *Corrugated roofing sheets*

For producing corrugated roofing sheets, coir fibre is soaked in mineralized water for 1–2 h[38]. The free water is drained off the fibre, which is then mixed with dry portland cement. A mat of suitable thickness is formed on a corrugated mould and pressed. It is held under pressure for 4–8 h. The sheet after demoulding is cured under shade and dried. A coat of waterproofing paint, based on cashew-nut shell liquid, is applied on one side of the sheet. Sheets of 2 × 1 m size and of 6–8 mm thickness have been produced. Their physical properties are listed in Table 7.6.

Experimental structures built with corrugated roofing sheets have shown no

Table 7.6 Physical properties of corrugated coir roofing sheet[38]

Properties	Coir roofing sheet	Asbestos sheet
1. Pitch of corrugation (mm)	146	146
2. Depth of corrugation (mm)	48	48
3. Length of the sheet (m)	1.5 to 2.0	1.5 to 3.0
4. Breadth of the sheet (m)	1.0	1.05
5. Weight (kg/m^2)	11.5 to 12	13.5
6. Breaking load at a span of 60 cm (kg/cm)	5.0	—
7. Breaking load at a span of 100 cm (kg/cm)	1.9	5.0 (minimum)
8. Thermal conductivity (kcal/cm/m^2)	0.09	0.24
9. Water permeability through finished surface (24 h)	almost nil	—
10. Acid resistance as per IS: 5913–1970 (g/m^2)	930	926

leakage of rain water or any visible sign of deterioration over the past 3 years. The expected life of these sheets is about 8–10 years. These sheets are suitable for use, in place of asbestos cement sheets, in low-cost housing projects in developing countries.

7.5 Concluding remarks

Natural fibres are at present used mainly to produce mats, mattresses, carpets, ropes, bags for packing cement, food grains, sugar, fertilizer, etc., cushions and cover seats for automobiles, railway carriages and house fixtures, hard brushes, fancy articles like purses, handbags, artificial hair and wigs, and threads and fabrics for decoration purposes. These fibres, which are abundantly available in most tropical and subtropical countries, can also be used to produce low-cost alternative building materials such as building boards; composite panels consisting of an inner core of rice straw, rice husk, etc., and an outer covering of fibre reinforced concrete; and corrugated roofing sheets for use in housing projects for the masses. The production technology is simple and does not require heavy capital investment on plant or equipment. It can be adopted on a cottage industry scale utilizing local skill and labour in developing countries.

References

1. *Proc. Conf., New Materials in Concrete Construction*, ed. S.P. Shah, University of Illinois at Chicago Circle, Chicago, 15–17 December 1972.
2. *Proc. Int. Symp., Fibre Reinforced Concrete*, ACI Publication SP-44, American Concrete Institute, Detroit, 1974.
3. *Proc. RILEM Symp.* 1975, *Fibre Reinforced Cement and Concrete*, 14–17 September 1975, ed. A. Neville, Construction Press, Lancaster.
4. ACI Committee 544 (1973) State-of-the-Art Report on Fibre Reinforced Concrete, *J. Amer. Concr. Inst.* **70**(11) 729–743.
5. Majumdar, A.J. and Ryder, J.F. (1968) Glass fibre reinforcement of cement products. *Glass Technology* **9**(3) 78–84.
6. Steele, B.R. (1972) Glass fibre reinforced cement. *Proc. Int. Building Conf.—Prospects for Fibre Reinforced Construction Materials*, London, November 1972, 29–39.
7. Hills, D.L. (1975) Premixed glass fibre reinforced cement. *Precast Concrete* **6**(5) 251–254.
8. Edgington, J., Hannant, D.J. and Williams, R.I.T. (1974) Steel fibre reinforced concrete. *BRE Current Paper* **69/74**, July, Building Research Establishment, 1–17.
9. Swamy, R.N. and Mangat, P.S. (1974) Influence of fibre geometry on the properties of steel fibre reinforced concrete. *Cement Concr. Res.* **4**(3) 451–465.
10. Mangat, P.S. and Swamy, R.N. (1974) Compactibility of steel fibre reinforced concrete. *Concrete* **8**(5) 34–35.
11. Swamy, R.N. and Lankard, D.R. (1974) Some practical applications of steel fibre reinforced concrete. Part I, Design and Construction. *Proc. Inst. Civil Engrs*, **56**, 235–256.
12. Gunasekaran, M. (1975) The strength and behaviour of lightweight concrete reinforced with metallic fibres of mixed aspect ratio. *Ind. Concr. J.* **49**(2) 48–49, 55.
13. Gram, H.-E., Persson, H. and Skarendahl, A. (1984) Natural fibre concrete. *SAREC Report R 2*, Swedish Agency for Research Cooperation with Developing Countries, 1-139.
14. Monfore, G.E. (1968) A review of fibre reinforcement of portland cement paste, mortar and concrete. *J. PCA Res. Devel. Lab.* **10**(3) 36–42.

15. Williamson, G.R. (1966) Response of fibrous reinforced concrete to explosive loading. *Technical Report* No. 2-48, Dept. of the Army, Ohio River Division Laboratories, Corps. of Engineers, Cincinnati.
16. Satyanaryana, K.G., Kulkarni, A.G. and Rohtagi, P.K. (1981) Natural fibres as valuable resource for materials in the future of Kerala. *Materials Science and Technology in the Future of Kerala*, eds. P.K. Rohatgi, K.G. Satyanaryana and S. Mohan, RRL Trivandrum. 127–178.
17. Singh, S.M. (1985) Alkali resistance of some vegetable fibres and their adhesion with portland cement *Research and Ind.* **30**(2) 121–126.
18. Satyanarayana, K.G., Sukumaran, K., Ravikumar, K.K., Bramhakumar, M., Pillai, S.K.G., Pavithram, C., Mukherjee, P.S. and Pai, B.C. (1984) Possibility of using natural fibre composites as building materials. *Proc. Int. Conf. on Low Cost Housing for Developing Countries*, CBRI, Roorkee, 12–17 November, 177–181.
19. Chand, N., Verma, K.K., Saxena, M., Khazanchi, A.C. and Rohatgi, P.K. (1984) Materials science of plant based materials of Madhya Pradesh. *Ibid.*, 191–201.
20. Aziz, M.A., Paramasivam, P. and Lee, S.L. (1984) Concrete reinforced with natural fibres. *New Reinforced Concretes*, ed. R.N. Swamy, Surrey University Press [Blackie], Glasgow and London, 106–140
21. Krishnamoorthy, S. and Ramaswamy, H.S. (1982) Fibre reinforced concrete with organic fibres. *Proc. Nat. Seminar on Building Materials – Their Science and Technology*, New Delhi, 15–16 April, vol. II(A) 15, 1–5.
22. Sridhara, S., Kumar, S. and Minare, M.A. (1971) Fibre reinforced concrete. *Ind. Concr. J.* **45**(10) 428–430, 442.
23. Siraskar, K.A. and Kumar, S. (1972) Fibre reinforcement for shatter proofness of concrete. *Proc. Symp. Modern Trends in Civil Engineering*, University of Roorkee, Roorkee, 254–259.
24. Velpari, V., Ramachandran, B.E., Bhaskaran, T.A., Pai, B.C. and Balasubramanian, N. (1980) Alkali resistance of fibres in cement. *J. Mater. Sci. Lett.* **15**, 1579–1584.
25. Bose, T.N. (1950) Use of bamboo as reinforcement in cement concrete *J. Assoc. Engs (India)*, **26**, April–June, 52–62.
26. Karamchandani, K.P. (1959) Role of bamboo as a constructional material. *Proc. Symp. on Timber and Allied Products*, NBO, New Delhi, 18–22 May, 430–435.
27. Narayana, S.K. and Rahman, P.M.A. (1962) Bamboo-concrete composite construction. *J. Instn. Engrs (India)* **42**, May, 426–440.
28. Gupchup, V.N., Jayaram, S. and Sukhadwalla, J.N. (1974) Suitability of bamboo strips as tensile reinforcement in concrete. Impact of Research on the Built Environment. Theme 11/3: The Impact of Research On Design – *New Techniques in Concrete and Reinforced Concrete Proc. CIB 6th Congr.*, Budapest, 3–10 October, 464–470.
29. Kalita, U.C., Khazanchi, A.C. and Thyagarajan, G., (1977) Bamboocrete low-cost houses for the masses, *Ind. Concr. J.* **51**(10) p. 309–319.
30. Kurian, N.P. and Kalam, A.K.A. (1977) Bamboo-reinforced soil-cement for rural use. *Ind. Concr. J.* **51**(12) 382–389.
31. Indian Institute of Technology, Madras (1978) Bamboo reinforced soil-cement. *Appropriate Technology*, **5**(1), 29.
32. Datye, K.R., Nagaraju, S.S. and Pandit, C.M. (1978) Engineering applications of bamboo. *Proc. Int. Conf. on Materials of Construction for Developing Countries*, Bangkok, August 1978, Vol. 1, 3–20.
33. Kalita, U.C., Khazanchi, A.C. and Thyagarajan, G. (1978) Bamboocrete wall panels and roofing elements for low cost housing. *Ibid.*, 21–25.
34. Soundarajan, P. (1979) A novel technique using bamboocrete as roofing material. *Appropriate Technology*, **6**(2) 20.
35. De. P.L., Gupta, T.N., Mangal, R.C., Mohan, D., Rai, M., Sharma, J.S. and Verma, N. (1980) Appropriate technologies and materials for housing and building in India. *UNIDO Monographs on Appropriate Industrial Technology* **12,** Appropriate Industrial Technology for Construction and Building Materials, United Nations, New York, 129–159.
36. George, J. and Shirsalkar, M.M. (1963) Particle boards from coconut husk. *Research and Ind.* **8**(5) 129–131.
37. George, J. (1970) Building materials from coconut husk and its byproducts. *Coir* **14**(2) 19–44.
38. Singh, S.M. (1978) Coconut husk—a versatile building material. *Proc. Int. Conf. on Materials of Construction for Developing Countries*, Bangkok, August 1978, 207–219.

39 Shirsalkar, M.M., Jain, R.K. and George, J. (1964) Fire resistant building boards from coconut pith. *Research and Ind.*, **9**(12) 359–360.

40. Jain, R.K. and George, J. (1970) Utilisation of coconut pith for thermal insulating concrete *Coir* **14**(3) 25–31.

41. Taneja, C.A. (1971) Fibrous gypsum plaster boards. *Building Materials Note* No. 11, CBRI, Roorkee, November, 12 pp.

42. Singh, M., Rehsi, S.S. and Rai, M. (1981) *Evaluation of phosphogypsum for the manufacture of gypsum plaster and plaster board*. CBRI, Roorkee, July, 1–60.

8 Natural fibre concrete roofing

HANS-ERIK GRAM

Abstract

Roofing material of natural fibre concrete is arousing considerable interest in many countries since it can be manufactured from domestic raw materials at a comparatively low cost.

The natural fibres increase the cohesion of the fresh cement mortar and this makes it easy to shape the roofing product. When the concrete has hardened, the fibres act as reinforcement, although they do not prevent the formation of cracks but merely help to limit the width of the cracks so that the fracture toughness increases. Many of the development problems of natural fibre concrete roofing have been identified and solved during the last decade. Industrial production of roofing sheets has been started on a small scale in Tanzania, and production of tiles at a village level has begun in about 30 countries all over the world. The production of tiles is increasing rapidly at present. This chapter is devoted to discussing the properties of natural fibre concrete, production methods, various products, loads to which we can expect the products to be subjected during their service life, and product cost. A worldwide inventory of current production is also presented. This indicates that sheets are more subject to cracking than are tiles. The load on sheets is greater, and since the natural fibres decompose with time, these roofing products must be regarded as unreinforced. It is therefore recommended that measures be taken to improve the durability of natural fibres when manufacturing sheets or large roofing products.

8.1 Introduction

The need for adequate roofing in developing countries is a vital problem for the great majority of the population. The solution of this problem is often linked to the import of iron sheets or asbestos fibres for asbestos-cement sheets. The increasing economic burden that many less developed countries have to carry makes it harder to meet vital needs, such as roofing, by means of imports. The efforts to find appropriate solutions based on locally available raw materials have thus become more and more important.

Cement mortar containing natural fibres, such as sisal, coir, or jute, can be formed into thin sheets or pantiles for roofing. This chapter is devoted to a discussion of the properties of natural fibre concrete, different production methods for natural fibre concrete roofing and experience gained from the production and use of these roofing products.

8.2 Historical background

In this century the most important example of a fibre cement composite has been asbestos-cement. Asbestos-cement contains about 10% by volume of asbestos fibres and the material has outstanding mechanical properties. Some of the asbestos fibre has, in certain cases, been replaced with cellulose fibre (pulp) for economic reasons or due to a lack of asbestos fibre. During the 1950s increased interest was shown in the use of other fibres for reinforcing the cement matrix. Steel fibres, alkali-resistant glass fibres and plastic fibres were produced as alternative reinforcement materials. Thin products of sprayed glass-fibre cement (gfc) could be provided with mechanical properties which were almost as good as those of asbestos-cement.

Towards the end of the 1960s, studies began on natural fibres as reinforcement in cement mortar. The properties of the material were investigated in a number of countries[1], and reports of positive properties began to be published during the first half of the 1970s[2–4]. Natural fibre concrete can constitute an economically interesting alternative to asbestos-cement in developing countries, since it is based on domestically available raw materials. In the mid-1970s, various proposals for roofing sheets of natural fibre concrete, usually with the corrugated asbestos-cement sheet as model, were presented[5–7]. Towards the end of the 1970s and at the beginning of the 1980s, appropriate methods for manufacturing roof sheets of natural fibre concrete or mortar were developed, and they rapidly spread to countries in Central America, Africa and Asia[1].

8.3 Properties of natural fibre concrete

8.3.1 *Properties in the fresh state*

The properties of fresh fibre concrete depend on the properties of the matrix, the type of fibre added, the fibre length and the fibre volume. The workability of a mix is changed when fibres are added. Figure 8.1 shows that more water must be added to the mix for increasing amounts of sisal fibres if the same workability (slump) is to be maintained.

As can be seen from Figure 8.1, there is no significant effect from the fibre length itself. The reason for this is probably that the sisal fibres are flexible and that long fibres curl easily when mixed, not causing any stiffening.

The other important aspect is the 'balling' of fibres. The extent to which balling may occur in a given mix is determined by the type, length and thickness of fibres used and by the fibre volume fraction. The thinner a fibre,

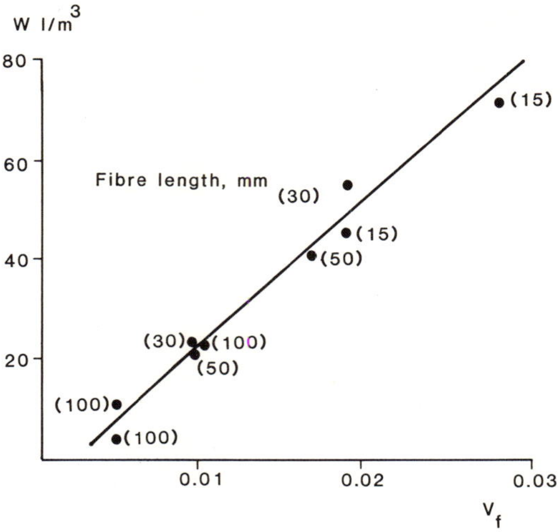

Figure 8.1 Extra water needed to maintain constant workability as a function of fibre volume (V_f). Sisal fibres of different lengths were used[8].

the shorter it must be to avoid balling. Balling should not be allowed to occur, as it has a detrimental effect on strength. Certain mixing methods can be employed to minimize the balling effect. Normally, the progressive addition of fibres at the end of the mixing process, after the other ingredients have been mixed, reduces the balling effect.

An example of how the tendency to ball is related to the volume of fibres and the fibre length is presented in Figure 8.2. The tendency to ball also depends, to a certain extent, on the type of mixer used.

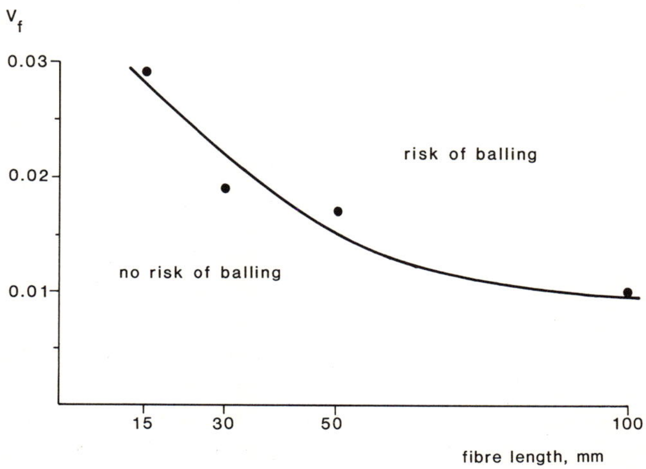

Figure 8.2 Evaluation, from tests with sisal fibres, of the tendency to ball, as visually examined. Limit values for mixes with a tendency to ball are indicated[8].

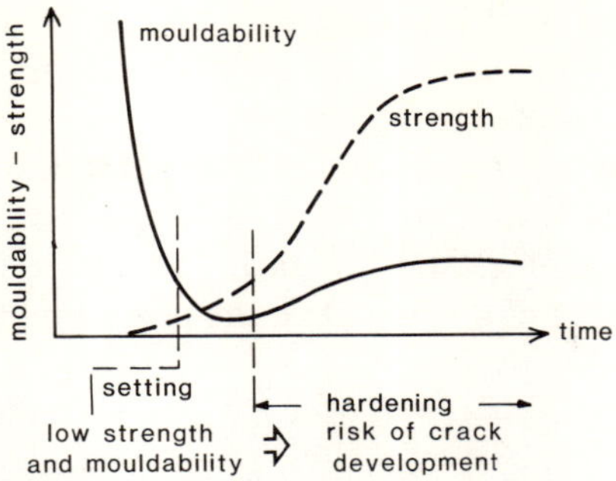

Figure 8.3 Relationships between mouldability and age, between strength and age[9].

The addition of fibres enhances the cohesion of the fresh concrete and makes it possible to mould a product in a simple manner.

8.3.2 *Properties during the change from fresh to hardened state*
As the fresh mix gradually changes over to the hardened state, the workability decreases and the strength of the matrix increases. Cracks can develop very easily during this phase (see Figure 8.3).

Cracks can be initiated by external or internal forces. One example of an internal force is plastic shrinkage caused by rapid evaporation. It is the same process that can occur when mud dries. It has been known for thousands of years that this cracking can be limited or completely avoided by reinforcing the material, for instance sun-dried bricks with straw. The incorporation of fibres in concrete can have a similar effect.

An example of an external force is the transfer of a thin fibre mortar mix, cast on a plastic sheet, to a corrugated mould. In this case, the fibre helps to hold together the fresh mortar. The time to setting is determined by a number of factors – evaporation from the mortar; cement quality and quantity; quantity of water; temperature, admixtures, porosity of the aggregate and, in certain cases, the cleanliness of the fibres. Natural fibres can contain substances that retard cement hydration. This extends setting time and has a negative effect on strength growth.

8.3.3 *Properties in the hardened state*

8.3.3.1 *General.* Properties in the hardened state are determined by the material composition selected. The strength of the material is determined by, above all, the proportion of cement to water and the percentage by volume of fibres. The properties are also affected by the manner in which the fibre concrete has been compacted during the casting, its curing and its age.

The main advantage of fibre reinforcement of concrete is the improvement of fracture toughness and of impact strength.

8.3.3.2 *Flexural strength.* When the percentage by volume of fibres used is below 5%, the behaviour of the natural fibre concrete in bending can be characterized as shown in Figure 8.4.

Of importance among the factors affecting the flexural strength are the modulus of elasticity, flexural and tensile strengths and deformability of the matrix; the modulus of elasticity, tensile strength, elongation at break, dimensions and volume of the fibre; and the phase boundaries governing the fibre–matrix bond.

As shown in Figure 8.4, the composite is elastic until the first crack develops. This point, the limit of proportionality, is not influenced, or is influenced to only a very limited extent, by the fibres. Consequently, the occurrence of the first crack is mainly determined by the properties of the matrix. When using chopped fibres, the flexural strength or modulus of rupture of the composite will be the same as the value obtained at the limit of proportionality, i.e. the flexural strength of the unreinforced concrete. However, the area under the stress–strain curve for the composite with chopped fibres in Figure 8.4 is larger than that for the unreinforced matrix. This means that the fracture toughness and impact strength increase with the addition of chopped fibres.

As can be seen in Figure 8.4, the modulus of rupture or the flexural strength of the composite reinforced with long fibres can exceed the limit of proportionality or flexural strength of the matrix. When the reinforcement consists of 4% long sisal fibres, the flexural strength exceeds the limit of proportionality by a factor of 4. Numerous cracks occur in the specimen during the flexural test. These cracks are comparatively small, with the exception of the failure crack. The deformation at failure is considerable, and this marked deformation is desirable, as it provides a visible warning that the material will fail. The fracture toughness (and impact strength) of the composite is considerably higher than in the case of chopped fibres (see Figure 8.4). This can be seen by comparing the area under the stress–strain curves.

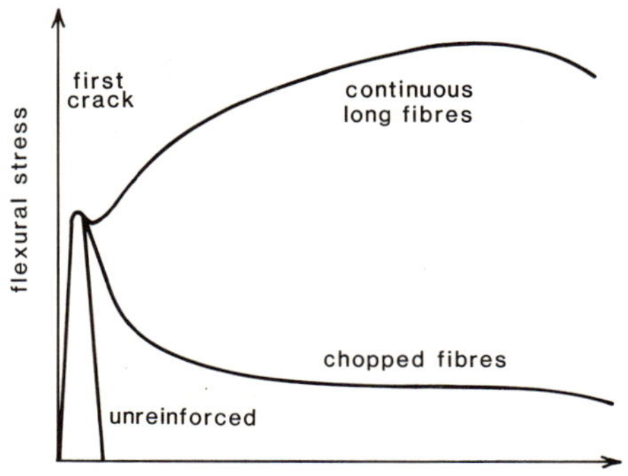

Figure 8.4 Stress–strain behaviour for unreinforced concrete, natural fibre concrete with chopped short fibres and continuous long fibres.

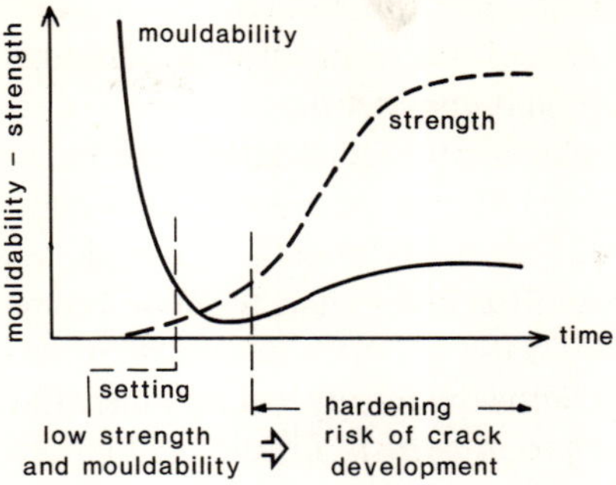

Figure 8.5 Stress–strain behaviour for natural fibre concrete with chopped short fibres and continuous long fibres as well as for asbestos cement.

8.3.3.3 *Tensile strength.* The effect of reinforcement in concrete is best shown in a direct tensile strength test. This test procedure is complicated, because it is not easy to apply a concentric load. An example of a stress–strain curve for natural fibre reinforced concrete is presented in Figure 8.5.

As can be seen in Figure 8.5, the natural fibres do not improve the tensile strength if they are added as chopped fibres. Lengthy fibres can improve the tensile strength of the composite, but only to a limited extent with the fibre volumes that are commonly used. The low modulus of elasticity of the natural fibre and the bond between the fibre and the matrix result in a fairly large strain at failure, concentrated on one crack when short fibres are used, but spreading to numerous cracks if long fibres and a high volume of fibres (4%) are used. The area under the stress–strain curve gives an idea of the toughness and impact strength of the material. As can be seen in the figure, the tensile strength of asbestos-cement is high, but it is not as tough as concrete reinforced with long sisal fibres.

8.3.3.4 *Compressive strength.* The compressive strength of a composite is not affected at all by the addition of natural fibres. If dirty fibres are used, the compressive strength can decrease. The incorporation of fibres may require the addition of extra water, which results in a higher water–cement ratio and therefore a lower compressive strength.

8.3.3.5 *Watertightness.* The watertightness of the composite is not affected by the incorporation of fibres. But, as is the case with the compressive strength, the watertightness can be influenced negatively if the water–cement ratio is increased to improve the workability of the fresh mortar mix.

8.3.3.6 *Shrinkage.* Drying shrinkage in the hardened composite is linked only to the water content of the fresh matrix. Fibres can give a better crack distribution–smaller crack widths, but a larger number of cracks. Again, it should be pointed out that the incorporation of short, chopped fibres may require more water in the mix to obtain good workability, which will increase the shrinkage.

8.3.3.7 *Other properties.* Properties like thermal conductivity, sound transmission, fire resistance, water absorption, linear expansion, moisture movements and temperature movements are not influenced by the incorporation of a few per cent by volume of natural fibres.

8.3.4 *Ageing, fibre deterioration and carbonation of the matrix*

Natural fibres are protected against microbiological decay in the matrix because of the high alkalinity of the pore water. Constituents of the pore water, for instance calcium hydroxide, penetrate the fibre and 'mineralize' it. It is not known what happens to a fibre if cracking occurs. Probably there is a crack width limit where the fibre is still protected against microbiological decay.

The high alkalinity of the pore water results, however, in chemical attack on the fibre and causes reduction or even nullifying of its strength. Fibres in carbonated concrete, where the alkalinity of the pore water is low, are not decomposed chemically, and also seem to be protected against microbiological decay.

Concrete carbonates through a reaction with the carbon dioxide in the air. The carbonation thus takes place first at the surface and moves further and further into the concrete with time.

Shrinkage occurs in connection with the carbonation of the matrix. This shrinkage is of the same magnitude as the drying shrinkage. Since the carbonation process is normally slow, concrete creep may contribute to preventing the occurrence of cracks due to shrinkage.

The carbonation front moves deeper into the concrete with time. A thin concrete product may be completely carbonated within a few years. Microcracking may have a negative influence on the flexural and tensile strengths of the composite. The addition of fibres to the fresh mortar may require more water in the mix to obtain satisfactory workability. This leads to a higher water–cement ratio, faster carbonation and a larger carbonation shrinkage.

8.4 Production methods*

8.4.1 *General*

The production of natural fibre concrete roofing sheets and other products can be subdivided into the following operations:

*(This section has been taken from reference (1) and has been written by H. Persson and Å. Skarendahl.)

(i) Preparation of the constituent materials
(ii) Mixing
(iii) Casting
(iv) Curing and storing
(v) Quality control.

Depending on whether chopped or continuous fibres are used, and on what is considered most practical for the process in question, the material flow into the production process is normally in accordance with one of the two methods shown in Figure 8.6. A combination is also possible.

For both the premix and the lamination methods, all steps in the process can be done manually, without the use of electricity, heat, pressure or vacuum.

The production of natural fibre concrete products is thus similar to the production of normal concrete products, and will likewise be suited for different degrees of mechanization to improve quality and productivity. The requirements for the production thus differ radically from the production process for asbestos cement, where advanced machinery is imperative.

8.4.2 *Preparation of the constituent materials*

Even if the price may vary within a wide range, cement is a readily available

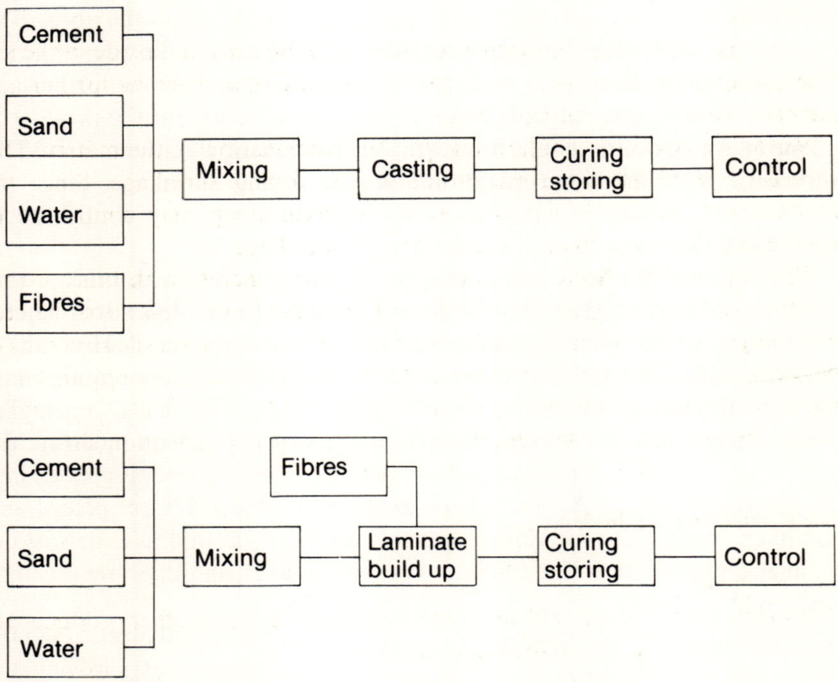

Figure 8.6 Two alternative production processes for natural fibre concrete products, premix and lamination[1].

material in most areas. Cement is most often delivered in bags. The suitability of the local cement for natural fibre concrete and the use of additives must be evaluated. Storing the cement in a proper way to keep it dry is, of course, essential.

The sand should have a rather fine grading, with a maximum aggregate size of approximately 2 mm. The sand should not contain any organic substances which can be harmful to the cement hydration. Sieving of the local aggregate will often be necessary, and simple procedures for the testing of contamination, grading, etc. should be set up.

The water can be of a rather varying quality, but it should not contain contaminations which make it unsuitable for use in cement-based materials. Whether the water is appropriate or not can easily be tested.

The natural fibres will be used either chopped or at full length in nets or in fabrics. The preparation of the fibres required in the concrete production process will depend on the type and form of fibres used. Possible effects on cement hydration from surface contamination of the fibre should be evaluated, and application of an impregnation to improve durability must be considered.

8.4.3 *Mixing*

Chopped fibres are mixed with cement, sand and water in a 'premix' process. When long fibres are used, the matrix (cement, sand and water) is mixed separately and then, in a 'lamination' process, worked in with the fibres. The mixing can be done with shovels or forks, or mechanically in pan or drum mixers. The workability of the mixture, temperature, and other factors must, of course, be chosen according to the method of production for the product in question. The mixture should be rather wet, but not too fluid, since a mixture which is too wet leads to separation and poor quality.

8.4.4 *Casting*

Premix natural fibre concrete is cast by simply pouring the mixture into a mould and levelling it to the desired thickness. Only slight tamping or hammering on the mould is required for sufficient compaction. However, a vibration table or the like will give a more efficient compaction and thus a more dense matrix.

In the lamination process, a thin layer of the concrete matrix is cast on the bottom of the mould, followed by a layer of fibres. The fibre layer is pressed into the matrix by rollers. A new matrix layer and a fibre layer are placed and compacted, and this procedure is repeated until the desired thickness of the lamination is obtained. The top layer is a matrix layer, and the surface can be smoothed over using simple tools.

The fibres give the concrete composite material a very good plasticity in the fresh state, i.e. normally during the first couple of hours after mixing the cement with water. This property makes it possible to reshape the material in various forms after it has been cast. It is, for example, possible to cast a flat sheet of natural fibre concrete on a plastic film and then alter the flat sheet to a corrugated shape by placing the film with the fibre concrete on another, already corrugated, product.

The plasticity of natural fibre concrete also makes possible the postforming of different shapes by placing the fresh fibre concrete on a product with the desired shape, thus making a replica of the mould. This method can be used for making buckets, bins, etc.

The complete casting process can be done manually, using simple hand tools. In small-scale industries, simple, basic machinery will probably be appropriate for fibre handling, matrix mixing, lamination build-up, compaction and postforming. The moulds can be made rather simply, but if high-quality surfaces, low tolerances and a high production rate are aimed at, the moulds should be fairly rigid, and preferably not of wood.

8.4.5 *Curing and storing*

All cement-based materials have to be water-cured after casting in order to prevent the water, needed for the hydration of the cement, from evaporating. The consequences of improper curing are cracking, warping due to uneven shrinkage, and reduced strength gain.

Curing can be performed with wet rags and the like by covering the product with plastic film or by keeping the product in a water basin. The time required for proper curing mainly depends on the type of cement used and the ambient temperature, and should be evaluated for each production situation. Demoulding can usually be done after one day, if carefully carried out.

The strength of concrete increases with time, but at a decreasing rate. If ordinary portland cement is used, a major part of the final strength will be present after four weeks. The product is thus not ready for delivery directly after the wet curing, but has to be kept in storage for some weeks. Handling during curing and storing must be done with care.

8.4.6 *Quality control*

Control measures must be taken for the various stages in the production process. As mentioned above, certain tests of the constituent materials should be carried out to verify that they are suited for natural fibre concrete production. Also, during the process itself, some testing should be done to verify that the process is under control. Tests regarding thickness, density and fibre content should be considered here.

A quality control procedure must be applied to the finished product to ensure acceptance. The tests might include strength, watertightness, etc. and the use of methods for testing described in appropriate standards should be encouraged. Proposals for quality controls are provided in references 1, 19, 34 and 35.

8.4.7 *Village-level production*

Natural fibre concrete can be produced using various levels of mechanization. When producing on what could be called a 'village level', it is essential that the demand for capital investment is very limited. On the other hand, production might very well be labour-intensive. The raw material cost should be low, the use of simple tools sufficient and only limited technical skills needed.

The production procedure for village-level production can be described as follows:

(i) Preparing sand by sieving by hand.
(ii) Chopping fibres in advised lengths, or preparing longer fibres or nets, as required.
(iii) Mixing by shovel of cement, sand, water and fibres (if chopped) using a simple mix design.
(iv) Pouring the concrete in a mould, if the premix technique is used. Otherwise, building up a lamination. Compaction by tamping and surface treatment with a trowel made of wood.
(v) Postforming by transferring the products by hand.
(vi) Curing under wet rags or the equivalent.

A complete village level production layout has been developed by ITDG Building Materials Workshop (see Figure 8.7). Short fibres are used. A production process where long fibres are used is described in a manual from the Building Research Unit in Tanzania, BRU[10]. Further descriptions are given in references 9 and 11–19.

8.4.8 *Small-scale industrial production*

The manufacture of natural fibre concrete products in small-scale industries is of interest in order to increase production rate, to improve quality and to be able to produce a wider variety of products. In this case, some capital investment for basic machinery will be needed. The production process should be mechanically designed but preferably not automatic, and should still lead to production utilizing the local labour force. The industry will cover a certain market area. Technical and economic skills on a higher level than those relevant for production on the village level are required to run the industry.

The process in a small-scale industrial production of natural fibre concrete can be described as follows:

(i) Preparing the sand by mechanical sieving or hand-sieving (depending on the raw material).
(ii) Preparing fibre fabrics or fibre nets, including impregnation for improved durability.
(iii) Mixing the concrete matrix in a mixer using a simple mix design.
(iv) Building up laminations in a casting machine. Compaction by pressing and mechanical hammering or vibration.

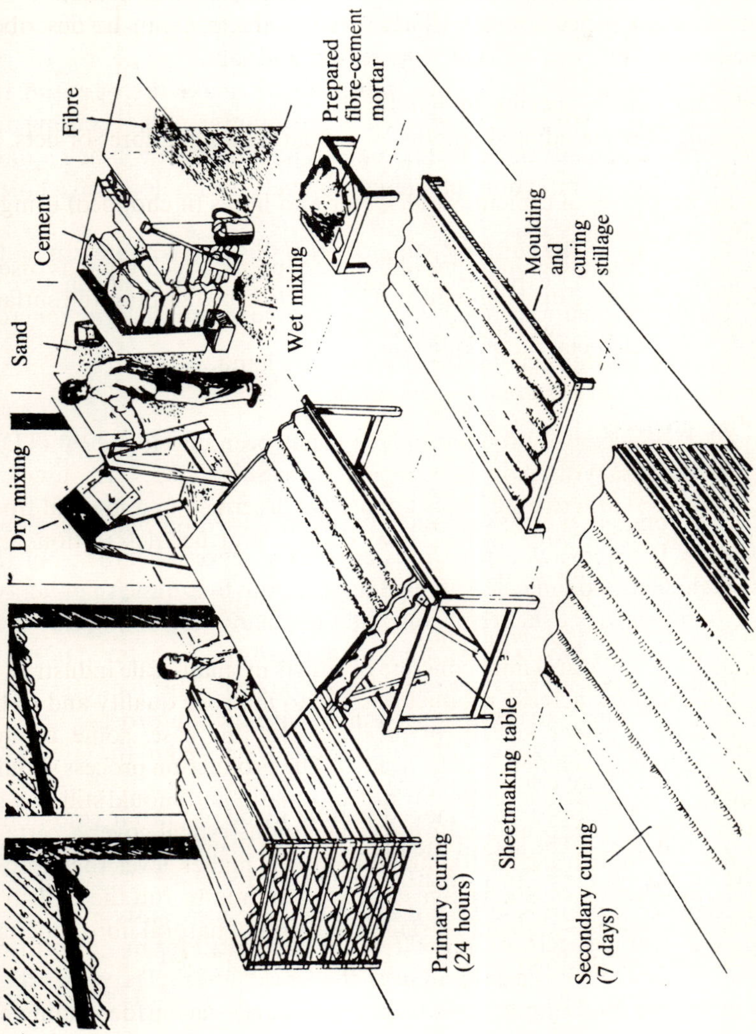

Figure 8.7 Process layout for a village-scale production of fibre concrete corrugated sheets using chopped fibres, according to Parry[7].

(v) Postforming on roller benches.
(vi) Curing in fog chambers, in water basins or under rags.
(vii) Quality control, storage and delivery.

8.4.9 *Automated production*

Natural fibre concrete can also be produced by highly-mechanized and automated industries in the same way as asbestos-cement. Compared to small-scale production, production rate is increased and labour is decreased, but more capital is required. The process will probably make use of automatic sieving, continuous mixing, pressing, vacuuming, pumping, steam curing and other well-known and advanced technologies. Qualified staff will be required for administration, production planning, concrete technology, machinery maintenance, sales, etc.

To the knowledge of the authors*, no process of this type is at present under development, but if the experience of small-scale industrial production is positive, highly automatic production processes for natural fibre concrete products should also emerge in due course.

8.5 Roofing products

8.5.1 *Tiles*

Unreinforced concrete tiles have been satisfactorily used on buildings for at least 80 years. Problems with frost resistance were reported earlier, but the products manufactured today have a high resistance to freeze and thaw action. There are many different shapes of concrete tiles, many resembling clay tiles. They are often coloured. Nowadays, they are normally manufactured of a very dry sand–cement mix with very low water–cement ratios. The thickness varies between 5 and 20 mm. Extrusion is used in the production process[20].

Roofing tiles reinforced with natural fibre were developed in the late 1970s in Kenya and elsewhere (see Figure 8.8).

Parry[15,16] has developed an S-shaped pantile made of mortar containing chopped natural fibres. The size of the tile is 500 × 250 mm and the thickness 6 mm. The weight per m^2 is 20 kg and the cement used 5 kg/m^2. The volume of timber used in the support is less for the tiles than for previously introduced fibre concrete roofing sheets. About 600 sets of equipment for manufacturing Parry roofing tiles have been sold all over the world (1987). The production capacity is about 50 roofing tiles (4 m^2) per worker per day. Skarendahl and Persson[1] present a roofing tile measuring 400 × 250 × 6 mm. The weight per m^2 is approximately 25 kg and the cement consumption approximately 6 kg/m^2.

Lola[22] reports a 'Perera extra-large tile' measuring 390 × 1000 × 10–15 mm. The cement consumption and weight per m^2 are not known.

* Persson and Skarendahl.

Figure 8.8 Flat, fibre-reinforced roofing tiles (top), folded tiles and corrugated sheets at Kenyatta University College[21].

Figure 8.9 Roofing tiles in a factory in Kenya (concept, IT Building Materials Workshop).

Figure 8.10 House in Kenya clad with roofing tiles on both walls and roof.

Ghavami *et al.*[23] report concrete tiles reinforced with coir fibres produced in a laboratory in Brazil with the aid of pressure, of dimensions 200 × 200 × 10–30 mm. The cement consumption and weight per m^2 are not known. Examples of the use of roofing tiles are shown in Figures 8.9 and 8.10.

8.5.2 *Sheets*

The literature contains many different suggestions for the appearance of a roofing sheet. Table 8.1 presents some data for sheets, with short or long fibres and with different dimensions. The trend has been towards smaller roofing sheets. In the late 1970s, efforts were often made to use the same dimensions as asbestos-cement sheets, i.e. 1.5 or 2 m long sheets, see for instance Figure 8.7. Due to cracking during handling, transport or application on the roof, the size was reduced to about 1 m^2 at the beginning of the 1980s. Today the size is still

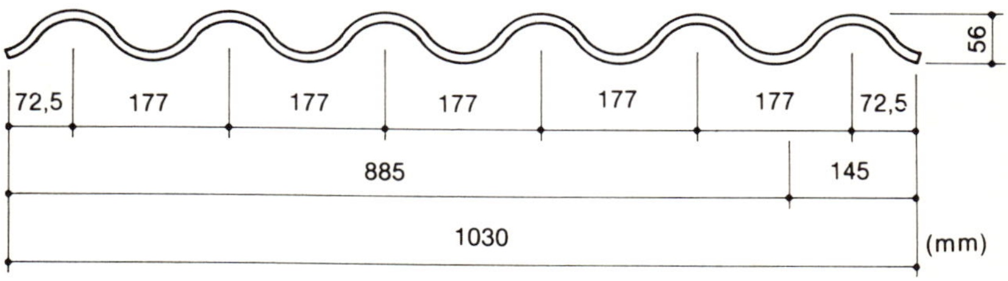

Figure 8.11 Corrugated roofing sheet according to ref. 1. Span 1200 mm; thickness 10 mm.

Table 8.1 Selected data on roofing sheets

Fibre used and fibre length (mm)	Dimensions of sheet (mm)	Flexural strength (MPa)	Equivalent load for 1 m span (kN)	Country	Reference
Elephant grass, 50	NA	10.5	—	Zambia	24
Coir	1500 × 1000 × 6	6.7	—	Thailand	25
Coir, 25	915 × 460 × 10	22.0	—	Singapore	26
Coir, 100	NA	10.0	—	Sweden	27
Coir/wood wool	1 500 × 850 × 10	6.5	—	India	28
Sisal, continuous	NA	30.0	—	Sweden	27
Sisal, continuous	NA	12.0	—	Brazil	29
Sisal, 30	NA	6.7	—	Brazil	29
Sisal, continuous	1200 × 1030 × 10	30.0	—	Sweden	1
Sisal, 10–15	1000 × 730 × 9	—	2000	Malawi	30
Sisal, 10–15	1200 × 920 × 7	—	1300	Zimbabwe	30
Sisal, 25 + continuous	1000 × 920 × 8	—	2300	Kenya	30
Sisal, 20 + continuous	1000 × 850 × 5	—	1200	Tanzania	30
Sisal, 12	980 × 730 × 8	—	2650	Guatemala	22
Sisal, 12	990 × 770 × 8	—	3000	Guatemala	22
Sisal, 12	1150 × 850 × 9	—	920	Dominican Rep.	22
Sisal, 12	1460 × 750 × 9	—	1400	Dominican Rep.	22
Sisal, 20	1200 × 840 × 12	—	2070	Colombia	22
Coir, 15	1000 × 390 × 12	—	1650	Sri Lanka	22
Coir, 15	1010 × 900 × 10	—	1190	Sri Lanka	22
Asbestos	1500 × 1050 × 6 (3000 × 1050 × 6)	17.8	—	Zambia	24, 25

Figure 8.12 Building in Nairobi covered with corrugated roofing sheets of sisal fibre concrete.

Figure 8.13 Building in Morogoro covered with trapezoidal roofing sheets developed at the Building Research Unit, Dar es Salaam, Tanzania.

Figure 8.14 Experimental roof at the Building Research Unit, Dar es Salaam.

falling, to around 0.7 × 1.0 m or even smaller. One example of suggested dimensions of a sheet is shown in Figure 8.11.

Figures 8.12 and 8.13 show examples of buildings covered with roofing sheets of natural fibre concrete. Figure 8.14 shows an experimental roof in Dar es Salaam.

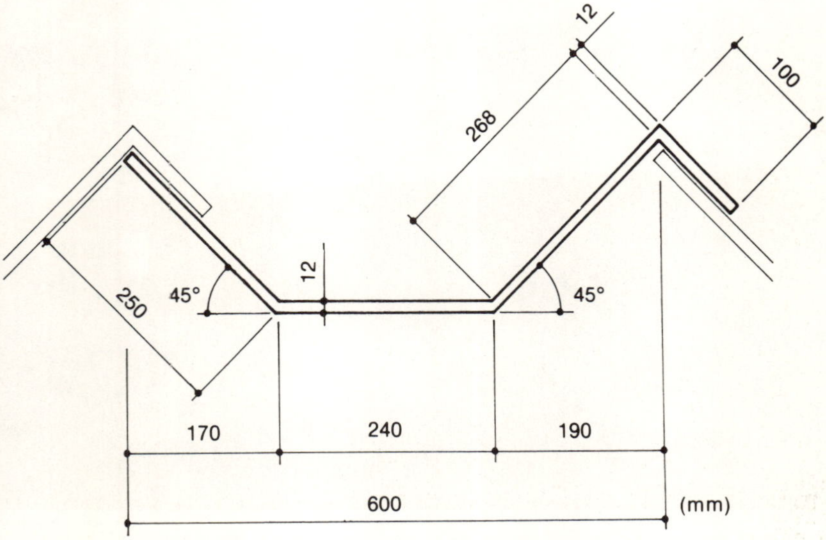

Figure 8.15 Dimensions for folded sheet, according to Persson and Skarendahl[1].

Figure 8.16 Experimental roof at the Building Research Unit, Dar es Salaam. Long, self--supporting roofing elements in the centre and trapezoidal roofing sheets to the right.

Figure 8.17 Test loading of folded sheet at the Building Research Unit, Dar es Salaam.

Proposals for designing roofing sheets are also provided by Berwanger and Rif[31].

8.5.3 *Larger, self-supporting elements*

Long, self-supporting roofing elements can be manufactured on condition that the natural fibre can be used as reinforcement in the cement matrix. Roofing elements of this type would eliminate the need for a supporting roof structure. An example of a roofing element with a length of 4.5 m is shown in Figures 8.15–8.17.

Since the natural fibre is broken down in the alkaline environment in the concrete, the use of natural fibre as reinforcement in concrete is not recommended. Neither, consequently, is the manufacture of self-supporting roofing elements until satisfactory solutions have been produced for the durability problems.

8.6 Loads on roofing products

A roofing element is subjected to many different types of load. During production and curing there is plastic shrinkage due to water evaporation. Later, the drying shrinkage, with a magnitude of perhaps 0.2%, may cause internal stresses and cracking. Cracking can also occur during handling, demoulding and transport (see Figure 8.18).

Storing in the wrong way (for example, stacking) may also result in stresses and cracking.

Figure 8.18 The sheet can crack during handling[19].

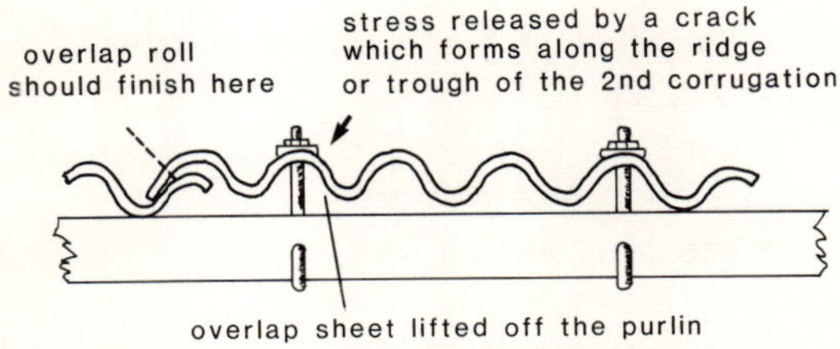

Figure 8.19 Poor fitting at side overlaps and overtightened hook bolts may cause longitudinal cracking[32].

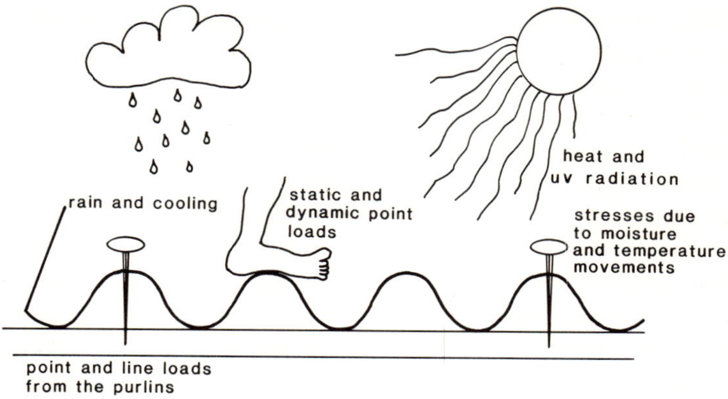

Figure 8.20 Different types of loads on a corrugated roofing sheet.

The element must also withstand static and dynamic loads during installation on the roof. Once on the roof, stresses may be concentrated to different points or lines on the roofing elements (see Figures 8.19 and 8.20). Deformations in the timber support may also cause stress concentrations in the element. Temperatures cycling from about 10°C at night to 70°C during the day result in a movement of 0.7 m/m. The moisture movements caused by rain and sun also initiate movements in the elements and these must be dealt with. The moisture movement of natural fibres is larger than that of the matrix and this may result in internal stresses, especially if the volume of fibres is high and the fibres are utilized as reinforcement (long fibres). If the fibres are decomposed due to chemical attack, new stress situations may also occur. Finally, the carbonation process in the matrix also leads to shrinkage and microcrack development.

8.7 Costs

When discussing the cost of a roofing material, the entire roof, in other words, both the roof covering material and the supporting structure, should be considered. The cost for a finished roof embraces a number of factors, some of which are listed below:

(i) Design of roof truss (available material, selected roof covering material, weights, wind load, etc.).
(ii) Cost of roof truss material.
(iii) Transport of material to construction site.
(iv) Labour cost when constructing roof.
(v) Cost of roof covering material.
(vi) Service life of roof covering material.
(vii) Maintenance cost.

The cost of the roof covering material depends on, among other things: capital investments in site, building, equipment and know-how; capital, labour, raw material and transport cost; interest; breakage; advertising and product information; sales organization and taxes.

The different costs vary from one country to another and within countries. According to Bijen[17], 1 m^2 of natural fibre cement board in 1986 cost US$ 2.75 produced at village scale (capacity 10 000 m^2/year), US$ 2.55 on a small home industry scale (100 000 m^2/year) and US$ 2.45 on a large industrial scale (4 000 000 m^2/year). According to Skarendahl and Persson (1984)[1] the costs for corrugated roofing sheets of natural fibre concrete (1984) are US$ 2.90/m^2 manufactured at village level and US$ 2.67/m^2 manufactured at small-scale industrial level. In reference 9 (1987), the cost per m^2 varies between US$ 2 and US$ 4.

Economies of scale obviously provide advantages with regard to the price of the product. The choice of scale for the production is guided, however, by factors such as the availability of capital, trained labour and infrastructure. Aid to developing countries is preferably given in the form of small-scale production plants so as to create a larger number of job opportunities. When investing in small-scale plants, the costs of training are often forgotten. The result is an unrealistic product price, or deficient or non-existent training with a low production capacity or poor quality.

8.8 Evaluation of worldwide field study

In 1985 SKAT[9] prepared a questionnaire which was sent out to all the manufacturers then known of roofing products of fibre concrete. Responses were received from 19 manufacturers in 17 countries all over the world. Tables 8.2–8.4 show the type of roofing product manufactured, the length of time production has been in progress, the number of units produced, the number of units installed, production capacity, material composition and rejection rates. A smaller evaluation of the manufacture of roofing sheets in four countries has been carried out by Lola[22,33].

Evaluation of the questionnaires shows that most producers are non-governmental and that the products are used mainly in rural areas. The typical production capacity is small, less than 80 m^2/day, in most cases less than 20 m^2. The most commonly used fibre has been the sisal fibre, but elephant grass, palm nut fibres, jute, coir and ijuk fibres have also been used. Fibre concrete is cheaper on the market than nearly all other roofing materials in most countries and under most circumstances.

Most sheet producers see many constraints to wider application of their product because of low durability, cracks and problems with transport. Tile producers, however, are very confident and see very few constraints.

Tiles are generally termed successful, whereas large sheets have led to failures in about 50% of the cases studied. Failures in these cases were due to

Table 8.2 Basic data on FC roofing

Country	Product width, length, thickness (mm)	Produced since	Number of elements produced	Number of elements installed in roofing	Production capacity (m^2/day)
Africa					
Mozambique	240 × 500 × 7 ITW-tile	1985	1500	1000	4
Zambia	670 × 1000 × 6	1984	2000	2000	13
Tanzania	750 × 750 × 7.5	1978	5000	5000	20
Kenya	Parry ITW-tile	1983	100 000	48 000	18
Kenya	Parry ITW-tile	1985	20 000	1500	17
Zimbabwe	700 × 1000 × 6–7.5	1979	2000	200	15
Zambia	sheets (NA)	1982	—	—	10
Ghana	tiles (NA)	1985	500	150	NA
Malawi	760 × 1000 × 10–15	1980	34 000	21 000	15–60
Asia					
Bangladesh	795 × 1000 × 6.5	1983	880	530	9
Bangladesh	790 × 1000 × NA	1983	1070	834	8
Bangladesh	Parry ITW-tile	1983	80	—	—
India	650 × 1000 × 10	1982	10 000	9500	21
India	1000 × 2000 × 7	NA	FEW	FEW	—
Indonesia	920 × 1150 × 7	1982	2000	1800	—
Sri Lanka	390 × 1000 × 10–15	1984	15 000	14 000	NA
America and Australia					
Dominican Republic	750 × 1350 × 6–9	1984	3000	1500	16
Dominican Republic	1000 × 1450 × 10–15	1982	1000	900	—
Nicaragua	840 × 1000 × 6–9	1984	14 000	8000	80
Solomon Islands	Parry ITW-tile	1984	1000	1000	4
Haiti	Parry ITW-tile	1984	42 000	34 000	40
Guatemala	770 × 990 × 6–9	1981	1500	1500	NA
Colombia	610 × 1140 × 10	1982	300	200	NA

NA = No Answer

Table 8.3 Composition of matrix, type of fibre, and compaction method used.

Country	Type of element[1]	Cement–sand ratio	Water–cement ratio	Type of fibre and length (mm)	Fibre volume (%)	Type of compaction[2]
Africa						
Mozambique	T	1:2	0.52	Sisal, 12	1.7	v
Zambia	S	1:1	—	Sisal, 10–20	1.1	t
Tanzania	S	1:1	0.56	Sisal, 20–50	3.0	v
Kenya	T	1:3	—	Sisal, 12	1.1	v
Zimbabwe	S	1:1	—	Sisal, 15–20	0.9	t
Zambia	S	NA	—	Sisal, 20–50	NA	NA
Zambia	S	NA	—	Elephant gr, 20–50	NA	NA
Ghana	T	NA	—	Palm nut, 60	NA	NA
Malawi	S	1:2	0.6	Sisal, 20–30	1.0	t
Asia						
Bangladesh	S	1:0.8	—	Jute, 18–25	1.1	t
Bangladesh	S	1:0.8	—	Jute, 15	1.5	NA
Bangladesh	T	—	—	—		—
India	S	1:1	—	Coir, 12	1.9	t
India	S	1:0	—	Coir, 100–200	16	v (+pressure)
Indonesia	S	1:0.67	—	Ijuk, 30	3.4	t
Sri Lanka	S	NA	—	Coir, 20–30	NA	v
America and Australia						
Dominican Republic	S	1:1	—	Sisal, 25	1.2	t
Dominican Republic	S	1:1	—	Sisal, 50–75	NA	t
Nicaragua	S	1:1	—	Sisal, 25	3.5	t
Solomon Islands	T	1:3	—	Coir, 20	1.1	v
Haiti	T	1:2.5	—	Sisal, 20	NA	v
Guatemala	S	NA	—	Sisal, 20–30	NA	v
Colombia	S	1:1.2	—	Sisal, 20–30	0.6	v

NA = No Answer
1. T, tile; S, sheet
2. v, vibration; t, 'tamping'

Table 8.4 Rejection rate before shipment, damage after shipment, mode of fixing, type of observed cracking and expected durability.

Country	Type of element[1]	Rejection rate before shipment (%)	Damage rate after shipment (%)	Way of fixing	Crack pattern[2]	Expected durability (years)
Africa						
Mozambique	T	0.5	0	Wire loops	none	—
Zambia	S	4.5	5.5	Wire loops	l	5–10
Tanzania	S	1.0	1.0	Nail, screw	l	30
Kenya	T	4.0	1.0	Wire loops	l	20–25
Kenya	T	0	2.0	Wire loops	none	15
Zimbabwe	S	5	10	Wire loops	l + t	20
Zambia	S	20	30	Wire loops	NA	7–10
Ghana	T	2	0	Wire loops	NA	—
Malawi	S	2–20	2–10	Wire loops	l	10–70
Asia						
Bangladesh	S	5.0	5.0	Wire loops	l + t	15
Bangladesh	S	6	2	J-hooks + wire loops	l + t	10
India	S	5.0	—	Wire loops	l	10
India	S	5.0	5.0	J-hooks	l + t	30
Indonesia	S	—[3]	—[3]	Wire loops	l + t	—
Sri Lanka	S	—	2	J-hooks	l + t	4–6
America and Australia						
Dominican Republic	S	3.0	2.0	Screws	l + t	8
Dominican Republic	S	—	—	Nails	l + t	5–10
Nicaragua	S	6.0	8.0	J-hooks	l + t	10
Solomon Islands	T	1.0	1.0	Nails	l	10
Haiti	T	1.0	NA	Wire loops	corners	—
Guatemala	S	—	20	J-hooks	l + t	4–6
Colombia	S	—	53	J-hooks	l + t	—

NA = No Answer
1. T, tile; S, sheet
2. l, longitudinal; t, transverse
3. Figure not given, but the roofing structure classed as failure because of many problems with cracks.

errors in production and installation by operatives who had no proper skill or training.

Sheets seem to have both longitudinal and transverse cracks. Two out of four tile producers report no cracking. The reasons for cracks are said to be poor curing, lack of homogeneity in the mortar mix, unevenly distributed fibres, unsatisfactory properties of the cement, sand or fibres, excessively high water–cement ratio or poor compaction during the production.

In one case, 45% of installed roofing sheets were cracked after only a few years of service. Tiles have only been in use for two years (1985) and so far have had less tendency to crack compared with sheets.

Four of the ten sheet producers wish to change to tile production. The four tile producers are satisfied with their product.

Sheets are usually produced with very simple equipment—masonry tools such as shovels, sieves, trowels and moulds made of wood or asbestos-cement. Compaction is by tamping. Tiles are made on a vibrating table. Quality control is rare—only two questionnaires report on daily quality control.

Generally there seems to be a lack of information and awareness about costs and economics. The costs of raw materials are well known, but other cost factors, such as investment costs, interest on loans, depreciation of capital, equipment and training costs, i.e. fixed costs, are not taken fully into account. Ignorance concerning these costs is mostly due to the way in which fibre concrete roofing projects are supported.

Fibre concrete roofing products are not yet commercial products. In only very few cases have these products entered the building materials market profitably. Among the advantages of fibre concrete roofing production are the low initial investment cost, no direct import of raw materials, on-site production, and low energy input. The disadvantages are scarcity of cement, lack of energy input for cement production, need for foreign currency to buy equipment, costs for know-how, and low productivity. Another problem is a lack of knowledge about the lifespan of the products, and a lack of marketing experience.

In many cases the crucial point has been to obtain enough money to purchase the hardware for production. Training is generally treated as a lower-priority extra. There exists a lack of continued know-how transfer, mainly concerning management, marketing, installation and up-to-date results of research, even if workers have experience in meeting problems that occur in the production and application process. Most of the information on the production of roofing sheets is based on manuals or articles from the Intermediate Technology Workshop, based in the UK.

From the answers in the questionnaires it can be concluded that marketing in most cases has not been planned, either because the elements were produced to meet local demand only or because marketing was considered a problem of secondary priority.

A serious constraint on a wider application of the technology was shown to be a general lack of know-how. Acceptance by the market is not a problem of first priority for the respondents. The lack of good raw materials, mainly good sand and clean fibres, was a problem in several cases.

The questionnaires also showed that a technical manual or training course should be as comprehensive as possible. The manual or course should mainly cover technical aspects of fibre concrete roofing production, emphasizing management and marketing aspects as a second priority. According to the questionnaires, the following aspects of fibre concrete roofing technology were ranked in order of decreasing importance: durability, acceptance, manufacturing method, economics, quality control, raw material, know-how transfer/professional training, limits of application, installation of fibre concrete roofing elements, research and development, transport and handling, supporting structure/maintenance and marketing. Many of these aspects are linked to know-how transfer.

Since 1979, 165 000 tiles and 90 000 sheets have been produced by the participants in the present survey. About 85 000 tiles and 67 000 sheets have been set up on roofs. The interest in tiles is increasing very strongly. The following comparison of sheets and tiles has been made.

Sheets	**Tiles**
Sheets will be compared to asbestos or corrugated iron sheets; expectations will be difficult to meet.	Tiles will be compared to clay tiles; expectations are easier to meet.
Only a good work team will be able to guarantee a good overall aesthetic appearance of a finished roof.	Good aesthetic appearance is easy to achieve, as minor errors do not show easily.
Mouldings need a very well coordinated team of at least two workers.	One qualified worker can do the moulding.
Tamping or vibrating are difficult to standardize.	Vibration is standardized.
Production is possible without electricity and with very simple equipment.	Production process needs electrical or manual vibration.
Handling of the fresh product is difficult; risk of breakage.	Easy handling; almost no breakage of one day-old tiles.
Curing in water tanks is difficult because of the considerable size of sheets.	Curing in many kinds of water tank is easily possible.

Transport is rather difficult and risky.	Transport is less difficult.
Lower costs for equipment.	Higher investment for production equipment.
Higher cement content (1:1); therefore more cement per square metre and higher material costs.	Lower cement content (1:2); therefore less cement per square metre and lower material costs.
Heavier roofing elements and less overlaps will withstand stronger wind forces.	Lighter roofing elements; more easily moved by wind if not tied down.
Handling of the sheets on the roof is difficult and may lead to breakage.	Easy handling on the roof.
Traditional craftsmen (used to corrugated iron sheet and clay tile roof laying) need special training for installing sheets.	Where clay tiles are common, most craftsmen will be able to install tiles.
Poor supporting structure or warping of timber often leads to leakage.	Uneven parts of the roof structure can be absorbed to a certain extent by the tiles.
Replacing sheets on roofs is difficult; other sheets may break in the process of replacement.	Replacing tiles is easier, though not without problems.
Training is essential and must be done by a qualified instructor. Minimum training time 2 to 3 weeks.	Initial training can be achieved in a shorter time but follow-up is required at initial stage of technology.
The stresses in the sheets are higher than in the tiles. Sheets are therefore more susceptible to cracking.	Due to its smaller size the stress within a tile is smaller, and less cracking occurs.

By way of conclusion, we can note that the failures which have occurred so far seem to be the result of insufficient knowledge, the use of faulty production methods and incorrect installation of the fibre concrete roofing. The general conclusion may be summarized in the statement that the fibre concrete roofing technology is promising, provided that its know-how is adequately and comprehensively communicated to all users and is followed up in some organized way.

8.9 Future prospects

Roofing products of natural fibre concrete can potentially meet the need for cheap roofs which occurs all over the world. This is particularly true since the use of asbestos will probably decrease in roofing sheets due to the health hazards of the asbestos fibre.

The production of tiles is increasing continuously while the production of sheets manufactured on a small scale appears to be decreasing. A small-scale industrial production of sheets has been started in Tanzania and the manufacture of these sheets includes measures taken to improve their durability. The size of the sheets has been reduced and may be further reduced in the future for technical reasons.

Small-scale production of tiles will, in all certainty, be rendered more efficient through improved know-how transfer. Improved know-how transfer can be achieved by establishing regional centres, which can be responsible for training and external quality control, for example. The size of the tiles may perhaps increase, and productivity may be raised by minimizing and perhaps even excluding the fibre content.

References

1. Gram, H.-E., Persson, H. and Skarendahl, Å. (1984) Natural fibre concrete. *Sarec report R2:1984* (Swedish agency for research cooperation with developing countries), Stockholm, 139 pp.
2. Marotta, P. (1973) Tests on panels of coconut fibre impregnated with portland cement. *Report 2*, Division of Building Research, CSIRO Victoria, Australia, 18 pp.
3. Nilsson, L. (1973) Reinforcement of concrete with sisal and other vegetable fibres. Thesis for the Bachelor of Engineering Degree, The Department of Building Construction, Royal Institute of Technology, Stockholm, 55 pp (in Swedish). See also Swedish Council for Building Research, Document D14:1975, Svensk Byggtjänst, Stockholm, 68 pp. (in English).
4. Sridhara, S., Kumar, S. and Sinare, M.A. (1971) Fibre reinforced concrete. *Ind. Concr. J.* **45**, 428–430, 442.
5. Singh, S.M. (1975) Corrugated roofing sheet and building panel from coir wastes. *Coir*, April, 3–6.
6. Persson, H. and Skarendahl, Å. (1978) Sisal fibre concrete for roofing sheets and other purposes. *Forum on Appropriate Industrial Technology*, UNIDO, New Delhi, Swedish International Development Authority, Stockholm, 57 pp. See also UNIDO, *Monographs on Appropriate Industrial Technology* **12**, 1980, Appropriate Industrial Technology for Construction and Building Materials, United Nations, New York, 105–128.
7. Parry, J.P.M. (1980) Building materials and components. UNIDO, *Monographs on Appropriate Industrial Technology* **12**, Appropriate Industrial Technology for Construction and Building Materials, United Nations, New York, 27–45.
8. Gram, H.-E. and Skarendahl, Å. (1978) Sisal reinforced concrete – Study No 1 on materials. *Report* **7822**, Swedish Cement and Concrete Research Institute, Consultant Section, Stockholm, 15 pp.
9. Gram, H.-E. *et al.* (1987) *FCR – Fibre Concrete Roofing*, SKAT (Swiss Center for Appropriate Technology) St Gallen, Switzerland, Intermediate Technology Publications Ltd, London, 185 pp.
10. Building Research Unit (BRU) (1979) Roof sheets made of sisal reinforced concrete. *BRU Data Sheet* **F 8.1**, Dar es Salaam.
11. Parry, J.P.M. (n.d.) *The Development of Fibre Cement Technologies for Roofing – A Handbook for the Ultimate User.* IT Workshops. Cradley Heath, UK, 88 pp. (Probably 1984.)

12. UNCHS-HABITAT (1981) Construction with sisal cement. *Technical Notes* **1**, Nairobi, Kenya, 4 pp.
13. Skarendahl, Å. (1982) Natural fibre concrete for roofing. *Seminar on Building Materials*, National Construction Council, Dar es Salaam, Tanzania, 17 pp.
14. Paamand, K. (1984) *Sisal Cement Roof Sheets Manual.* Department of Appropriate Technology, Hlekweni Friends Rural Service Centre, Bulawayo, Zimbabwe, 31 pp.
15. Parry, J.P.M. (1987) *Fibre Concrete Tiles—Users' Manual.* IT Workshops, Cradley Heath, England, 42 pp.
16. S-shaped pantiles from simple flat-bed machine. *Concrete Plant and Production* **3**(8)(1985) 307–310.
17. Bijen, J. (1986) Vegetable fibre cement board. *Intron Report* **86076**, Maastricht, Netherlands, 56 pp.
18. Swift, D.G. and Smith, R.B.L. (1979) Sisal cement composites as low-cost construction materials. *Appropriate Technol.* **6** (3), 6–8.
19. Parry, J.P.M. (1984) *Fibre Cement Roofing.* The technical record of the development and overseas field trials between 1977–84 of the low cost building material of the future. IT Workshops, Cradley Heath, UK, 73 pp.
20. Watson, J.A. (1979) *Roofing Systems* – Materials and Application. Reston Publishing Co. Reston, Virginia, 127–148.
21. Persson, H. and Skarendahl Å. (1979) Implementation of a building material—Sisal fibre concrete roofing sheets. Dept. of Building Engineering, *Rapport* **20**, Royal Inst. of Technology, Stockholm, 26 pp.
22. Lola, C.R. (1985) Fibre reinforced concrete roofing sheets. *Technology Appraisal Report.* AT International, Save the Children Federation, Washington DC, 28 pp.
23. Ghavami, K. and Van Hombeeck, R. (1984) Application of coconut husk as a low cost construction material. *Low-Cost and Energy Saving Construction Materials*, Vol 1, Rio de Janeiro, July 1984, Envo Publ. Co., 53–60.
24. Lewis, G., Mirihagalia, P. (1979) A low cost roofing material for developing countries. *Building and Environment* **14** (2), 131–134.
25. Balaguru, P.N. and Shah, S.P. (1985) Alternative reinforcing materials for less developed countries. *Int. J. Devel. Technol.* **3**, 87–105.
26. Paramasivam, P., Nathan, G.K. and Das Gupta, N.C. (1984) Coconut fibre reinforced corrugated slabs. *Int. J. Cement Comp. and Lightweight Concr.* **6** (1), 19–27.
27. Gram, H.-E. (1983) Durability of natural fibres in concrete. *CBI Research fo. 1.83*, Swedish Cement and Concrete Research Institute, Stockholm, 255 pp.
28. Coir waste-wood wool corrugated roofing sheets. AGR 19, Central Building Research Institute, Roorkee, India, 23–24.
29. da Silva Guimaraes, S. (1984) Experimental mixing and moulding with vegetable fibre reinforced cement composites. *Low-Cost and Energy Saving Construction Materials*, Vol. 1, Rio de Janeiro, July 1984, Envo Publ. Co., 37–51.
30. Sakula, J.H. (1982) Sisal-cement roofing in east and southern Africa. Evaluation of sisal-cement roofing sheet technologies in Malawi, Zimbabwe, Zambia, Tanzania and Kenya. Intermediate Technology Industrial Services, Development Group Ltd, Rugby, England, 186 pp.
31. Berwanger, C. and Rif, J. (1985) Analyses of corrugated roofing sheets for optimum cross-sections using computer programs. *Report* **1**, Civil Engineering Department, University of Ottawa, 27 pp.
32. Parry, J.P.M. (1985) *Fibre Cement Roofing. Position report.* IT Workshops, Cradley Heath, UK, 19 pp.
33. Lola, C.R. (1986) Fibre reinforced concrete roofing sheets technology appraisal report. *Third Int. Symp. on Developments in Fibre Reinforced Cement and Concrete*, Sheffield, July 1986, Vol 1, 5.
34. Johansson, L. (1984) Corrugated sheets of natural fibre reinforced concrete. Draft standard, Document D3:1984, Swedish Council for Building Research, 37 pp.
35. Parry, J.P.M. (n.d.) *Production and Installation of Handmade FRC Roof Cladding Components. Basic Operating Manual.* IT Workshops, Cradley Heath, UK, 26 pp. (Probably 1984.)

Index

ACCOUNTING IN EUROPE

To my wonderful wife Luisa,

who helped me write this book